More information about this series at http://www.springer.com/series/4622

Power Systems

Sumedha Rajakaruna · Farhad Shahnia
Arindam Ghosh
Editors

Plug In Electric Vehicles in Smart Grids

Energy Management

Springer

Editors
Sumedha Rajakaruna
Electrical and Computer Engineering
Curtin University
Perth, WA
Australia

Arindam Ghosh
Electrical and Computer Engineering
Curtin University
Perth, WA
Australia

Farhad Shahnia
Electrical and Computer Engineering
Curtin University
Perth, WA
Australia

ISSN 1612-1287 ISSN 1860-4676 (electronic)
Power Systems
ISBN 978-981-287-301-9 ISBN 978-981-287-302-6 (eBook)
DOI 10.1007/978-981-287-302-6

Library of Congress Control Number: 2014957146

Springer Singapore Heidelberg New York Dordrecht London

Printed on acid-free paper

Springer Science+Business Media Singapore Pte Ltd. is part of Springer Science+Business Media (www.springer.com)

Preface

Plug in Electric Vehicles (PEVs) use energy storages usually in the form of battery banks that are designed to be recharged using utility grid power. One category of PEVs are Electric Vehicles (EVs) without an Internal-Combustion (IC) engine where the energy stored in the battery bank is the only source of power to drive the vehicle. These are also referred as Battery Electric Vehicles (BEVs). The second category of PEVs, which is more commercialized than the EVs, is Plug in Hybrid Electric Vehicles (PHEVs) where the role of the energy storage is to supplement the power produced by the IC engine. These two types of PEVs are predicted to dominate the automobile market by 2030. Widespread adoption of PEVs allows the world to reduce carbon emissions in transportation needs significantly. Therefore, it is vital to the success of a collective global effort in meeting the climate energy targets and to reduce the dependence on increasingly scarce fossil fuels. However, significant challenges are thrust upon the utility grid operators on how best to manage the power demand arising due to the charging of PEVs by the grid (G2V) and the power supply due to the Vehicle to Grid (V2G) discharging of energy storages in PEVs.

This book covers the recent research advancements in the area of energy management that can be employed to accommodate the anticipated high deployment of Plug-in Electric Vehicles (PEVs) in smart grids. The topics that are covered in this book include smart coordination based on real-time pricing, decentralized demand side management, optimal and distributed control of both G2V and V2G modes of PEVs, minimizing the energy procurement cost and financial risks in an energy hub, voltage droop controller with an event-driven control strategy for the coordination of charging PEVs, Additive Increase Multiplicative Decrease to control both the active and reactive power consumption and injection in the smart grid, an optimal controller to maximize profit based on a novel dynamic model of the system, aggregator bidding into the day-ahead electricity market with the objective of minimizing charging costs, optimal operation of plug-in vehicle fleets in a microgrid characterized by the presence of other distributed resources and a mixed integer linear programming energy management optimization model to schedule the charging and discharging times of PEVs. Hence, this book introduces many new strategies proposed recently by researchers around the world to address the energy management of smart grids with high penetration of PEVs. The book is aimed at engineers, system planners, energy market operators, researchers, and graduate students, who are interested in the latest developments in this field of research.

Contents

1 **Overview of Plug-in Electric Vehicles Technologies** 1
Antonio Carlos Zambroni de Souza, Denisson Queiroz Oliveira and Paulo Fernando Ribeiro

2 **Smart Coordination Approach for Power Management and Loss Minimization in Distribution Networks with PEV Penetration Based on Real Time Pricing** 25
Bhuvana Ramachandran and Ashley Geng

3 **Plug-in Electric Vehicles Management in Smart Distribution Systems** 59
Antonio Carlos Zambroni de Souza and Denisson Queiroz Oliveira

4 **An Optimal and Distributed Control Strategy for Charging Plug-in Electrical Vehicles in the Future Smart Grid** 79
Zhao Tan, Peng Yang and Arye Nehorai

5 **Risk Averse Energy Hub Management Considering Plug-in Electric Vehicles Using Information Gap Decision Theory** 107
Alireza Soroudi and Andrew Keane

6 **Integration of Distribution Grid Constraints in an Event-Driven Control Strategy for Plug-in Electric Vehicles in a Multi-Aggregator Setting** 129
Klaas De Craemer, Stijn Vandael, Bert Claessens and Geert Deconinck

7 **Distributed Load Management Using Additive Increase Multiplicative Decrease Based Techniques** 173
Sonja Stüdli, Emanuele Crisostomi, Richard Middleton, Julio Braslavsky and Robert Shorten

8 Towards a Business Case for Vehicle-to-Grid—Maximizing Profits in Ancillary Service Markets 203
David Ciechanowicz, Alois Knoll, Patrick Osswald and Dominik Pelzer

9 Integration of PEVs into Power Markets: A Bidding Strategy for a Fleet Aggregator 233
Marina González Vayá, Luis Baringo and Göran Andersson

10 Optimal Control of Plug-in Vehicles Fleets in Microgrids 261
Guido Carpinelli, Fabio Mottola and Daniela Proto

11 Energy Management in Microgrids with Plug-in Electric Vehicles, Distributed Energy Resources and Smart Home Appliances 291
Okan Arslan and Oya Ekin Karaşan

Reviewers

K. Ramalingam, Indian Institute of Technology, India
Alexander Schuller, Karlsruhe Institute of Technology, Germany
Ahmed M.A. Haidar, University of Wollongong, Australia
Ricardo Rüther, Universidade Federal de Santa Catarina, Brazil
Charalampos Marmaras, Cardiff University, UK
Alessandro Di Giorgio, University of Rome, Italy
Abdul Motin Howlader, University of the Ryukyus, Japan
Alireza Soroudi, University College Dublin, Ireland
Marina González Vayá, ETH, Switzerland
Minggao Ouyang, Tsinghua University, China
Linni Jian, South University of Science and Technology of China, China
Sonja Stüdli, University of Newcastle, Australia
Klaas De Craemer, KU Leuven, Belgium
Filipe J. Soares, INESC Technology and Science, Portugal
Mahmud Fotuhi-Firuzabad, Sharif University of Technology, Iran
Zechun Hu, Tsinghua University, China
Suryanarayana Doolla, Indian Institute of Technology Bombay, India
Erotokritos Xydas, Cardiff University, UK

Chapter 1
Overview of Plug-in Electric Vehicles Technologies

Antonio Carlos Zambroni de Souza, Denisson Queiroz Oliveira and Paulo Fernando Ribeiro

Abstract The advent of renewable energy is about to change power systems around the world. In this sense, operating a power system may become an even more complex task, with implications on the system security, reliability and market. The role played by the utilities may not be diminished, since they must be able to provide energy when intermittent sources are not available. In this near future, the plug-in electric vehicles will also have an important role for power distribution systems. But, at the same time, they have a big potential to help on integration of the renewable power generation in existing power systems. This chapter presents some of the existing plug-in electric vehicles technologies and discusses a few implication of this new scenario in the system operation.

Keywords Plug-in electric vehicles · Energy storage systems · Electric vehicle technology

1.1 Introduction

Bulk power systems are operated in an interconnected mode in order to guarantee reliability and robustness to final users. In this sense, many studies, like dynamic and voltage stability, reliability, protection and power quality are carried out during the planning and operating scenarios. This is because a problem in part of the system must be managed in such a way that it is not spread to neighbor areas. Considering natural catastrophes, like tsunamis, makes the problem even worse, since electricity is one of the infrastructures to be preserved. This has gained increased attention from researchers because of the penetration of renewable generation. Preserving a system after a catastrophe takes place brings the concept of a

A.C. Zambroni de Souza (✉) · D.Q. Oliveira · P.F. Ribeiro
Federal University of Itajuba, Itajuba, Brazil
e-mail: zambroni@unifei.edu.br

S. Rajakaruna et al. (eds.), *Plug In Electric Vehicles in Smart Grids*,
Power Systems, DOI 10.1007/978-981-287-302-6_1

smart grid, theoretically capable of playing a self-healing process, in order to preserve the system.

Micro-grids fit in smart grids concepts. However, unlike shipboard power generation systems, which always work on an isolated mode, micro-grids may work in an islanded manner or connected to a bulk system, though in an autonomous way. A further characteristic lies on the fact that it may work in both ways simultaneously, i.e., though connected to a system, it is self-sufficient to provide its own load. The power generation portfolio of micro-grids has conventional and renewable generation-based options. In this sense, inverters are necessary to provide an alternate current (AC) to the network in a uniform manner, so the voltage is locally controlled and a single frequency is guaranteed. This increases the complexity of the system, since new components are now necessary to be modelled and new interfaces must be understood.

The components above tend to change the way a generation scheme is managed, since a large number of tiny players may be involved. This problem becomes even more dramatic by considering the penetration of plug-in electric vehicles (PEVs). Currently, several PEVs are being traded and are characterized by higher efficiency, fuel saving and low noise. On the other hand, PEVs still have greater prices compared to conventional models, due to their battery energy storage system.

As these vehicles increase their market share, another problem deserves attention from electricity companies. Their inclusion in power systems represents a large increase on load demand, causing many problems as voltage violation, power quality degradation, power losses increase, thermal limits violation on distribution transformers, harmonics and fault currents increase. This subject deserved the attention of some previous works that aimed to propose an efficient charging policy for the system, so the voltage profile is kept within an acceptable range while PEVs are charged [1–4]. Another interesting question is posed if PEVs are considered to supply power to the system in emergency conditions. This changes the paradigm of power systems' operating conditions, since a reserve of power generation may be considered in blackouts or catastrophic conditions.

This chapter presents an overview of PEV's technologies available in the literature. The power systems in the future will certainly require a change of paradigm in the way it is faced by societies in general. In this sense, the final costumer may also be a producer, creating the term "prosumers". The utilities are not to be neglected. Rather than that, they will play an important role on the system reliability, since the intermittence nature of prosumers may jeopardize the continuity of supply. In this sense, two basic assumptions are recognized on energy planning philosophies.

One philosophy takes economic progress and economic profits as a starting point where the main values are: free market, no regulation, maximization of profits, and self-regulation.

The other system takes sustainability in the present economic system as a starting point in which the main values are responsibility over nature and human life, sustainability, long term view, and maximizing added value for society. In this approach electricity companies, governments, etc. join responsibility to develop a sustainable policy. This is the scenario where PEVs are considered. When considering the inclusion of plug-in vehicles in the grid, the following philosophy issues should be analyzed:

- Interoperability of self-sustainable micro-grids and their reliability.
- The feasibility of computational tools in this scenario.
- Governmental rules of safety. How do the utilities are faced in this scenario?
- One could be tempted to disregard the utility role in the new system operating structure. Two reasons, however, inhibit this path:
- Renewable energy systems may be obtained as a summation of small domestic units, creating a big amount of local energy in the system. This is particularly important for domestic consumers, however, the industry still claims for a huge amount of the total energy of the system.
- Renewable energy systems may also be associated to large blocks of generation, like big wind or solar farms. In this case, operating and maintaining such structures depend on the utility abilities.

Because this chapter focuses on technical issues of PEVs, these topics are not directly addressed here. However, the concepts are placed in such a way that they are compromised by the implementation of any of the items presented above. Besides, because plug-in electric vehicles are to be a part of smart grids, a brief discussion on the philosophy of these systems is also presented. This next section addresses the basics of the PEV's technologies. This is accomplished by a brief history of energy production in the world followed by a technical discussion of each technology considered.

1.2 PEV Technologies

Unlike many people believe, the electric vehicle technology is not a recent innovation. In fact, the first electric vehicles were built during the 19th century after the development of lead acid batteries. In addition, research on nickel-iron and nickel-zinc batteries, and the development of recharging process helped on electric vehicles popularization. Beyond the batteries, two new technologies collaborated to improve electric vehicles performance: the regenerative breaking systems and the hybrid vehicles [5]. In 1900, 28 % of the American automobile fleet was composed of electric vehicles.

The combustion gasoline engine was first presented by Daimler and Benz in 1885, while Rudolf Diesel presented the diesel engine in 1892. Until the first years of 20th century, three different propulsion technologies competed with each other in automotive market: steam, electric and gasoline propulsion. But, only after Henry Ford presented the Ford Model T and the assembly line, in 1908, the combustion engine vehicles really overcame the concurrence. The following aspects can explain this outcome:

- The Ford's assembly line decreased the vehicle production cost and time. The Ford Model T price was about $850, a half of an electric vehicle price. Besides, the car was robust, safe, easy to drive and reliable.
- In 1912, the electric starting was developed, eliminating the need of a crank to start the gasoline engine.
- The battery autonomy is low, impeding from traveling long distances without recharging the batteries.
- The existence of oil reserves and an infrastructure to supply fossil fuel to the transportation sector.

Only during the 1970s energy crisis, when the oil shortage from Middle East caused a raise on prices, the electric vehicles became an attractive option to reduce oil addiction from transportation sector. Also on 1970s, the concerns on environmental aspects raised the debate on energy generation and consumption. Following this trend, the actual Plug-in Electric Vehicles (PEVs) may be considered the result of efforts made by automotive industry to develop a product that reflects the worry about the environment and the future.

At first sight, a PEV looks like any other vehicle. But, the vehicle components are different from those of conventional cars. First, the PEVs have an Energy Storage System (ESS), which can be composed of a set of batteries, or by a fuel cell, or by other storage technology. The ESS is charged through a battery charger, which is connected to a conventional outlet or to a special fast recharge point.

The ESS role is to supply electricity to an electric motor, which provides traction power through a mechanical transmission. The electric motor drive is another component of PEVs, which should have an adequate performance for vehicular applications. This means it should have a good performance over a wide speed and torque range, low cost, weight and volume, high robustness and reliability. The electric energy flows from the battery through an electronic power converter to feed the motor drive. Regarding the power electronics converters, these devices should be reliable, robust and meet the packaging requirements for vehicular applications. These requirements include a better performance on severe conditions of temperature, harsh environment and mechanical vibrations.

The state of art of PEV technologies is presented in the following sections.

1.2.1 Energy Storage Systems

Energy Storage Systems (ESSs) are systems which store energy in various forms such as electrochemical, kinetic, pressure, potential, electromagnetic, chemical and thermal, using, e.g. fuel cells, batteries, capacitors, flywheels, compressed air, pumped hydro, super magnets, hydrogen etc. The principal criteria of an ESS required for a specific application are [6, 7]:

- The amount of energy in terms of specific energy and energy density.
- The electrical power.

- The volume and mass.
- Reliability.
- Durability.
- Safety.
- Cost.
- Recyclability.
- Environmental impact.

For PEV application, some ESSs features are desired, such as:

- Specific power.
- Storage capacity.
- Specific energy.
- Response time.
- Efficiency.
- Self-discharge rate/charging cycles.
- Sensitivity to heat.
- Charge-discharge rate lifetime.
- Environmental effects.
- Capital/operating cost and maintenance.

Many of the technical advantages and limitations of batteries, fuel cells and capacitors are rooted in the fundamental electrochemical mechanisms they employ. It is unlikely that the performance of these technologies will ever converge and so their suitability for particular applications will always remain different [6].

Upfront cost is a key factor for public acceptance and untakes, and is proving to be one of the steepest challenges for electrochemical energy systems. Energy density is often cited as the largest problem for electrochemical storage devices.

The specific energy and energy density of batteries and capacitors are unlikely to ever compete with liquid hydrocarbons. This is, however, an unfair and misleading comparison. While the fuel in a conventional vehicle weighs relatively little by itself, plenty of additional equipment is required to convert it into motion. In contrast, the batteries of a PEV comprise the majority of the powertrain mass. A second issue is that of conversion efficiency. The value of hydrocarbons, hydrogen and electrons lies in how much of their energy can be converted into a useful form of energy [6].

Within the last few decades the automobile industry has undergone a revolution in overall vehicle reliability. With the large number of components and the number of potential failures, it has become necessary for ESSs to provide a high reliability over its operational life, typically assumed to be 10 years of 240,000 km, in order to satisfy customer expectations [7].

Cyclic State of Charge (SoC) usage has historically been a dominating battery failure mode for batteries in heavy-duty applications. Recently, some factors have tended to increase the cyclic wear rate of batteries: the increasing number of on-board electronic control units, and the power consumption of comfort devices not matched by the alternator output. In both cases, the discharge/charge cycles are typically very shallow, but the accumulated Ah turnover with time may be significant. The ESS performance on partial SoC operation is other point of concern [7].

1.2.1.1 Batteries

A battery is an electrochemical cell, or Galvanic cell, that transforms chemical energy into electrical energy. It consists of an anode and a cathode, separated by an electrolyte, which is an ionic conductor and also an insulating medium. Electrons are generated at the anode and flow towards the cathode through the external circuit while, at the same time, electroneutrality is ensured by ion transport across the electrolyte. The two main types of battery used in PEVs are nickel metal-hydride (NiMH) and lithium-ion (Li-ion) batteries [6].

NiMH batteries are usually used in hybrid PEVs. They use an alkaline solution as the electrolyte. The batteries are composed of nickel hydroxide on the positive electrode, and the negative electrode consists of an engineered alloy of vanadium, titanium, nickel and other metals. The components of the NiMH battery are harmless to the environment and the batteries can be recycled. NiMH batteries offer significantly higher-cycle life and energy density, which is twice that of the lead-acid battery [8]. The components of NiMH batteries include an anode of hydrogen absorbing alloys (MH), a cathode of nickel hydroxide ($Ni(OH)_2$) and a potassium hydroxide (KOH) electrolyte. The general electrochemical reactions are as follows [6]:

$$\begin{array}{l} M + e^- + H_2O \rightarrow MH + OH^- \\ \underline{Ni(OH)_2 + OH^- \rightarrow NiO(OH) + H_2O + e^-} \\ M + Ni(OH)_2 \leftrightarrow MH + NiO(OH) \end{array} \tag{1.1}$$

The major advantage from a manufacturing point of view is the safety of this type of battery at high voltages compared to other kinds, as Li-ion. NiMH batteries are preferred in industrial and consumer applications due to their design flexibility (from 30 mAh to 250 Ah), environmental acceptability, low maintenance, high power and energy densities, cost, safety, long life cycle, storing volumetric energy and power, wide operation temperature ranges and resistance to overcharge and discharge. Their current cost is at $250–$1,500/kWh [6, 8].

On the other hand, if repeatedly discharged at high load currents, the life of NiMH is reduced to about 200–300 cycles. The best operation performance is achieved when discharged 20–50 % of the rated capacity. Other technological issues are primarily their limitations at extreme temperatures and memory effect [7, 8].

Lithium-ion batteries are light, compact and operate with a cell voltage of approximately 4 V with a specific energy in the range of 100–180 Wh/kg. In these type of batteries both the graphite anode and lithium metal oxide cathode are materials into which, and from which, lithium ions migrate through the electrolyte in an organic solvent, then is inserted or extracted into the electrodes [6].

Thus, when a lithium ion battery is discharging, Li is extracted from the anode and inserted into the cathode and when it is charging, the reverse process occurs according to the following reactions:

$$\begin{aligned} &LiCoO_2 \rightarrow Li_{1-x}CoO_2 + xLi + xe^- \\ &xLi^+ + xe^- + 6C \rightarrow Li_xC_6 \\ \hline &xLi^+ + xe^- + LiCoO_2 \rightarrow Li_2O + CoO \end{aligned} \tag{1.2}$$

Li-ion batteries store more energy and have lower memory effect than NiMH; however they suffer from major issues as cost, approximately $1,000/kWh, wide operational temperature ranges, materials availability, environmental impact and safety. It is often observed that these batteries suffer from electrolyte decomposition leading to the formulation of oxide films on the anode, thus blocking extraction sites of lithium, and severe oxidative processes at the cathode due to overcharging, in turn causing dissolution of protective films on the cathode and excess and continuous oxidation of the electrolyte.

Li-ion batteries manufacturing process has progressed much since their introduction to the market in the early 1990s. In fact, manufacturing has scaled so well that the cost of standardized cells fell to 1/9 of the initial value [9]. Breakthroughs in battery technology are urgently required, with innovative, performing and durable material chemistries for both the electrodes and the electrolyte sub-components. The major objective is to identify materials exhibiting higher performance and durability than those currently offered.

The calendar life of Li-ion batteries is still a problem, as the rate capacity loss has not been improved. Lifetimes of these batteries are in the order of 2,000 cycles to 80 % depth of discharge (DoD) before 20 % of power is lost. The number of cycles is approximately reciprocal with the DoD, meaning that approximately 4,000 cycles to 40 % DoD can be expected.

However, differently from most previous measurements indicate, the Li-ion battery capacity does not decrease only as a result of cycling and DoD. Some investigations are conducted on [10], and they show that the cycle life versus DoD/cycle feature is different from the data given from manufacturers. Some results show that deep discharges of 95 % do not affect severely the battery cycle life. At the same time, shallower DoD values do not appear to increase cycle life significantly. These results suggest that a greater portion of the cell capacity could be used during each cycle and the DoD does not have a great effect on lifetime.

Moreover, the results from [10] contradict the literature claiming that there is little or no relationship between DoD and capacity fade and the electrolyte decomposition occurs at the same rate regardless of SoC and degree of graphitic lithiation. More investigation is necessary to determine the most important parameters for battery degradation.

At present Li-ion batteries are expensive but it is anticipated that the price will decline rapidly and that they will be the cheapest rechargeable batteries in 10 years. Scarcity of lithium was once thought of as a looming concern for the electrification of PEV fleet. However, it should be noted that only around 1 % of a Li-ion battery is Li by weight, implying around 0.08 kg Li/kWh of storage capacity. There also is the possibility to recycle the batteries in the future.

The lithium/air battery is attractive because lithium has the highest theoretical voltage and electrochemical equivalence of any metal anode considered for a practical battery system. Lithium metal, atmospheric oxygen, and water are consumed during the discharge, and excess LiOH is generated. The cell discharge reaction is depicted below.

$$2Li + H_2O + {}^1\!/_2O \rightarrow 2LiOH \tag{1.3}$$

The cell can operate at high coulombic efficiencies because of the formation of a protective film on the metal that retards fast corrosion after formation. On open-circuit and low-drain discharge, the self-discharge of the lithium metal is rapid due to the parasitic corrosion reaction. This reaction degrades the anode coulombic efficiency and must be controlled if the full potential of the lithium anode is to be realized.

The main advantage of the lithium/air battery is its higher cell voltage, which translates into higher power and specific energy. However, in view of their availability, cost and safety advantages, the development of metal/air batteries has concentrated primarily on zinc and aluminum [9].

In Lead-Acid Batteries, the spongy lead works as the negative active material of the battery, lead oxide is the positive active material and diluted sulfuric acid is the electrolyte. For discharging, both positive and negative materials are transformed into lead sulfate. As the cell discharges, both electrodes are converted into lead sulfate. The process reverses on charge as depicted on (1.4).

$$\begin{array}{l} Pb \leftrightarrow Pb^{2+} + 2e \\ Pb^{2+} + SO_4^{2-} \leftrightarrow PbSO_4 \\ PbO_2 + 4H^+ + 2e \leftrightarrow Pb^{2+}2H_2O \\ Pb^{2+} + SO_4^{2-} \leftrightarrow PbSO \\ \hline Pb + PbO_2 + 2H_2SO_4 \leftrightarrow 2PbSO_4 + 2H_2O \end{array} \tag{1.4}$$

As advantages for PEV applications, the lead-acid batteries have good high-rate performance, moderate performance for a wide range of temperature, high cell voltage, easy SoC indication, good performance for intermittent charge applications. They are available in production volumes today, yielding a comparatively low cost power source. In addition, the lead-acid battery technology is a mature technology [8, 9].

However, the lead-acid battery is not suitable for discharges over 20 % of its rated capacity. When operated at a deep rate of SoC, the battery would have a limited life cycle. The energy and power density of the battery is low due to the weight of lead collectors [8]. Valve-regulated lead acid batteries have been shown to withstand Ah turnover at least three times higher than conventional batteries. Further improvement can be expected for high-rate partial SoC operation [7].

Due to their unrivalled low cost, improved flooded batteries will continue serving as the primary energy storage system for automotive applications where the charge turnover is not critical. These advanced aqueous-electrolyte battery systems have the advantage of operating close to ambient temperature. Nevertheless, complex system design and circulation of electrolyte are needed to meet performance objectives. In order to meet the vehicular application requirements, these batteries would need significant improvements in shallow-cycle life and dynamic charge acceptance. Furthermore, the inherent tendency to build up acid stratification needs to be addressed because it aggravates sulfation during partial SoC operation as well as cyclic wear [7, 9].

Regarding the flow batteries, the zinc-bromine is an attractive technology for both utility-energy storage and vehicular applications. The concept of a battery based on zinc-bromine couple is old, but development to a commercial battery was not possible due to the tendency of zinc to form dendrites upon deposition and the high solubility of bromine in the aqueous zinc bromine electrolyte. Dendritic zinc deposits could easily short-circuit the cell, and the high solubility of bromine allows diffusion and direct reaction with the zinc electrode, resulting in self-discharge of the cell.

The major advantages of the zinc-bromine battery are the good specific energy and efficiency, low-cost and readily available materials, low environmental impact materials, recyclability, ambient temperature operation, adequate power density, and possibility of deep discharge and fast recharge. By the other side, the main drawbacks are the high self-discharge rate when shut down while being charged, safety and the need of cooling and temperature control system. The electrochemical reactions in a zinc-bromine battery can be simply represented as follows [9].

$$\begin{array}{l} Zn^{2+} + 2e \leftrightarrow Zn^0 \\ 2Br^- \leftrightarrow Br_2(aq) + 2e \\ \hline ZnBr_2(aq) \leftrightarrow Zn^0 + Br(aq) \end{array} \tag{1.5}$$

The Nickel-Zinc (Ni-Zn) battery is an alkaline rechargeable system. It is a combination of the nickel electrode, as used in other batteries and the zinc electrode. Currently, the Ni-Zn system is capable of delivering about 50–60 Wh/kg. They have high energy and power density, low-cost materials, and deep cycle capability (500 cycles at 100 % DoD) and are environmentally friendly. The operation of Ni-Zn batteries ranges from −10 to 50 °C, which means they can be used under severe working circumstances. However, they suffer from poor life cycles due to the fast growth of dendrites, which prevents the development of commercial Ni-Zn batteries for vehicular applications [8, 9].

The electrochemistry on zinc in alkaline solution is quite complex than other battery technologies, so the reactions presented here are for illustration purpose only of discharge (1.6), charge (1.7) and overcharge (1.8) modes [9].

$$\begin{array}{l} 2NiOOH + 2H_2O + 2e^- \rightarrow 2Ni(OH)_2 + 2OH^- \\ Zn + 2OH^- \rightarrow Zn(OH)_2 + 2e^- \\ \hline 2NiOOH + 2H_2O + Zn \rightarrow 2Ni(OH)_2 + Zn(OH)_2 \end{array} \tag{1.6}$$

$$2Ni(OH)_2 + Zn(OH)_2 \rightarrow 2NiOOH + 2H_2O + Zn \tag{1.7}$$

$$\begin{array}{l} 2OH \rightarrow {}^1\!/_2O_2 + H_2O + 2e^- \\ 2H_2O + 2e^- \rightarrow H_2 + 2OH^- \\ Zn + {}^1\!/_2O_2 \rightarrow ZnO \\ \hline H_2O \rightarrow H_2 + {}^1\!/_2O_2 \end{array} \tag{1.8}$$

Nickel-Cadmium (Ni-Cd) batteries are being considered for use in electric buses. They offer a good power density, maintenance-free operation over a wide temperature range, long cycle life, a relatively acceptable self-discharge rate, and can be fully discharged without damage. Another advantage is their capability for fast charging. The specific energy of Ni-Cd batteries is around 55 Wh/kg. These batteries can be recycled, but cadmium is a kind of heavy material that could cause pollution if not properly disposed of. Another important drawback is the high cost, as most nickel batteries.

Their longer cycle life may offset some of this cost on a life cycle basis. New electrode developments such as plastic-bonded and nickel foam electrodes promise to improve performance and reduce costs. Some kinds of Ni-Cd batteries are the vented pocket-plate, the sintered-plate and the portable sealed. The pocket-plate battery can stand both severe mechanical and electrical operational conditions, as overcharging, reversal and short-circuiting. The sintered plate Ni-Cd battery can be constructed in a much thinner form and has 50 % higher energy density than the vented pocket-plate model and the cell has a much lower internal resistance and superior high-rate and low-temperature performance. Sealed batteries incorporate specific design features to prevent a buildup of pressure in the device caused by gassing during overcharge. The overall reaction for Ni-Cd batteries is depicted in (1.9) [8, 9].

$$\begin{array}{l} Ni(OH)_2 + OH^- \rightarrow NiOOH + H_2O + e^- \\ Cd(OH)_2 + 2e^- \rightarrow Cd + 2OH \\ \hline Cd + 2NiOOH + 2H_2O \leftrightarrow Cd(OH)_2 + 2Ni(OH)_2 \end{array} \tag{1.9}$$

1.2.1.2 Fuel Cells

The fuel cell was invented by Sir William Grove, in 1839. It is an electrochemical device operating at various temperatures (up to 1,000 °C) that transforms the chemical energy of a fuel (hydrogen, methanol, natural gas etc.) and an oxidant in the presence of a catalyst into water, heat and electricity. Furthermore, the power generated by a fuel cell depends largely upon the catalytic electrodes and materials used. It is considered to be an efficient and non-polluting power source offering much higher energy densities and energy efficiencies than any other ESSs. Fuel cells are considered to be promising energy devices for transport, mobile and stationary sectors.

The essential difference between a fuel cell and a battery is the manner for supplying the source of energy. In a fuel cell, the fuel and the oxidant are supplied continuously from an external source when power is desired. The fuel cell can produce electrical energy as long as the active materials are fed to the electrodes. In a battery, the fuel and oxidant (except for metal/air batteries) are an integral part of the device. The battery must be replaced or recharged [9]. Fuel cell technology can be classified into two categories [9]:

- Direct systems where fuels, such as hydrogen, methanol and hydrazine, can react directly in the fuel cell.
- Indirect systems in which the fuel, such as natural gas or other fossil fuel, is first converted by reforming to a hydrogen-rich gas which is then fed into the fuel cell.

The heart of a fuel cell consists of a non-conductive electrolyte material sandwiched between two electrodes—the anode and the cathode. The fuel and the oxidant are fed continuously to the anode and the cathode sides, respectively. At the anode side the fuel is decomposed into ions and electrons. The insulator electrolyte material allows only ions to flow from both the anode to the cathode side through an external electrical circuit. The recombination of the ions with the oxidant occurs at the cathode to form water. The electrodes materials of the fuel cell are inert in that they are not consumed during the cell reaction, but have catalytic properties which enhance the electroreduction or electrooxidation of the reactants [6, 9].

Fuel cell systems can take a number of configurations depending on the combinations of fuel and oxidant, the type of electrolyte, the temperature of operation, and the application. There are currently six main groups of fuel cells available: proton exchange membrane fuel cell (PEMFC); alkaline fuel cell (AFC); phosphoric acid fuel cell (PAFC); molten carbonate fuel cell (MCFC); solid oxide fuel cell (SOFC); and microbial fuel cell (MFC) [11].

Regarding the fuels, the hydrogen seems to be the most promising. It has the highest energy content by weight, but very low energy content by volume. This makes storage and distribution to the point of use costly. However, the problem with low volumetric energy density can be decreased by storing the hydrogen either under increased pressure, at extremely low temperatures as a liquid or in metal-hydride systems. Hydrogen storage on board the vehicle is the key factor for achieving market success. The methods of hydrogen storage in vehicle are:

- Liquid hydrogen, which demands very low temperatures (−253 °C in ambient pressure) and causes a loss of chemical energy.
- Compressed hydrogen, which also consumes energy in compression step.
- Metal hydride, which is the safest method, but very heavy and require a lot of time to store the hydrogen and has an insufficient release rate.
- Carbon nanotubes and Metal organic frameworks.

There are a few challenges related to hydrogen generation, storage and utilization. These include:

- The design and development of low-cost and efficient hydrogen production system using novel technologies, with a particular emphasis on the production of hydrogen from renewable sources.
- Novel technologies associated with low carbon emission hydrogen production and utilization from fossil fuels and distributed hydrogen technologies.
- The development of novel materials, systems and solutions for hydrogen storage and transportation, with low costs and high energy efficiency.

In recent years, the PEMFC has been extensively demonstrated worldwide in many applications fields and is now on the verge of commercialization. A single cell PEMFC consists of a membrane electrode assembly sandwiched between two flow field plates. Each cell produces approximately 1.1 V, so to obtain the required cell voltage the single cells are combined to produce a PEMFC stack. The membrane is typically made of a proton exchange membrane material designed to conduct protons H^+; anode and cathode electrodes or catalyst layers made of an electrocatalyst supported on carbon and proton exchange materials, and; gas diffusion layers, which allows reactants to diffuse to the active sites on the electrocatalyst and facilitating water management by allowing water vapor to diffuse out, and the liquid water produced on the cathode side to flow out of the fuel cell.

In a PEMFC, the following electrochemical reactions occur at the anode and at the cathode, respectively.

$$\begin{array}{l} H_2 \rightarrow 2H^+ + 2e^- \\ \underline{O_2 + 4H^+ + 4e^- \rightarrow 2H^+ + 2e^-} \\ H_2 + {}^1\!/_2 O_2 \rightarrow H_2O \end{array} \tag{1.10}$$

The oxygen reduction reaction at the cathode is a kinetically slow process, which has a more dominant effect on the performance of the PEMFC than the hydrogen oxygen reaction. Consequently, developing active catalysts is the main focus of research on PEMFC.

As all fuel cells are normally distinguished by the materials used, although the manufacturing of the fuel cell electrodes is also different in each case, the main objective in fuel cell technologies is to develop low-cost, high-performance and durable materials. As well as reducing the cost, the main target in automotive PEMFC is to operate the system above 100 °C with low humidification of reactant hydrogen and air. Large-scale deployment of PEMFCs for the transportation sector demands the development of low-cost and high-performance membrane electrode assemblies.

Currently, fuel cells are too expensive and not durable. Platinum is the key catalytically active component in the most fuel cell types. Although its degradation process is well understood, there are a few solutions under investigation to prevent this effect. Platinum price is also high and contributes with one third of a stack cost. However, from an electrochemical point of view, platinum is still the best electrocatalyst for PEMFC as it is stable, durable and very active towards the electrochemical reactions. Although promising, surrogate technologies, as nanomaterials, have shown poor durability and stability in acidic and aggressive environments and cell voltage cycling under real operating conditions.

In addition to the high material cost, platinum is extremely sensitive to poisoning by CO, H_2S, NH_3, organic sulfur-carbon and carbon-hydrogen compounds in the H_2 stream and NO_X and SO_X in air. It is also prone to dissolution and/or agglomeration resulting in performance degradation. Moreover, platinum is rare and is mined in a limited number of countries, with more than 80 % in South Africa. Thus, the political aspect is also a determining factor for supply [6]. Fuel cell lifetimes are assessed by the number of hours until 10 % rated power is lost.

1.2.1.3 Other Technologies

Electrochemical capacitors are high power density and low energy density devices. There are two energy storage mechanisms for capacitors: (i) electrochemical double layer capacitors i.e., double-layer capacitance arising from the charge separation at the electrode/electrolyte interfaces and (ii), pseudo-capacitors i.e., pseudo-capacitance arising from fast, reversible faradaic reactions occurring at or near the solid electrode surfaces. In all cases, electrochemical capacitors rely upon the separation of chemically charged species at an electrified interface between a solid electrode and an electrolyte. The electrolyte between the anode and the cathode is ionic, usually a salt in an appropriate solvent. The operating cell voltage is controlled by the breakdown voltages of the solvents with aqueous and organic electrolytes [6].

Electrochemical capacitors are currently being proposed by many automotive manufacturers due to their load level demand, quickly inject or absorb power to help minimize voltage fluctuations in the electronic systems, and provide pulse power well over 1,000 W/kg with a cycle life reaching more than 500,000 cycles. In contrast to batteries and fuel cells, the lifetime of electrochemical capacitors is longer and in some cases their energy efficiency rarely falls below 90 % provided they are kept within their design limits. Finally, it is speculated that the next generations of capacitors are expected to come close to Li-ion battery technology in energy density while maintaining their high power density [6]. The prospects for capacitors are excellent regarding to durability and degradation, with calendar times measured in decades and charge/discharge cycles in millions.

1.2.2 Electric Motors

An electric motor drive forms the central core of a PEV. In research and development works worldwide, significant consideration is given to the use of high-performance and high-efficiency electric motors at high speed in the range of 12,000–15,000 rpm [12]. The electric motor in its normal mode can provide constant rated torque up to its base or rated speed. At this speed, the motor reaches its rated power limit. The operation beyond the base speed up to the maximum speed is limited to a constant power region. The range of the constant power operation depends primarily on the particular motor type and its control strategy [13].

In an industrial point of view, the major categories of electric motor drives adopted or under consideration for PEVs are: permanent magnet motor (PM) drives, which could be AC or DC, induction motor (IM) drives, the DC motor, and the switched reluctance motor [12–14].

Choosing a motor for PEV traction is challenging due to the following reasons and it should be emphasized that the motor design is a part of the system tradeoff [12, 14, 15]:

- Wide speed range, including constant-torque and constant-power regions.
- Adequate torque capability for passing and overtaking at high speed, and high torque at low speeds for starting and climbing.
- A high instant power and a high power density.
- Limited DC voltage level since battery favors low voltage.
- Very limited packaging space.
- Highest possible efficiency for travel range.
- Lowest possible cost.
- Reliability and robustness for various vehicle-operating conditions.
- Driver expectations.
- Energy Source.
- Market acceptance degree of each motor type.
- High controllability, steady-state accuracy and good transient performance.

In the context of PEV design, the motors can be classified according to their maximum speed. Low speed motors have a maximum speed of 6,000 rpm. From 6,000–10,000 rpm, the motors can be classified as medium speed. And, above 10,000 rpm the motor drives are classified as high speed motors. The maximum motor speed affects the gear size and has an effect on the rated torque of the motor, but it does not affect the power requirement. Low speed motors with extended constant power speed range have a much higher rated shaft torque. Consequently, they need more iron to support these currents. This will also impact the power converter silicon size and conduction losses. Extended speed range, however, is necessary for initial acceleration as well as for cruising intervals of operation. Therefore, the rated motor shaft torque can only be reduced through picking a high speed motor, but this affects the gear ratio. A good design is the result of a tradeoff between maximum motor speed and the gear size [13].

In addition to the regular central traction motors, numerous wheel-hub motor concepts have been developed, but no commercial vehicles have been developed using these concepts. One of the main technical challenges of wheel motors is the unsprung mass [16]. In this section, the main motor drives technologies and their features are presented.

1.2.2.1 Permanent Magnet Motor Drives

The Permanent Magnet (PM) machine drives include the DC motor drives, the synchronous motor and the brushless DC motor, which is a type of electronically commutated PM synchronous machine. The PM DC motor usually needs two or three gear ratios due to their limited speed range, heavy weight and large size, while the PM synchronous machine and brushless PM DC motors provide a good performance over 12,000 rpm without any need of gear ratios due to their magnetic materials [12].

The PM motors have several distinct advantages, namely, high efficiency, high power factor and relatively higher stability. Smooth brushless operation and simple rotor construction of PM synchronous motors offer additional advantages, particularly for high-speed applications. Furthermore, a high number of pole pairs reduce the weight and material content. Absence of rotor cooper losses at synchronous speed makes it highly efficient. The stator and the drive electronics are similar to those for the IM drives. Most of the ac motors for PEVs are liquid-cooled, as this reduces size and weight. It also keeps the magnet temperature down, which is beneficial to the magnetic material [12].

However, there are also a few disadvantages. PM motor drives do not have self-starting torque. If cage windings are included, they can be operated at asynchronous mode during starting. The cost of magnetic materials is quite high. Magnetic corrosion and relatively low temperature tolerance are potential hazards and fixed flux level gives low speed range at constant power [12].

The PM brushless DC motor is specifically known for its high efficiency, high power density due to reduced overall weight and volume for a given output power and efficient heat dissipation. The high power factor of the PM brushless DC motor also reduces the volt-ampere rating of the converter. However, this kind of motor drive suffers from a rather limited field weakening capability. This is due to the presence of the PM field which can only be weakened through production of a stator field component which opposes the rotor magnetic field. Nevertheless, extended constant power operation is possible through the advancing of the commutation angle [13, 14].

There are several combinations of PM brushless motors. Depending on the arrangement of the PM, basically, they can be classified as surface-magnet mounted or buried-magnet mounted, with the latter being the more rugged. PM hybrid motors offer a wide speed range and a higher overall efficiency [14].

By observing the major vehicle manufacturers, it is possible to perceive that almost the entire light-duty PEV industry has shifted from IM to PM motor drives in order to meet the increasing power density and efficiency requirements. In the future, it is expected that PM machines start to extend into medium and heavy-duty markets. In the long run, there is a lot of concern about the prices and availability of rare PM materials and the development of other materials are very important [16].

1.2.2.2 Induction Motor Drives

AC induction motors are preferred over DC motors for their high reliability and maintenance-free operation. IM drives are standard technology with existing manufacturing infrastructure. Cage IMs are potential candidates for electric propulsion of PEVs, owing to their reliability, ruggedness, low maintenance, low cost, and ability to operate in hostile environments [14, 15]. Differently from the light-duty fleet, for medium and heavy-duty vehicles the IM motor drives are still the main preference [16].

However, when frequency changes, the motor impedance will also change and the variation of stator impedances will lead to the change of air gap flux and, thus, the output torque. The nonlinearity of its dynamic model with coupling between the direct and quadrature axes also makes the control more complex. Moreover, IM motors have an inherent disadvantage of slip-dependent rotor copper loss, and there is the problem of heat extraction from the machine core. The motors are either two or four-pole types, with large end windings and considerable back iron yoke, causing their weight to increase [12].

The presence of a breakdown torque limits its extended constant-power operation. At critical speed, the breakdown torque is reached. Generally, for a conventional IM, the critical speed is around two times the synchronous one. Any attempt to operate the motor at the maximum current beyond this speed will stall the motor. Moreover, efficiency at high-speed range may suffer in addition to the fact that IMs efficiency is inherently lower than of PM motors, due to the absence of rotor winding and rotor copper losses [14].

Field orientation control of IM can decouple its torque control from field control. This allows the motor to behave in the same manner as a separately excited DC motor. This motor, however, does not suffer from the same speed limitations as the other. Extended speed range operation is accomplished by flux weakening, once the motor has reached its rated power capability, but it is limited by the presence of a breakdown torque. It may be mentioned here that the torque control in induction motor is achieved by PWM control of current. In order to retain the current control capability in the extended speed constant power range, the motor is required to enter the field weakening range before reaching the base speed, so that it has adequate voltage margin to control the current [13].

In general, IM motor drives are facing a number of drawbacks that pushed them out of the race of PEVs propulsion. These drawbacks are mainly high loss, low efficiency, low power factor and low inverter-usage factor, which are more serious for the high speed, large power motor. These problems should be taken into account in the design step of IM motor drives for PEVs [14].

1.2.2.3 DC Motors Drive

Traditional DC commutated motors have been prominent for PEV propulsion systems because their torque-speed characteristics suit the traction requirement. This kind of motor drive is inherently suited for field weakened operation, due to its decoupled torque and flux control features. Extended constant power operation is possible with this motor through its separate field weakening. Control of these motors is very simple. However, the presence of mechanical commutators imposes a severe restriction on the maximum speed of the DC motors. This low speed and extended constant power operation would necessitate higher motor shaft torque. Consequently, more iron is needed, which causes the motor to be bulky, heavy and expensive. It also requires considerable maintenance of its brushes and commutators. It is also restricted by the sparks, which come from the mechanical commutator system. These problems make them less reliable and unsuitable for maintenance-free operation [12–14].

Moreover, the development of rugged solid-state power semiconductors made it increasingly practical to introduce the IM and PM synchronous motor drives that are mature to replace the DC motor drive in traction applications. In fact, the commutatorless motors are attractive, as high reliability and maintenance-free operation are prime considerations for electric propulsion. Nevertheless, with regard to the cost of the inverter, AC drives are used generally just for higher power. At low power ratings, the DC motor is still more than an alternative [14].

1.2.2.4 Switched Reluctance Motor Drive

Due to the simple construction, the absence of rotor conductors, low inertia, fault-tolerant operation, simple control and outstanding torque-speed characteristics, the switched reluctance motor (SRM) is a potential motor drive for PEV applications. In a SRM, the phase windings on the stator in turn set up a magnetic dipole between stator and rotor poles. The resulting tendency is to reduce the air-gap reluctance. It results in that the rotor poles moves toward an aligned position with the excited stator pole. This operational feature is much different than the electromechanical energy conversion that takes place in other types of motor drives.

The SRM can reach extremely high speeds with a long constant power range. Operation in constant power is made possible in this motor by the phase advancing of stator current conduction angle until overlapping between the successive phases occurs. The SRM drive has a high starting torque and high torque-inertia ratio. The absence of magnetic sources makes it relatively easy to cool and insensitive to high temperatures, which are desired features for vehicular applications. Furthermore, it typically has a low-cost construction due to the absence of windings and permanent magnets on the rotor structure.

As drawbacks, the SRM motor has a low power factor which can penalize the converter, torque ripple, excessive bus ripple, acoustic noise and electromagnetic-interference noise generation. But, the existence of a long tail of natural mode of operation, beyond the constant power range, can offset some of this disadvantage by making the motor power rating smaller [13, 14].

1.2.2.5 Other Motor Drive Types

The hysteresis motor (HM) has not only simple constructional features with conventional polyphase stator windings and a solid rotor hysteresis ring, but also high built-in self-starting torque during the run-up and synchronization period. It has no rotor slots and, thus, it has a low noise during operation. These advantages make the hysteresis motor especially suitable for applications in which constant torque, constant speed, and quiet operation are required. The HM motor also has high magnetizing current and high parasitic losses. When it reaches synchronous speed, the rotor flux ceases to sweep around it and the residual flux density of hysteresis material on the rotor is relatively fixed. As the eddy current torque disappears at synchronous speed, the motor behaves as a temporary PM motor [12].

The combination of permanent magnet and hysteresis materials in the rotor constitutes a hybrid design motor drive with many advantages over conventional PM or hysteresis motors. This hybrid motor in which the permanent magnets are inserted into the slots at the inner surface of the hysteresis ring is called the hybrid permanent magnet hysteresis synchronous (HPMHS) motor. During asynchronous speed, the motor torque consists of the hysteresis torque, eddy current torque and permanent magnet brake torque. While in synchronous speed, the motor torque is comprised of the hysteresis and permanent magnet torques. So, it combines the advantageous of both machine drives. The negative effect of the magnet brake torque of a conventional PM motor is ideally compensated by the high eddy current and hysteresis torque, particularly at the initial run-up period [12].

The choice of the motor drive for PEVs is mainly determined by three factors. Reference [12] provides a detailed description of this. The main factors are weight, efficiency and cost, so the PEV motor selection may carry out taking into consideration technical and economic aspects.

1.2.3 Power Electronic Converters

The power switching devices, electric motors, and associated control systems and components play a key role in bringing PEVs to market with reliability and affordability. The power electronic system should be efficient to improve the range and efficiency of PEVs. The selection of power semiconductor devices, converters/ inverters, control and switching strategies, packaging and system integration is crucial to the development of efficient and high performance vehicles [17].

To meet the requirements of the automotive environment, several technical challenges need to be overcome, and new developments are necessary from the device level to the system level, as enumerated below.

- The development of a power device that combines the MOS gate control characteristics of thyristor-type structure whose forward voltage drop, even at higher currents above 400 A, must be less than 2 V and, at the same time, can be operated at switching frequencies higher than 10 kHz.
- The design of a new power diode with superior dynamic characteristics, such as MOS-controlled diode.
- The research on silicon carbide needs to be accelerated to make possible their application to high-power switching devices at higher operating temperatures.
- The devices and the rest of the components need to withstand thermal cycling and extreme vibrations.

The technologies related to device packaging need to be investigated by the semiconductor industry to develop a power switch for vehicular applications. Wire bonding, device interconnections, etc., are the barriers to the development of high-current-density power units. The power electronic systems available in the market are still bulky and difficult to package for automotive applications [17].

Although the technology of power semiconductor devices has advanced, it still needs to be improved. The cooling method needs to be adequate to quickly take away the heat from the devices. In addition, the impact of current intensiveness in a system on lower efficiency, larger passive components such as inductors and capacitors, and a thicker wiring harness among the components should be properly taken into consideration at the design stage [17].

However soft-inverters have the advantage of lower switching losses and low electromagnetic interference, they need more components, higher operating voltage devices, and more complicated control compared to hard-switched inverters. There is a need to develop an inverter topology that achieves the performance of a soft-switched inverter but with less components and simplified control. Fault-tolerant and control techniques also need further investigation.

The technology development effort needs to be focused on the sensorless operation of electric machine drives and the reduction or elimination of current sensors on inverters. The development of low-cost high temperature magnets would lead to the widespread use of PM motors, which have higher efficiency and need lower current to obtain the same torque as other machines [17].

Considering the technologies described above create a new scenario for electric systems of the future. Some of the implications are discussed next.

1.3 New Power System Operation Philosophies

The advent of renewable sources brings along a quite important discussion about the system reliability and cooperation. These issues are important because reliability is increased as renewable sources are connected to the grid. On the other hand, the intermittent nature of these sources requires the grid to be able to supply the load in adverse weather conditions, in case of a huge penetration of wind and solar generation.

The supply of sustainable energy is one of the greatest challenges of our modern society. Governments, universities and industries must cooperate intensively to develop sustainable energy sources that meet future requirements.

A great advantage of electric energy supply is that it can be transported easily over large distances. It is widely recognized that the development and integration of sustainable sources also requires an innovation of the electrical grid and associated technologies for generation, transmission, distribution and energy storage systems. The present electrical system is based on a model of large and centralized electricity generators (large-scale plants based on fossil fuels) whereas the future electrical system will be based on a large amount of smaller, local generators (solar panels, wind turbines). The changes in renewable energy generation will induce large changes in the management and distribution of electrical energy due to their unpredictability. The present electrical system has to be made smarter in order to accommodate and balance demand and supply on local, regional, national, and transnational level.

The long-term global prospects continue to improve for generation from renewable energy sources making the fastest-growing sources of electricity with annual increases averaging 2.8 % per year from 2010 to 2040. In particular, non-hydropower renewable resources are the fastest-growing sources of new generation in the outlook, in both OECD and non-OECD regions. Non-hydropower intermittent renewables, which accounted for 4 percent of the generation market in 2010, could increase their share of the market to 9 % in 2040 [18].

Sustainable energy systems need innovation in three basic areas:

(a) development of reliable sustainable energy sources;
(b) development of smart grids to accommodate production and consumption of energy under market signal incentives;
(c) development of models to understand the non-technological aspects of the production and consumption of energy, e.g., social and ethical questions. In addition, these non-technological aspects have to be integrated in the design of sustainable sources and smart grids.

Creating this environment demands a huge effort from system designers, which includes a different philosophical point of view. A model to deal with this problem has been given in [19], as shown in Fig. 1.1. This model states that designers have to use three different perspectives to specify new technologies: integrality, inclusiveness and ideal-drivenness. The idea of integrality refers to the different aspects that have to be taken into account, the idea of inclusiveness to the different stakeholders whose interests are at issue, and the idea of ideal-drivenness to the ideals, value systems, or basic believes that underlay the development of smart grids. This model is based on the ontology as developed by the philosopher Dooyeweerd [20] and the practice model developed by the philosophers Hoogland, Jochemsen, Glas, Verkerk and others [21].

The first 'I' refers to the different aspects that have to be analyzed. In total fifteen different aspects are identified, varying from the numerical, physical, social, economic, juridical to the moral dimension, see Table 1.1. Each aspect has its own

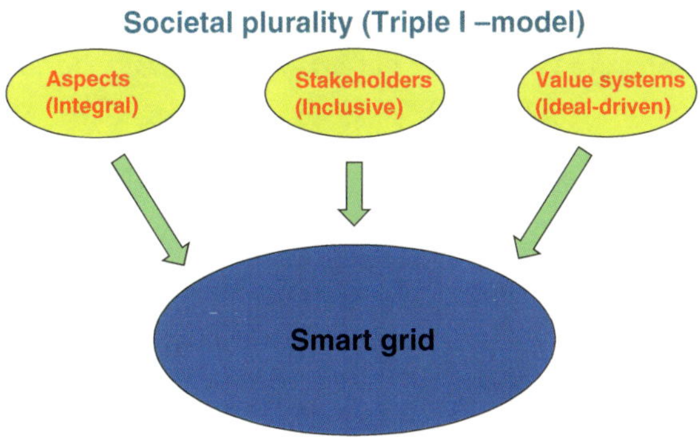

Fig. 1.1 General model proposed

Table 1.1 Overview of different aspects of design

Aspects	Sustainable energy generation and smart grids
Arithmetic	Measurable quantities like voltage, current and power
Spatial	Spatial location of power generation plants, overground or underground transmission and distribution networks
Kinematic	Rotating generators, energy, flow, water flow (hydro energy)
Physical	Different physical laws that determine the generation and transport of electric energy
Biotic	Influence of energy generation and electromagnetic fields on life (plants, animal, human)
Physic	Influence of energy generation and electromagnetic fields on the emotions of human depression, feelings of uncertainty and so on
Analytical	Distinction between different types of sustainable energy generation and distinction between different types of grid
Formative	Control of power generation and consumption, influence that citizens has (on the use of) smart grids in their home
Lingual	What meaning do citizens give to words like "smart grids"? Threatening? Promising development? How to name sustainable energy to promote the use of these sources?
Social	The influence of micro grids on the social behavior of citizens. How "to seduce" customers to adapt their activities to balance supply and demand (nudging)?
Economic	Price differentiation depending on momentary supply and demand
Esthetic	Esthetic design of wind mills and solar panels. Integration of sustainable energy sources in architectural designs
Judicial	Liability for safety of micro, smart and super grids
Moral	How to design micro and smart grids so that they really support and care about human life? How to prevent that human life is harmed?
Pistic	Do people trust smart meters? Or are they afraid that their privacy will be threatened? Do people trust smart systems in their house? Do they welcome these technologies? How to design for trust?

nature, dynamics and normativity. Consequently, these different aspects cannot be reduced to each other but every aspect has to be analyzed in detail.

The second 'I' refers to the different stakeholders and their justified interests. Based on a philosophical approach, Ref. [19] argues that the interests of stakeholders are different. For example, this comes to the fore when we analyze how different stakeholders will cope with breakdown of widespread blackout of the electrical system. Industrial enterprises will balance the risks, potential losses and prevention costs on economic grounds, hospitals will always choose for back-up installations to prevent harming patients, and citizens will accept the risks as long as their normal life is not hampered strongly. So, 'inclusiveness' requires the analyses of the interests of all different stakeholders. In this analysis the lists of aspects will be very helpful.

The third 'I' refers to the ideals, values and basic beliefs that underlay the search for sustainable sources and the design of the energy system of the future. It has to be noted that in Western culture different value systems are present. Some people believe that economic considerations have to be dominant (neoliberal approach), others believe that the present system can be adapted to meet environmental and sustainability requirements ('swallow ecology'), and again other state that do not only need technological innovations but also radical societal reforms ('deep ecology') [22]. It is important to make this third 'I' explicit in order to discuss the 'why' of sustainable energy and smart grids and to prevent that these fundamental questions are suppressed by technological and economical perspectives.

The approach of [19] is summarized in Fig. 1.2. It shows that for every (sub-) technology an extensive analysis of the three I's is required. On the one hand, it is hard to do this kind of analyses. Especially, because this demands from engineers an additional research. On the other hand, failures in this field are so costly that no organization or institution can permit itself big failures.

Combining the existing technologies with a new philosophy of system designing is certainly a great challenge for engineers and scientists. The problems to be overcome are complex and require knowledge not usually taught in the engineering

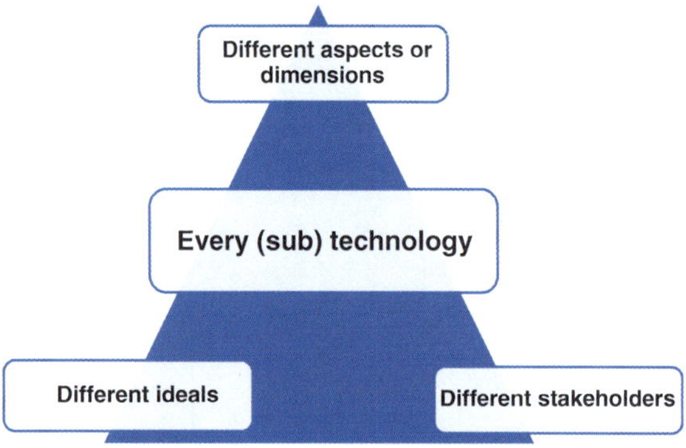

Fig. 1.2 Overview of the proposed approach

curricula. Ignoring this problem, however, may create a system susceptible to failure and ethical problems of operation. Understanding the existing technology and proposing a model to integrate the available devices into a new paradigm of system designer are the challenge to face.

References

1. Clement-Nyns K, Haesen E, Driesen J (2010) The impact of charging plug-in hybrid electric vehicles on a residential distribution grid. IEEE Trans Power Syst 25:371–380
2. Sortomme E, Hindi MM, James MacPherson SD, Venkata SS (2011) Coordinated charging of plug-in hybrid electric vehicles to minimize distribution system losses. IEEE Trans Smart Grid 1(1):198–205
3. Deilami S, Masoum AS, Moses PS, Masoum MAS (2011) Real-time coordination of plug-in electric vehicle charging in smart grids to minimize power losses and improve voltage profile. IEEE Trans Smart Grid 3:456–467
4. Oliveira DQ, Zambroni de Souza AC, Delboni LFN (2013) Optimal plug-in hybrid electric vehicles recharge in distribution power systems. Electric Power Syst Res 98:77–85
5. Hoyer KG (2008) The history of alternative fuels in transportation: the case of electric and hybrid cars. Util Policy 16(2):63–71
6. Pollet BG, Staffell I, Shang JL (2012) Current status of hybrid, battery and fuel cell electric vehicles: from electrochemistry to market prospects. Electrochim Acta 84(1):235–249
7. Karden E, Ploumen S, Fricke B, Miller T, Snyder K (2007) Energy storage devices for future hybrid electric vehicles. J Power Sources 168(1):2–11
8. Khaligh A, Li Z (2010) Battery, ultracapacitor, fuel cell, and hybrid energy storage systems for electric, hybrid electric, fuel cell, and plug-in hybrid electric vehicles: state of the art. IEEE Trans Veh Technol 59(6):2806–2814
9. Linden D, Reddy TB (2001) Handbook of batteries, 3rd edn. McGraw-Hill, New York
10. Peterson SB, Apt J, Whitacre JF (2010) Lithium-ion battery degradation resulting from realistic vehicle and vehicle-to-grid utilization. J Power Sources 195(8):2385–2392
11. Sibley JT (2000) Fuel cells: clean, reliable energy for the future from CT companies. Case reports. Bull Conn Acad Sci Eng 15(1):9–12
12. Rahman M Azizur, Qin Ruifeng (1997) A permanent magnet hysteresis hybrid synchronous motor for electric vehicles. IEEE Trans Industr Electron 44(1):46–53
13. Rahman KM, Ehsani M (1996) Performance analysis of electric motor drives for electric and hybrid electric vehicle applications. In: IEEE power electronics in transportation, pp 49–56
14. Zeraoulia M, Benbouzid MEH, Diallo D (2006) Electric motor drive selection issues for HEV propulsion systems: a comparative study. IEEE Trans Veh Technol 55(6):1756–1764
15. Williamson SS, Emadi A, Rajashekara K (2007) comprehensive efficiency modeling of electric traction motor drives for hybrid electric vehicle propulsion applications. IEEE Trans Veh Technol 56(4):1561–1572
16. El-Refaie AM (2013) Motors/generators for traction/propulsion applications: a review. IEEE Veh Technol Mag 8(1):90–99
17. Emadi A, Joo Lee Y, Rajashekara K (2008) Power electronics and motor drives in electric, hybrid electric, and plug-in hybrid electric vehicles. IEEE Trans Industr Electron 55 (6):2237–2245
18. EIA DOE (2009) Annual energy review, US Energy Information Administration on http://www.eia.gov/forecasts/aeo/pdf/0383%282013%29.pdf, seen on 22 July 2014

19. Ribeiro PF, Polinder H, Verkerk MJ (2012) Planning and designing smart grids: philosophical considerations. IEEE Technol Soc Mag 31:34–43
20. Dooyeweerd H (1969) A new critique of theoretical thought. The Presbyterian and Reformed Publishing Company, Phillipsburg
21. Jochemsen H (2006) Normative practices as an intermediate between theoretical ethics and morality. Philosophia Reformata 71:96–112
22. Naess A (1973) The shallow and the deep. long-range ecology movement. A summary. Inquiry 16:95–100

Chapter 2
Smart Coordination Approach for Power Management and Loss Minimization in Distribution Networks with PEV Penetration Based on Real Time Pricing

Bhuvana Ramachandran and Ashley Geng

Abstract The impact of Plug in Electric Vehicles (PEV) will be most significantly felt by the electric power distribution networks, and specifically by distribution transformers that exist on each neighborhood block and cul-de-sac as customers charge their PEVs. That impact is unlikely to be positive. Since PEV adoption is initially expected to cluster in neighborhoods where demand for PEVs is strongest, the new load may overload transformers, sap much-needed distribution capacity and also increase distribution network losses. Hence, the national goal of putting one million PEVs on the road by 2015 could easily impose a severe burden on the distribution network. Whether PEVs will help or hinder electricity provision will depend on how frequently and at what times the customers charge their vehicles. This behavior will be driven in part by the rate structures that are offered by utilities, as well as the price responsiveness of PEV owners to those rate structures. In this chapter, we propose a method to optimally charge the PEVs in order to minimize the system distribution network losses and to maximize energy transferred to PEVs. A novel short term prediction unit consisting of a receding time horizon method is proposed to forecast the PEV load and a multi objective bacterial foraging algorithm is used as an optimization tool. Also it is interesting to study the manner in which distribution network losses vary with PEV charging behavior. Hence the purpose of this chapter is to demonstrate a power management strategy using smart coordination approach to (a) design a charging and discharging infrastructure for the PEVs that maximizes energy delivered to PEV batteries and (b) reduce the distribution network losses to avoid overloading of the grid.

B. Ramachandran (✉) · A. Geng
Department of Electrical and Computer Engineering, University of West Florida, Pensacola, FL, USA
e-mail: bramachandran@uwf.edu

A. Geng
e-mail: xgeng@uwf.edu

S. Rajakaruna et al. (eds.), *Plug In Electric Vehicles in Smart Grids*, Power Systems, DOI 10.1007/978-981-287-302-6_2

Keywords Plug in electric vehicles • Smart charging • Power management • Time of use rates • Distribution network loss minimization

2.1 Introduction

The growing use of electricity increases grid loading, power losses, and the risk of congestion. However, employing electricity for heating and transportation, also introduce a significant level of flexibility to the traditional consumption pattern [1]. Over the past 5 years, transportation sector has been revolutionized due to the advent of Plug-in Electric Vehicles (PEV). The growing societal awareness of environmental issues as well as ongoing concerns about reducing dependence on foreign oil or petroleum have made the concept of PEV very popular during the past few years [2]. Preliminary studies indicate that PEVs will dominate the electricity industry in the near future as pollution-free alternatives to the conventional petroleum based transportation and they will populate residential feeders, especially in USA and Australia. Due to the high penetration levels of PEVs, significant impacts will be felt especially at the distribution level [3–10]. In the absence of proper coordination, it is most likely that these PEVs will charge and discharge during the overall peak load period [3] causing severe branch congestions, unpredictable system peak demands, unaccepted voltage deviations, significant increase in losses and poor power quality. Some studies conducted by authors in [7, 9] have observed that the existing distribution system infrastructure would only support a very low PEV penetration level without grid operation procedure changes or additional grid infrastructure investments. To overcome these problems, several PEV coordination approaches have been suggested in literature [5, 11–17].

The charging and discharging process of PEVs can be controlled so that energy will be transferred from grid to vehicle (G2V) or from vehicle to grid (V2G) respectively. Several PEV coordination techniques based on deterministic and stochastic dynamic programing were discussed [5]. Several other authors have adopted prediction of PEV charging profiles and vehicle range reliability using recorded vehicle usage data and also designs a minimum cost load scheduling algorithm based on the forecasted electricity price and PEV power demands.

Many countries have ventured into smart metering and smart appliances to improve the system load profile and to reduce peak demand so that demand side management (DSM) can be implemented for load control and power management in the electrical grid [18–22]. PEVs can be utilized to provide ancillary services including energy storage and frequency regulation [16, 17, 23]. These added benefits of PEVs enable electric grids to rapidly heal and self-regulate under conditions of emergency thereby improving system security and reliability and efficiently manage energy delivery and consumption [24–30].

Majority of existing strategies on load control and power management treat loads as individual entities, even for loads sharing the same load characteristics. With such an approach, either computational complexity (for centralized schemes) or communicational effort (for decentralized schemes) would grow significantly as the number of loads in the network increases. In this chapter, we consider groups of loads rather than individual loads, by categorizing loads into a relatively small number of load types. With this scheme, the size of the proposed optimization problem does not change as the load population increases, which is a valuable feature for large-scale load management.

To accomplish these objectives, this chapter proposes a novel real time smart coordination approach using a receding time horizon method to coordinate multiple charging and discharging of PEVs while reducing system stresses that can severely impact grid reliability, security and performance [24]. Real time charging control issues were addressed by very few authors such as [12, 31] where it is very challenging to obtain performance guarantees. The proposed PEV charging algorithm developed for smart coordination consists of a forecasting module and an optimization module which will improve power system resource utilization. The forecasting module sends information about the number of PEVs in the parking garage and also their arrival and departure rates.

The module then calculates and forecasts the number of PEVs that will be present at the same time in the parking garage for the next time interval. The heuristic multi-objective optimization module takes in the present and future power demands for all loads including PEV's over a finite time interval. The aim of this optimization module is to maximize the energy delivered to PEV batteries and satisfy the SOC criteria for the PEV while including constraints related to the power grid and customer demands. The optimization module is also designed to minimize distribution network losses considering charging time zone priorities specified by PEV owners.

To validate the power management infrastructure and distribution network loss minimization, the smart coordination strategy is implemented on a IEEE 13 node test feeder and a 38 bus power system consisting of a mix of residential, commercial and industrial customers penetrated with PEVs. To estimate the economics of charging, simulation results will be presented for uncoordinated and coordinated charging scenarios for three different Time of Use (TOU) rates and different PEV penetrations.

2.2 Research on Smart PEV Charging Coordination

Literature review of research carried out in the area of coordination of charging and discharging of PEVs throws light on the two categories of work so far. One category of research was focused on charging and discharging decisions based only on the present information about the state of the grid. The second category is the one which is based on forecasted estimates of the state of the grid and future power demands in the grid are considered while making decisions about charging or discharging. In [12], a real time coordinated PEV charging approach was proposed in which the time varying energy process was accounted for and along with charging time and zone preferred by the PEV owner. A DSM based charge control was proposed in [32] where the objective was to provide dynamically configurable dispersed energy storage during peak demand and outage conditions. An optimal PEV charging model that responds to the time-of-use price in a regulated market is proposed in [33]. In these papers, the impact of present and future PEV charging and discharging decisions on the grid were not considered. This means that the charging of PEVs would not result in a target state of charge level for the PEVs and hence would affect the reliability of the power system.

Several other authors have proposed probabilistic models and charging coordination strategies considering day ahead or real time markets [16, 34, 35]. The optimization model used could be either single objective optimization (to optimize cost or losses) or multi-objective optimization (to optimize operating cost with losses). The ultimate objective of this research is to develop a smart coordinated charging and discharging framework for smart grids based on TOU rates which would improve the system reliability and security.

2.3 Electric Vehicles and Distribution System

If PEV owners were to simultaneously charge their vehicles in a small geographical area, the increased demand would cause severe problems for the utility that must serve the load reliably. If PEV owners were to simultaneously charge their vehicles in a small geographic area, the increased demand caused due to charging could cause major problems for the utility that must reliably serve that area. While simultaneous charging of PEVs at system peak could result in supply shortages or create a need for large new investments in expanding generating capacity and setting up new generation plants, the most serious concern due to simultaneous PEV charging will be the congestion problem at distribution level for most utilities (Fig. 2.1).

First, consider the effect of PEV adoption on system peak demand. Assume that one in every four homes owns a PEV, or roughly 250,000 residential customers with an electric vehicle in the example utility considered. Assume that half of these customers are simultaneously charging their vehicles at the time of the system peak (other owners may not yet be home from work or could already have a full charge). Assuming a charging demand of 3.3 kW per vehicle, the resulting increase in peak demand would be roughly 400 megawatts (MW) (calculation: 250,000 customers × 50 % peak-coincident charging × 3.3 kW). While not an insignificant number, a mid-sized utility with, for instance, 5,000–10,000 MW of existing load would have the capability to address this load growth over a long-term forecast horizon.

Now, consider what could happen at the distribution level. There is evidence to suggest that adoption of PEVs will be geographically "clustered." Assume that of the residents living on a street that is served by a single transformer and in a "green" neighborhood, half own a PEV. A charging demand of 3.3 kW could double the daily demand of these homes. As a result, if the PEV owners were all to plug in their vehicles when returning home from work in the evening, the load on that street's transformer could increase by 50 % (calculation: 50 % PEV ownership × 100 %

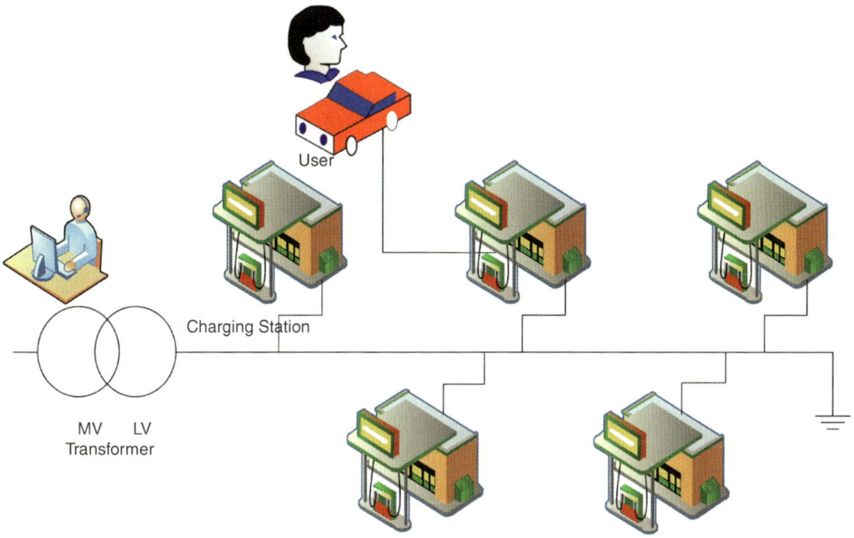

Fig. 2.1 Distribution network with PEV charging stations

increase in load per PEV owner). If the transformer was already being loaded at 70 % of capacity, then this increase would be enough to overload the transformer and create severe havoc in the distribution system. Dynamic pricing schemes, such as reduced rates for nighttime charging allow drivers to choose how to respond to change in prices. These pricing schemes allow users to choose their charging time and it does offer some relief to the grid in terms of motivating the user to charge during off peak periods by offering low tariff at those times. Such a smart grid can accommodate PEV charging according to schedule determined/chosen by the user.

Certainly, PEV adoption rates will vary from one service territory to the next, and the vehicles will be charged at varying rates and at different times of day. However, it is becoming clear that the existing generation resources will be in a much better position to accommodate future PEV market penetration than our distribution systems. Hence it is the distribution system infrastructure that needs to be restructured to accommodate high penetration of PEVs in communities.

2.4 Electric Vehicles and TOU Rates

The numerous potential benefits of widespread adoption of PEVs have been rated very high [36]. PEV are capable of reducing the greenhouse gas emissions due to reductions in the amount of gasoline burned by the vehicles internal combustion engines. Also since the price of gasoline is escalating, fueling with electricity is a least expensive option to the PEV owners. In a Smart Grid environment, if the owners decide to charge their vehicles late into the night, the vehicles represent an ideal off peak load that would complement new intermittent renewable energy resources such as wind and solar power.

The time and period of charging of PEVs could have a negative impact on the grid. Contrary to many expectations, PEVs will not result in unmanageable demands on generation resources. The real challenge would be at the distribution level. If all the residents of a small community purchased PEVs and they all charged at the same time, there would be a heavy spike in demand that could overload the transformers feeding those houses and would result in a severe damage to the distribution system. This could happen in reality if several of the PEV owners cluster in specific neighborhoods. Hence the utilities are trying hard to encourage off peak charging by allowing customers who own PEVs to take all or part of their electric service on some form of TOU pricing, often at higher voltages to facilitate faster charging. Many have approved TOU tariffs specially dedicated to PEVs. Several of the utilities offer different rates depending on whether the metering is done for the whole house or separately for the electric vehicle. It is somewhat common for utilities not to have created an EV-specific TOU rate, but to recommend that EV owners enroll in an existing residential TOU rate. TOU pricing is to encourage trend for charging PEVs efficiently since their owners can lower their electric bills by charging during off-peak hours.

PEV owners have the option to choose between charging based on convenience or only during those times when electricity costs are lowest. Saving money would motivate some owners to charge when the cost of charging is less. In the absence of incentives and benefits, PEV owners may not plug in their car for charging when they come back home at 6 pm and charge it to full capacity so that the vehicle is ready for them the very next morning. However, there are customers who might find it more convenient to charge their vehicle whenever they want to depending on their work schedule, availability for charging stations (Fig. 2.2) outside their home, extent of their tolerance to a less than fully charged battery and the regularity of their driving among other factors.

Fig. 2.2 Charging stations for PEVs

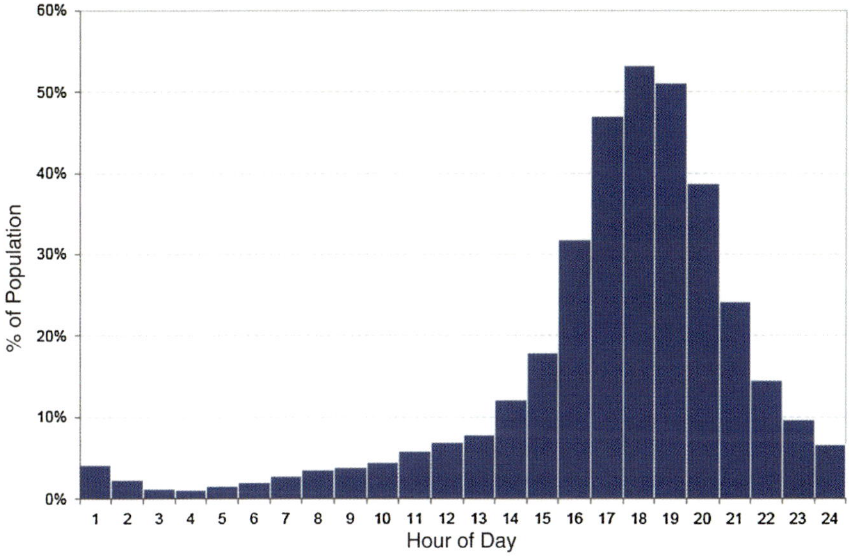

Fig. 2.3 Charging profile of PEV owners

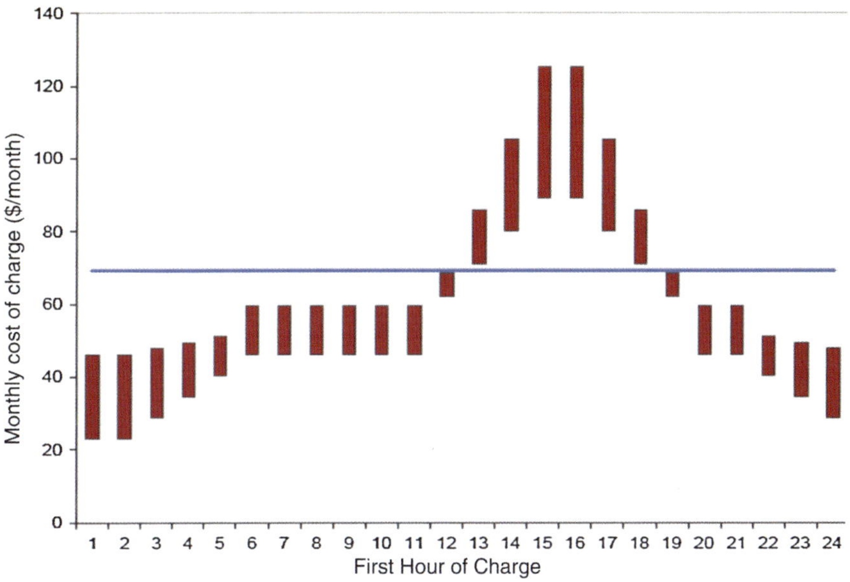

Fig. 2.4 Charging costs across TOU rates by time of day

An aggregated charging profile for the PEVs is given below in Fig. 2.3. Figure 2.4 shows the charging costs across TOU rates by time of day. A driver who is on the low TOU rate has the least incentive to charge during the cheapest periods, since their cost exposure is much less than that of an owner on either the medium or high TOU rates. A priori, one would expect drivers on the high TOU rate to display the largest price responsiveness and drivers on the low TOU rate to display the least.

Even in the high TOU rate case, the savings are modest. The difference between charging at 6 pm and 1 am is about $60 a month. Now the question arises as to whether a PEV owner will pay much attention to saving this sum of money. Research with other dynamic pricing and TOU pricing pilots suggests that despite the modest savings that accrue to customers on such pricing designs, people do move their load profiles in response to higher prices. Drawing upon empirical evidence from more than 100 tests with dynamic pricing, we would expect a peak-to-off-peak price ratio of 8:1 to produce a drop in peak load of around 15 %. The implied arc elasticity is fairly small (around −0.04) but is still capable of producing significant demand response with a potent rate design. Hence in this chapter we have implemented a real time pricing scheme for charge coordination of PEVs.

2.4.1 PEV Owners' Price Response and Distribution Transformer Overload

To conclude without any doubts that price responsiveness would alleviate any distribution transformer overload and loss issues, based on the TOU rates already established and the aggregate charging profile for the case study under consideration, price elasticity of demand of −0.04 is made use of. The percentage of customers charging during peak period would drop from 60 to 55 %. This is not beneficial to the grid operators trying to mitigate the adverse impact on the distribution system. Authors in [36] have tried various different price elasticities to effectively eliminate peak time charging.

2.4.2 Prediction of Charging Behavior

To predict charging behavior of PEV owners, a large number of volunteers were surveyed to study their charging behavior under various TOU rates. These volunteers were then randomly allotted to control groups and treatment groups where the control group members continue to drive their existing vehicles throughout the day whereas the treatment group members were supplied with a PEV. Both the control and treatment group's driving behavior was observed over a period of several months before and after the treatment group was supplied with PEVs. Results from the study carried out by [36] have shown that TOU rates may help reduce future grid reliability issues as PEVs penetrate the vehicle market. However, the extent to which properly designed rates would assist in maintaining grid reliability was not explored because of lack of information about the PEV owners' price responsiveness.

2.5 Coordinated and Uncoordinated Charging

To find applicable solutions to the problem of distribution transformer overloading, two general PEV coordination schemes have been considered in the literature.

- Centralized Coordinated PEV Charging—The system operator as a central controller sends commands through the smart grid communication network to each individual PEV to set the charging start time and rate. The decisions can be made based on several factors such as system capacity, system loss minimization, node voltage profiles, final state of charge, budget, etc. Therefore, a stable and more secure network can be achieved. However, centralized architectures with few central data stores require customer information and may lead to unscalable systems and costly initial infrastructure investments.
- Decentralized Coordinated PEV Charging—Each PEV is allowed to determine its own charging pattern. The decision can be made on the base of system capacity and conditions. The consequence of a decentralized approach may or may not be optimal, depending on the information and methods used to determine local charging patterns. Indeed, this approach does not require substantial knowledge of individual customers.

A comparison of both approaches is given in Table 2.1.

The phrase "decentralized" implies the ability of individual PEVs to make their own charging decisions. Most PEV charging algorithms have a centralized philosophy and structure, with all PEVs to be controlled from a central dispatch center. That is, PEV chargers cannot make any individual decisions on the starting time, rate and duration of their charging process. On the other hand, there are a few recently proposed

Table 2.1 Comparison of PEV coordination approaches

	Centralized PEV charging	Decentralized PEV charging
Idea	The system operator acts as central controller and sends commands through the smart grid communication network to each individual PEV to set its charging start time and rate. The decisions can be made based on several factors such as system capacity, system loss minimization, node voltage profiles, final state of charge, budget, etc.	Each PEV is allowed to determine its own charging pattern. The decision can be made on the bases of system capacity and conditions
Advantages and disadvantages	• More stable and secure network	• Easy to implement
	• Optimal coordination	• Preserves individual authority
	Centralized architectures with few central data stores may lead to unscalable systems and costly initial infrastructure investments	• Independent operations of PEV chargers
		• More dynamic and flexible system
	• Relies on customer information. Hard to implement	The results of a decentralized coordination approach may or may not be optimal

decentralized PEV coordinated charging algorithms, which rely on smart meter information and make their own individual decisions on charge time, rate and duration.

This chapter will first show the detrimental effects of uncoordinated charging of PEVs on distribution network and then introduces a new real time smart coordinated charging of PEVs in unbalanced residential network to control the distribution network losses and energy transferred to the PEVs. Detailed simulations are performed and presented to demonstrate the abilities of the proposed PEV charging algorithm. The main research goals are to formulate the optimal PEV coordination problem, define the objective function and select appropriate constraints such that the following requirements are fulfilled within a 24 h period:

1. Grid losses are minimized over the 24 h.
2. Each PEV charger operates independently and only relies on the information available at its own smart meter.
3. The distribution transformer loading is kept within its designated rated level to prevent possible damages to the equipment.
4. Finally, coordination is performed such that the system losses are minimized and energy transferred to the PEVs is maximized as a result of PEV charging activities.

The model developed in this chapter for smart coordination consists of a short term forecasting module and an optimization module. The short term forecasting module sends information about the number of PEVs in the parking garage and also their arrival and departure rates. The module then calculates and forecasts the number of PEVs that will be present at the same time in the parking garage for the next time interval. The heuristic multi-objective optimization module takes in the present and future power demands for all loads including PEV's over a finite time interval. The aims of this optimization module is to maximize the energy delivered to PEV batteries and satisfy the SOC criteria for the PEV while including constraints related to the power grid and customer demands.

The optimization module is also designed minimize distribution network losses considering charging time zone priorities specified by PEV owners. To validate the power management infrastructure and distribution network loss minimization, the smart coordination strategy is implemented on IEEE 13 node test feeder and a 38 bus power system consisting of a mix of residential, commercial and industrial customers penetrated with PEVs. To estimate the economics of charging, simulation results will be presented for uncoordinated and coordinated (centralized and decentralized) charging scenarios for different PEV penetrations.

2.6 Power Management

Electricity demand varies both by day and by year and since it is difficult to store electricity in large quantities it is produced at the same time as it is consumed. Hence, the variations in demand result in variations in the electricity generation and generation capacity must be designed to handle the peak demand. Similarly, the transmission capacity in the grid must be designed to handle the peak power in the system. The variation in electricity demand leads to increased cost of electricity since it requires a higher transmission capacity in the electric grid and since the electricity consumed during the peak is usually produced by generation plants with high production cost.

Power and Energy Management (PEM) can be performed on the supply side or demand side. On the supply side, PEM is undertaken when:

- There is a growing demand (demand requirement is higher than supply)
- There is a lack of resources (finance, energy) and PEM helps to postpone the construction of a new power plant.

On the demand side, energy management is used to reduce the cost of purchasing electrical energy and the associated penalties. The techniques used for PEM are aimed at achieving valley filling, peak clipping and strategic conservation of electrical systems. There are techniques that are used to decrease the need for additional capacity and the costs involved by increased fuel on the supply side. The implementation of the techniques leads to improving off-peak valley-hours and the load factor of the system. The common load management techniques to supply side or demand side are presented as load shedding and restoring. There are also more exotic means such as power wheeling, the installation of energy efficient processes and equipment, the use of energy storage devices, co-generation, use of renewable energy and reactive power control. Implementation of these techniques has found a steady increase in application and meets demand side management (DSM) objectives.

2.6.1 Power and Energy Management: Techniques

Energy management embodies engineering, design, applications, utilization, and to some extent, the operation and maintenance of electric power systems for the provision of the optimal use of electrical energy without violating other international standards. Load management in utility industries is the planning and implementation of the utility activities, which are designed to influence customers to use electricity in such a way, that it produces a desired change in the utility load shape. Different load management techniques have been proposed and used, e.g. time-of-use-tariffs, interruptible load tariffs, critical peak pricing, real-time pricing (RTP) and distribution system loss reduction [37, 38]. As stated in [38] different techniques can have differential impact on the electric grid.

Direct load control (DLC) This is the program designed to interrupt consumers' loads during the peak time by direct control of the utility power supply to individual appliances on a consumer premises. The control usually involves residential consumers. The cost benefit of DLC includes:

- Power system production cost savings.
- Power system generating capacity cost savings.
- Power system loss reduction.

The various control options for DLC are

- Direct load control, utility can switch off the load directly when required.
- Interruptible load control—the utility provides advance notice to the customer for switching off their loads.
- TOU tariffs, where utility rate structure is designed according to the time.

Mohamed and Khan [39] developed methods for classification of customers loads according to the size of load. Telephone, radio signal and power line were used to produce a signal that interrupted large industrial consumers. In this system, customers were required to reduce their electric demand to an emergency service load for only 10 min upon request. Under frequency, the relay was installed in the customer's loads, which responded very fast in the under frequency regime. Tools for evaluation of end-use monitoring DLC programs were described by [40] namely a duty cycle model (DCM) and demand side planning. The PC-based workstation had proven to be a viable and cost effective means of analyzing the voluminous data used in the program. The duty cycle model offered an integrated approach to DLC impact analysis. This is given by:

$$t = Average\,Load/Connected\,Load \tag{2.1}$$

In the case of PEM based on time dependent tariffs, load management is carried out by the influence of tariffs setting. The total cost of generating and delivering of electricity to consumers was being broken into four fundamental categories of services:

- Customer services,
- Distribution services,
- Transmission services,
- Generation services.

Integrated utilities in regulated states set the rates to cover the costs of all services. The electric consumers are billed as:

- Flat rate tariffs/two part tariff
- Time of use tariff
- Spot price

In a flat rate tariff, a customer pays the same amount for electricity at any time of day. In the TOU based method, the utility provides transparent information on the electricity price at different periods to the customers to encourage off peak and discourage peak period consumption by varying price of electricity. Time of use rates provide variation of the cost of energy by season or time of day. Rates are higher during peak demand periods and lower during off-peak periods. Some utilities have made TOU rates mandatory for large customers. Savings from time of use rates vary depending on the size of the peak/off-peak price differential and the length of the peak period. Another type of tariff setting for LM is spot price. The message is sent to customers to indicate the price of electricity for an instant of time. A spot price scheme is appreciable if electricity price fluctuation is high and if the consumer can anticipate the price behavior as well as being able to respond quickly when the electricity price is high or low.

2.7 Proposed Smart Coordination of PEV Charging Using Real Time Pricing

In the proposed approach, PEV owners are allowed to select one of the charging time periods and rates. Each PEV owner will provide to the system his charging tag number, required state of charge and parking duration. The command center receives and processes this vehicle data. The forecasting algorithm predicts the number of PEVs in the system during that time period. Forecasting algorithm will then be used to predict the number of PEVs that would be in the system during the next time interval. This forecasted data along with the actual data would then be sent to the centralized command center who will then operate the optimization module to schedule PEV charging until maximum energy is delivered to the batteries and distribution losses are minimized. This chapter explains in detail how PEVs can be scheduled using real time costs thereby reducing the burden on the local distribution networks. For online coordination of PEVs, a smaller optimization period should be chosen to start charging the PEVs as fast as possible.

2.7.1 Forecast Module for Predicting PHEV Owners Charging/Discharging Behavior/Schedule in a Smart Grid

A smart grid is a power grid with information transfer allowing agents on the grid to communicate and make decisions regarding load connections. One major advantage of a smart grid is the opportunity to more efficiently utilize the power that is generated. When considering a conventional power grid that uses load forecasting to predict power demands, it is possible to account for activities that have been exhibited for many years such as the cycle of the modern family to work or school and back home again. When an additional element outside of the historical forecasted data is added to the power demands it can be difficult to compensate. Such a scenario could present itself with the emergence of plug-in electric vehicles (PEVs). Not only is the additional load associated with PEVs uncertain due to their adoption rate, but it could also prove difficult to quantify because of the stochastic nature of vehicle use. This topic investigates the use of smart coordinated PEV charging on a smart grid allowing for a more manageable overall use of power. The information transfer is used by the PEV's control strategy to make efficient charging decisions.

Figure 2.5 presents a flowchart of the proposed approach. The first step is to gather the data needed for the study. From the data, four key parameters can be processed for further analysis: (i) the locations of the vehicles (i.e., where they can be charged); (ii) when they are parked (i.e. when they can be charged), (iii) the number of PEVs that are being charged and (iv) real time price during that interval. The second step is to formulate and implement the optimization models for different control strategies. The final step is to use the data in the models developed to

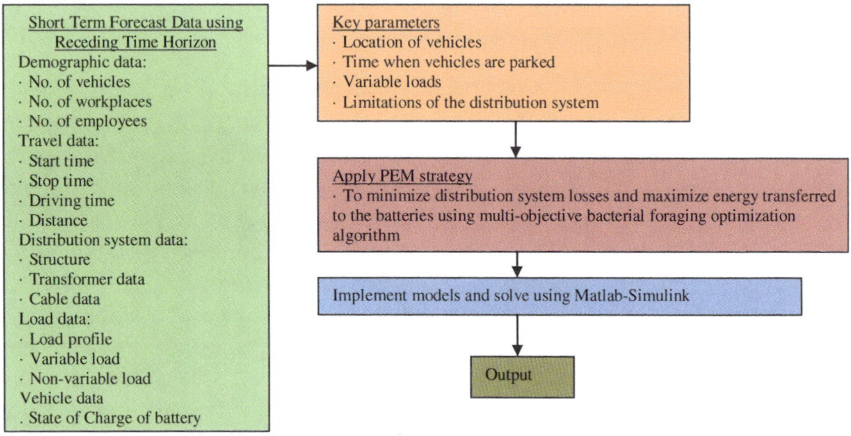

Fig. 2.5 Flowchart of smart coordination approach with forecasting module and optimization modules

evaluate the impacts of different control strategies on the distribution system. The optimization model is based on an AC optimal power flow framework which is described in [41], with the objective function being: maximization of energy transferred to the PEVs and minimization of distribution network losses. This model was developed using Matlab-Simulink.

2.7.1.1 Methodology

Uncontrolled PEV charging in a modern day grid with renewable energy resources may cause several local grid problem including additional extra power losses, voltage swings and power quality disturbances. Uncontrolled charging is the current practice for PEV charging and is expected to persist in the near future to enable a transition period for the PEV penetration to be significant, hence it paves the way for the coordinated charging, which is the second expected scenario. For this scenario, a coordinated charging system should be developed under the smart grid paradigm. This system must be able to deal with real-time measurements and parking lot dynamics through the utilization of the two-way smart grid communication infrastructure. The primary target of such a coordinated charging system is the best use of smart grid generation resources so that the PEV load can be shifted to optimal periods during PEV parking duration in order to maximize customer satisfaction without jeopardizing system equipment.

Smart coordination refers to coordinated charging and it has been shown that coordinated charging of PEVs can lower power losses and voltage deviation by flattening out peak power and improve the load profile. In the proposed approach, smart charging and discharging coordination architecture consists of two main modules: a prediction module, and an optimization module. The prediction module consists of a data collection and storage module which governs the collection of information related to current PEV power demands, the current state-of-charge (SOC) of PEV batteries, and the power demand of regular loads. In most cases, an aggregator is assumed to be in place to deal with PEV data collection and storage. The role of the aggregator is to collect information from the PEVs and send it to the grid operator, and to send charging/discharging decisions from the operator to the chargers. The short term prediction/forecasting module should provide accurate forecasts of future PEV power demands and regular loads in the power system. Based on this information, the optimization module should then make optimal coordinated charging and discharging decisions that guarantee maximum energy transferred to the customers PEVs and minimum distribution network losses (Fig. 2.5).

Accurately estimating the impact of PEV charging on electric power system components requires both component models and good estimates of the magnitude and timing of demand increases due to PEV charging. Early PEV research assumed very simple charging profiles, such as assuming that vehicles will charge daily starting at 17:00, 18:00 or 19:00 h, with batteries fully depleted at the start of each charge cycle. However actual PEV charging loads will depend highly on travel patterns, which vary tremendously from driver to driver and day to day. To better capture this variability in driving behavior, researchers have used either detailed GPS data for small groups of drivers, or survey data from larger populations. Authors have used data from 9 drivers to estimate variability in daily miles driven, but with fixed evening arrival times. Another study used GPS data from 76 vehicles to derive a stochastic model of miles driven and arrival/departure times. Other authors have used a larger set of GPS data to develop a Monte Carlo model that is similar to the one presented here, but the data are not used to model the miles driven, which is necessary to estimate the battery state-of charge on arrival.

In this chapter, the problem is formulated as an optimization problem with the objective function being a sum of convex and strictly increasing functions. This power scheduling problem is solved in a static fashion, that is, the optimization is performed only once before or at the beginning of the scheduling horizon. To take dynamic changes of loads into consideration, this chapter studies a real time implementation of the power scheduling. Our approach is to reformulate the optimization problem so that it is solved in the fashion of receding horizon. Generally, it works by solving optimization over the next T time steps, executes the first time step decision, and resolves the optimization problem for the next T time steps by incorporating new information available at the moment. Since the power management problem is not formulated in a traditional way, the receding horizon operation needs to be carefully designed. The main challenge comes from the fact that load groups may need to be reorganized during the execution.

The short term forecasting method implemented is the receding horizon formulation of power management problem discussed in [43]. Along this direction, two strategies are available: one is based on the conventional receding horizon idea, and the other is a reformed scheme with the merit of reducing online computational load. We assume individual EV charging loads (either residential or commercial) are connected to the power grid through a control unit. Each control unit monitors status of an EV battery, connect/disconnect load from the grid, and wirelessly communicates with a remote aggregator. The aggregator acts as a central scheduler and commander to communicate with the PEV owner/driver and to regulate the charging process of each load. Once a vehicle is plugged-in, the corresponding aggregator (e.g., parking deck operator) receives the battery state information (e.g., state-of-charge, state-of-health, voltage, and current) as well as customer information (e.g., customer identification, customer preference, and billing information) and sends in the real time pricing rate to the customer. Multiple aggregators serve as middleware between the central controller (e.g., distribution Company, and microgrid operator) and individual vehicles. Given the real-time information from multiple aggregators, the central controller performs the energy scheduling (optimization of losses and energy transferred) and sends back control signals periodically.

The load population is assumed to be large, by taking into consideration the anticipated high penetration level of PEVs. This requires our solution to the power management scalable and computational tractable. To this end, the PEV charging loads are classified into groups with the following definitions given in [43]:

Definition 1 *A load type* l^r is defined by $l^r = \{a^r, b^r, \tau^r, p^r\}$, $r = 1, 2, \ldots, m$, where a^r is the (earliest) charging start time, b^r the (latest) charging completion time, τ^r is the required total charging period, and p^r indicates the desired power level required by the EV charging system which is assumed to be time-varying.

Definition 2 A *family of load requests* is defined as $\mathcal{F} = \{(l^1, N^1), (l^2, N^2), \ldots, (l^m, N^m)\}$., where l^r, $r = 1, \ldots, m$, is the r-th load type, and N^r is the total number of requests from customers for the type-r load.

Here, the scheduling horizon is discrete and consists of T time steps, which is denoted as $[1, T]$. For each time step, the power level p^r desired by type r loads may be different; therefore, we introduce the notation of *charging stages* below.

Definition 3 For each time step j of the charging process for type-r power loads, denote the demanded power level by $p^r(j), j = 1, 2, \ldots, \tau^r$. It is said that the type-r load requires τ^r *charging stages*, and the jth *stage power level* is $p^r(j)$. With the above definition, the power of type-r loads can be expressed as a power vector $p^r = [p^r(1), p^r(2), \ldots, p^r(\tau^r)]$. This is an ordered vector for which the stage $p^r(i)$ must be completed before the $p^r(j)$ stage starts, for any $i<j$. The completion of charging could be intermittent, i.e., charging stages could be discontinuous in time.

2.7.2 *Optimization Module to Minimize Distribution Network Power Losses for Power Management*

A common approach to deal with power management is to assign each individual load a vector of binary numbers (1 or 0); each number is used to indicate the on/off state of the power at one discrete time step. As the load population increases, the number of decision variables increase proportionally, and the searching space grows exponentially. Therefore, such approach is not scalable generally for centralized strategies.

This chapter focuses on groups of loads rather than individuals. With the definitions of load type and charging stage, the decision variables are chosen to be the total number of loads for each load type to be powered on for a certain charging stage at any discrete time instant. Symbolically, the decision variable is denoted by $\beta_k^r(j)$, representing the total number of type-r loads being charged at charging stage j at time step k, where $k \in [1, T]$ is the time index, $j \in \{1, \ldots, \tau^r\}$ the charging stage index, and $r \in [1, \ldots, m]$ the load type index. Then, the total power consumption of the entire power system at time k is:

$$L_k = \sum_{r=1}^{m} \sum_{j=1}^{\tau^r} \beta_k^r(j) p^r(j) \tag{2.2}$$

The objective of power management is to minimize the total power losses and maximize energy transferred to the PEVs over time duration $[1, T]$. To this end, a multi-objective function

$$\text{Min} \sum_{k=1}^{T} C(L_k) + \text{Max} \sum E_D \tag{2.3}$$

is chosen where function $C(\cdot)$ is strictly increasing and convex. Convexity of the above cost function causes heavier penalty on larger instantaneous power losses, which is important in alleviating power loss values. More advantages of choosing such a cost function are discussed in [42]. E_D is the energy delivered to a PEV battery during the time interval [1, T]. Based on power flow constraints, bus voltages and generated real and reactive powers are specified. The decision variables are the voltage magnitudes and angles at all buses except slack bus and real and reactive power generated at slack bus. During each iteration of the optimization algorithm, voltage limits are checked to see if there are any violations.

The optimization problem of EV charging power management to minimize distribution network losses and maximize energy transferred is summarized below, and the detail can be found in [43] along with a two-layer strategy to reduce computation burden of the optimization. The total real and reactive power generated at each bus can be calculated based on current measurements and predicted data. The total real power consumed by load will be the sum of real power consumed by all other regular loads added to the real power consumed by 67 PEV load. Decision to charge or discharge is made based on the state of charge (SOC) of the PEV battery and is limited by the capacity of charger. Energy transferred to the battery is calculated as product of battery capacity and the difference between final SOC and initial SOC at a particular time interval. State of charge of the battery is limited by the desired SOC by the user. But also the incoming PEVs are expected to require a final SOC of 100 % when they leave and to arrive with a minimum SOC.

Problem P_p ***(EV Power Management Problem)***
Find $\beta_k^r(j)$ to

$$\text{Min} \sum_{k=1}^{T} C\left(\sum_{r=1}^{m}\sum_{j=1}^{\tau^r} \beta_k^r(j) p^r(j)\right) + \sum_{k=1}^{T} E_D \tag{2.4}$$

subject to the following constraints:

(a) $\beta_k^r(j) \in \mathcal{Z}^+$ for $k = 1, \ldots T, j = 1, \ldots \tau^r$, and $r = 1, \ldots, m$.
(b) $\beta_k^r(j) = 0$ for any $k < a^r$ or $k > b^r$.
(c) $\sum_{j=1}^{\tau^r} \beta_k^r(j) \leq N^r$, for $k = 1, \ldots T, r = 1, \ldots, m$.
(d) $\sum_{k=1}^{T} \beta_k^r(j) = N^r$, for any $j = 1, \ldots, \tau^r, r = 1, \ldots, m$.
(e) $\sum_{k=1}^{n+1} \beta_k^r(j+1) \leq \sum_{k=1}^{n} \beta_k^r(j)$, for all $n = 1, \ldots, T-1$, $j = 1, \ldots, \tau^r$, $r = 1, \ldots, m$.

Note that decision variables of problem P_p are number of EV charging loads which are integers and are not preferable by numerical optimization solvers. Assuming a large network with high population of EV loads, we can rewrite the problem with a set of new decision variables as defined below.

Definition 4 Given decision variables $\beta_k^r(j)$ of problem P_P, we define a *new set of decision variables* $\gamma_k^r(j)$ as the percentage of type-r load requests being switched on, i.e.,

$$\gamma_k^r(j) = \frac{\beta_k^r(j)}{N^r}, \; r = 1, 2, \ldots, m \tag{2.5}$$

where N^r is the total number of the type-r loads. Note that the above new decision variables, $\gamma_k^r(j)$, are rational numbers, which can be easily used to replace $\beta_k^r(j)$ in problem P_P. Once the optimization problem P_P is solved, the aggregator can plan out a more specific schedule on power allocation, by indicating which exact load needs to be powered on/off for any charging stage at any time step. One approach to such power allocation is described in Algorithm 1 below. More specifically, the solution to Problem P_P produces number of requests $\beta_k^r(j)$, where k is time index, j is charging stage index, and r is load type index. In compact form, we write $\beta_k^r(j)$ as matrix B^r, for each load type r, as follows:

$$B^r = \underbrace{\begin{bmatrix} \beta_1^r(1) & \beta_2^r(1) & \cdots & \beta_T^r(1) \\ \beta_1^r(2) & \beta_2^r(2) & \cdots & \beta_T^r(2) \\ \vdots & \vdots & \vdots & \vdots \\ \beta_1^r(\tau^r) & \beta_2^r(\tau^r) & \cdots & \beta_T^r(\tau^r) \end{bmatrix}}_{\text{Time step} \;\rightarrow\rightarrow} \begin{matrix} \text{Charging} \\ \text{Stage } j \\ \downarrow \\ \downarrow \end{matrix} \tag{2.6}$$

The problem P_P generates m of such matrix, $B^1, \ldots, B^m$, one for each load type. For each B^r, Algorithm 1 yields another matrix Λ^r with dimension $N^r \times T$:

$$\Lambda^r = \underbrace{\begin{bmatrix} \lambda_{1,1}^r & \lambda_{1,2}^r & \cdots & \lambda_{1,T}^r \\ \lambda_{2,1}^r & \lambda_{2,2}^r & \cdots & \lambda_{2,T}^r \\ \vdots & \vdots & \vdots & \vdots \\ \lambda_{N^r,1}^r & \lambda_{N^r,2}^r & \cdots & \lambda_{N^r,T}^r \end{bmatrix}}_{\text{Time step } \rightarrow\rightarrow} \begin{matrix} \text{Individual} \\ \text{loads} \\ \downarrow \\ \downarrow \end{matrix} \tag{2.7}$$

Each row of Λ^r corresponds to each individual load of type r and each column corresponds to one time step. Each element, $\lambda_{i,k}^r$, is the charging stage number of load i at time k. Note that we set $\lambda_{i,k}^r = 0$ when the load is not served. Thus, we have $\lambda_{i,k}^r \in \{0, 1, \ldots, \tau^r\}$. Below is the algorithm which creates Λ^r from B^r.

Algorithm 1 (*Individual power allocation*):

Given: $\beta_k^r(j)$, number of type-r loads being served with charging stage j at time step k, $j = 1, \ldots, \tau^r, k = 1, \ldots, T$.

Find: $\lambda_{i,k}^r$, charging stage index of type-r loads being served at stage $j = 1,2, \ldots, \tau^r$, at time step $k = 1,2, \ldots, T$.

Procedure:

```
For each row j of Λ^r, j = 1, ..., τ^r
    temp = 1;
    For each column k of Λ^r, k = 1, ..., T
        λ^r_{i,k} = 0;
        For i = temp to (temp + λ^r_{j,k})
            λ^r_{i,k} = j;
        End
        temp = temp + λ^r_{j,k};
    End
End
```

Here is an example to illustrate the algorithm. Suppose type-r loads have population $N^r = 10$, total number of charging stages $\tau^r = 3$, and scheduling horizon $T = 5$. The optimization problem P_p produces

$$B^r = \underbrace{\begin{bmatrix} 6 & 2 & 2 & 0 & 0 \\ 0 & 4 & 4 & 2 & 0 \\ 0 & 0 & 2 & 6 & 2 \end{bmatrix}}_{\textit{Time } 1 \rightarrow 5} \begin{matrix} \textit{Stage} \\ 1 \\ 2 \\ 3 \end{matrix} \tag{2.8}$$

Then, the outcome of Algorithm 1 generates

$$\Lambda^r = \underbrace{\begin{bmatrix} 1 & 2 & 3 & 0 & 0 \\ 1 & 2 & 3 & 0 & 0 \\ 1 & 2 & 0 & 3 & 0 \\ 1 & 2 & 0 & 3 & 0 \\ 1 & 0 & 2 & 3 & 0 \\ 1 & 0 & 2 & 3 & 0 \\ 0 & 1 & 2 & 3 & 0 \\ 0 & 1 & 2 & 3 & 0 \\ 0 & 0 & 1 & 2 & 3 \\ 0 & 0 & 1 & 2 & 3 \end{bmatrix}}_{Time\ 1\ \rightarrow\ 5} \begin{matrix} Loads \\ 1 \\ \vdots \\ \downarrow \\ \vdots \\ 10 \end{matrix} \tag{2.9}$$

This matrix indicates at any time step, which charging stage each load needs to be served. A zero in the matrix tells that the corresponding load needs to be turned off at that time step. For instance, load 3 will be turned on at time step 1, 2 and 4, for charging stage 1, 2 and 3, respectively. In summary, the static approach to deal with the power management problem is solving the optimization problem P_p followed by executing Algorithm 1. Algorithm 1 is extremely light in computation effort, with complexity linearly proportional to the load population size and independent of how P_p is solved; therefore, next we only need to focus on the receding horizon implementation of the optimization problem P_p. Receding horizon (RH) control provides a method to extend the above static optimization work to real-time power scheduling so that the dynamic changes of the power network can be taken into account and real time pricing rates could be applied to the customers. In general, RH works by solving optimization over the next T time steps, executes the first time step decision, and resolve the optimization problem for the next T time steps by incorporating new measurement data available at the moment. In short, RH scheme repeats the process of optimization, execution, and adaptation.

In this section, we present two schemes of receding horizon optimization: a global scheme and a local scheme, for our power management problem. Both these receding horizon formulations deal with the power management problem for a fixed time duration $[1, T]$. The process of optimization is illustrated in Fig. 2.6. At each

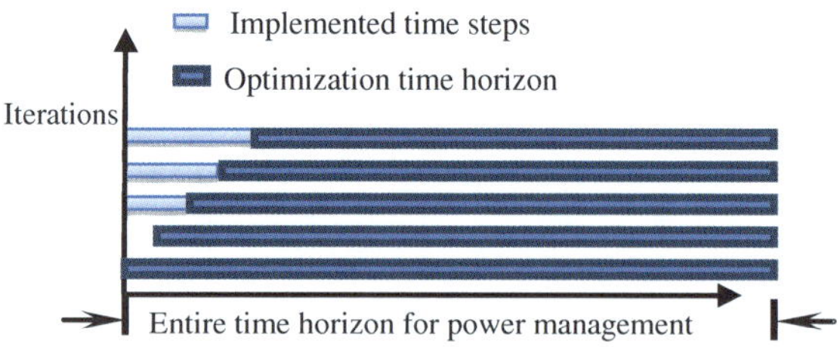

Fig. 2.6 Time horizons for receding horizon optimization

of iteration, the optimization problem will be updated by considering the changes of loads. For example, the entire time horizon is assumed to be $[1, 24]$ for 24-h period. The first iteration solve the optimization in time horizon $[1, 24]$, the second iteration considers optimization horizon $[2, 24]$, and so on. Our RH algorithm for power management problem follows the procedures listed in Algorithm 2. The formulation of receding horizon optimization problem mentioned in part c of Step 2 will be the focus of the rest of the section.

Algorithm 2 *(Receding Horizon Algorithm):*

Procedures:

Step 1. Set $k = 0$ *and solve optimization problem* P_p *followed by Algorithm 1.*

Step 2. At time step k:

a) Issue charging services scheduled at time step $k - 1$.
b) Collect updates in the network (dropouts and newcomers).
c) Solve the updated receding horizon optimization problem.

Step 3. Repeat *Step 2* procedure with $k = k + 1$ until $k = T$.

One major benefit of receding horizon approach is the ability to incorporate updated information; in EV charging power management, updates may include newly arrived charging requests as well as dropped outs. For example, a customer gets home late and plugs his EV to the grid hoping the charging to complete on time as usual. On the other hand, it is possible that existing requests are dropped from the system at time step k; for instance, some customers may unplug their devices in the middle of charging for an unplanned trip. The charging service requests come and go dynamically. The load controller equipped at the customer end monitors such condition and reports to the aggregator through a two-way communication link.

Complete Receding Horizon Optimization Consider the iteration at time step k. A few matters need our attention regarding the formulation of receding horizon optimization. First, decision variables here will be denoted by $\beta_i^r(j)$, instead of $\beta_k^r(j)$ as in problem P_p, to represent the total number of type-r loads being charged at charging stage j at time step i, where $i \in [1, T]$ is the time index, $j \in \{1, \ldots, \tau^r\}$ the charging stage index, and $r \in [1, \ldots, m]$ the load type index. The reason for this notation change is that only values of $\beta_i^r(j)$ at $i = k + 1$ will be implemented in this iteration, and the rest of values will be discarded. In addition, note that, at time step k, the loads being scheduled at past and present time steps, $1, 2, \ldots k$, have been implemented and cannot be rescheduled; hence we only need to solve for $\beta_i^r(j), i \in [k + 1, T]$. To keep the formulation consistent, we still consider the optimization horizon to be $[1, T]$ by constraining $\beta_i^r(j)$ to be zero for any $i \leq k$.

Secondly, due to the fact that problem P_p deals with groups of loads instead of individuals, the receding horizon optimization needs to be reformulated since individual loads in the same group may have undergone different charging services during previous execution stage. More specifically, for the same load type, at time steps $1, 2, \ldots, k$, some are powered on and others off, and for the ones being on, they may be at different charging stages. Therefore, at time step k, the optimization needs to keep track of and consider the existing state of the network due to all these executions.

Thirdly, before solving the optimization at time step k, the aggregator of the power system collects the messages sent from the loads about changes of the network. When a charging load withdraws its service request, a withdraw signal is triggered and sent by the load controller to the aggregator, similar to the case when a new charging request arrives. These changes are formulated below.

At time step k, let $D_k^r(j)$ denote the number of loads dropped out after they completed charging stage j and before their services are completed, where $j = 1, 2, \ldots, \tau^r - 1$. Note that a load is not counted as a dropout if its charging service is completed. Then, at time step k, the total number of type-r loads exiting abnormally from the network is

$$D_k^r = \sum_{j=1}^{\tau^r - 1} D_k^r(j) \tag{2.10}$$

Further, denote A_k^r the number of type-r loads newly arrived at the present time step k; thus, the total number of type-r loads currently in the network is

$$N_k^r = N_{k-1}^r - D_k^r + A_k^r \tag{2.11}$$

with $N_0^r = N^r$ and $k = 1, 2, \ldots, T - 1$.

Among the type-r loads in the network at time step k, we use $\theta_k^r(j)$ to denote the number of loads having completed stage j charging and waiting for stage $j + 1$, $j = 1, 2, \ldots, \tau^r - 1$. Then, it follows that,

$$\theta_k^r(j) = \theta_{k-1}^r(j) + \bar{\beta}_k^r(j) - \bar{\beta}_k^r(j+1) - D_k^r(j) \tag{2.12}$$

where $\bar{\beta}_k^r(j)$ and $\bar{\beta}_k^r(j+1)$ are calculated in the previous iteration and currently implemented, with $\bar{\beta}_k^r(j)$ being the number of loads served with stage j charging at the current time step, and $\bar{\beta}_k^r(j+1)$ being the number of loads served with stage $j + 1$ charging.

With the above notation, we present below the receding horizon optimization problem solved at each time step k in Algorithm 2.

Problem P_p^{CRH} (*Complete Receding Horizon Optimization*) Find $\beta_i^r(j)$ to

$$\text{Min} \sum \sum_{i=1}^{T} C\left(\sum_{r=1}^{m} \sum_{j=1}^{\tau^r} \beta_i^r(j) p^r(j)\right) + \text{Max} \sum_{i=1}^{T} E_D \tag{2.13}$$

subject to the following constraints:

(a) $\beta_i^r(j) \in \mathcal{Z}^+$ for $i = 1, \ldots T$, $j = 1, \ldots \tau^r$, and $r = 1, \ldots, m$.
(b) $\beta_i^r(j) = 0$ for any $i < \max(a^r, k + 1)$ or $k > b^r$.

(c) $\sum_{j=1}^{\tau^r} \beta_i^r(j) \leq N_k^r$, for $i = 1, \ldots T$, $r = 1, \ldots, m$.

(d) $\sum_{i=1}^{T} \beta_i^r(j) = N_k^r - \left[\theta_k^r(j) + \theta_k^r(j+1) \cdots + \theta_k^r(\tau^r)\right]$, for any $j = 1, \ldots, \tau^r$, $r = 1, \ldots, m$.

(e) $\sum_{i=1}^{n+1} \beta_i^r(j+1) \leq \sum_{i=1}^{n} \beta_i^r(j) + \theta_k^r(j)$, for all $n = 1, \ldots, T-1$, $j = 1, \ldots, \tau^r$, $r = 1, \ldots, m$.

The constraints of problem P_p^{RH} are explained in the following. Constraint (a) requires number of loads to be integers, and constraint (b) states that no type-r loads will be scheduled beyond its required horizon $[a^r, b^r]$, and no loads will be scheduled for present or past time steps. Note that due to constraint (b), the cost function satisfies

$$\begin{aligned} &\sum_{i=1}^{T} C\left(\sum_{r=1}^{m}\sum_{j=1}^{\tau^r} \beta_i^r(j) p^r(j)\right) + \sum_{i=1}^{T} E_D \\ &= \sum_{i=k+1}^{T} C\left(\sum_{r=1}^{m}\sum_{j=1}^{\tau^r} \beta_i^r(j) p^r(j)\right) + \sum_{i=k+1}^{T} E_D \end{aligned} \tag{2.14}$$

That is, as iteration advances, the lower bound of the optimization time horizon increases while the upper bound keeps the same, which is consistent with the optimization time horizon illustration in Fig. 2.7.

The last constraint (e) sets the sequencing requirement of charging stages, i.e., for any loads, charging stage $j+1$ cannot start before completion of stage j. The left side of the inequality is the total number of loads having been serving with stage $j+1$ charging up to time step $n+1$, and the first term on the right,

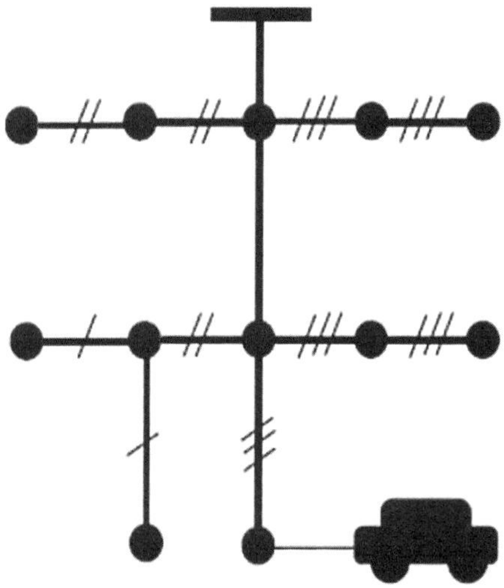

Fig. 2.7 IEEE 13 node system

$$\sum_{i=1}^{n} \beta_i^r(j) = \sum_{i=k+1}^{n} \beta_i^r(j), \tag{2.15}$$

represents the number of loads having gone through stage j during duration $[k+1, n]$, and the second term $\theta_k^r(j)$ includes the number of loads having gone through stage j before or at time step k. Two terms on the right together comprises the number of loads completed stage j prior to time step $n+1$.

The constraint (c) and (d) will be explained with the matrix

$$B^r = \underbrace{\begin{bmatrix} \beta_1^r(1) & \beta_2^r(1) & \cdots & \beta_T^r(1) \\ \beta_1^r(2) & \beta_2^r(2) & \cdots & \beta_T^r(2) \\ \vdots & \vdots & \vdots & \vdots \\ \beta_1^r(\tau^r) & \beta_2^r(\tau^r) & \cdots & \beta_T^r(\tau^r) \end{bmatrix}}_{\text{Time step } \rightarrow\rightarrow} \begin{matrix} \text{Charging} \\ \text{Stage}\, j \\ \downarrow \\ \downarrow \end{matrix} \tag{2.16}$$

Column sum of the matrix B^r is the total number of loads being powered on at one time step, and this number should be less than the total number of loads currently being present in the network; this explains constraint (c). Row sum of the matrix is handled in constraint (d), which is the total number of loads being allocated at each charging stage over time. This number should be no more than the number of loads currently in the network waiting for the service. This number is obtained by the total number of loads N_k^r in the network subtracting the sum, $\left[\theta_k^r(j) + \theta_k^r(j+1) \cdots + \theta_k^r(\tau^r)\right]$, which contains the loads which doesn't need stage j charging.

Overall, this complete receding horizon algorithm for power management is summarized in Algorithm 3.

Algorithm 3 (*Complete Receding Horizon Algorithm*)

Procedures:

Step 1. Set $k = 0, \theta_0^r(j) = 0$ and $N_0^r = N^r$. Solve optimization problem P_p^{CRH} to obtain $\beta_i^r(j)$ for $i = 1, \ldots, T, j = 1, \ldots, \tau^r$, and $r = 1, \ldots, m$.

Step 2. At time step k:

a) Issue charging services $\beta_k^r(j)$ for all j and r.
b) Let $\bar{\beta}_k^r(j) = \beta_k^r(j)$ for time step k, and obtain $D_k^r(j)$ and A_k^r from the aggregator. Update:

$$N_k^r = N_{k-1}^r - \sum_{j=1}^{\tau^r-1} D_k^r(j) + A_k^r \tag{2.17}$$

$$\theta_k^r(j) = \theta_{k-1}^r(j) + \bar{\beta}_k^r(j) - \bar{\beta}_k^r(j+1) - D_k^r(j) \tag{2.18}$$

c) Solve the updated receding horizon optimization problem P_p^{CRH}.

Step 3. Repeat *Step 2* procedure with $k = k + 1$ until $k = T$.

Partial Receding Horizon Optimization In the previous scheme of complete receding horizon optimization, the power management optimization is redone at each iteration for the entire network of EV charging. Here, we present a scheme by optimizing only a small part of the network.

Let us consider the formulation of the optimization problem for time step k. New EV charging requests may arrive at time step k, and we denote them as a new-arrival family $\tilde{F}_k = \{(l^1, \tilde{N}_k^1), \ldots, (l^m, \tilde{N}_k^m)\}$ with $l^r = \{a^r, b^r, \tau^r, p^r\}$ representing the type-r load characteristics, $r = 1, 2, \ldots, m$. Note that the only difference between $\tilde{F}_k$ and F (existing loads) is the population size of each load type. At this time step, previous approach is to execute the scheduled action for time step k, update the status of the existing requests, and combine the existing and new-arrived requests as a new family of power tasks. Here we look at a somewhat different action.

Now we consider the existing condition of the network at time step k. From the scheduling of the last iteration, $\beta_i^r(j)$ indicates the number of type-r loads being served with charging stage j at time step $i = k, k+1, \ldots, T$. With the same notation as in the complete receding horizon scheme, let $D_k^r(j)$ denote number of loads dropped out after they finish charging stage j at time step k. For convenience of formulation, define

$$D_i^r(j) = \begin{cases} D_k^r(j) & i = k \\ 0 & i \neq k \end{cases}. \tag{2.19}$$

Without considering newcomers, based on the previous iterative optimization, the power load at time step i, for $i = k+1, \ldots, T$, is

$$\bar{L}_i = \sum_{r=1}^{m} \sum_{j=1}^{\tau^r} \left[\beta_i^r(j) - D_i^r(j)\right] p^r(j) \tag{2.20}$$

Again, we allow $i = 1, 2, \ldots, T$ by constraining $\beta_i^r(j)$ to be zero for $i = 1, 2, \ldots, k$, since the charging services have been conducted at these time steps and rescheduling is no longer necessary.

At time step k, the optimization problem here will keep the schedule from the previous decision, and allocates only the new arrivals. For this reason, the first part of $\bar{L}_i$ in 0 never changes from iteration to iteration, so $\bar{L}_i$ can be obtained recursively by

$$\bar{L}_i = \bar{L}_i - \sum_{r=1}^{m} \sum_{j=1}^{\tau^r} D_i^r(j) p^r(j) \tag{2.21}$$

The optimization is formulated below as problem P_p^{PRH}.

Problem P_p^{PRH} (*Partial Receding Horizon Optimization*)

Find $\tilde{\beta}_i^r(j)$ to

$$\text{Min} \sum_{i=1}^{T} C\left(\sum_{r=1}^{m} \sum_{j=1}^{\tau^r} \tilde{\beta}_i^r(j) p^r(j) + \bar{L}_i\right) + \text{Max} \sum_{i=1}^{T} E_D \tag{2.22}$$

subject to the following constraints:

(a) $\tilde{\beta}_i^r(j) \in \mathbf{Z}^+$ for $i = 1, \ldots T, j = 1, \ldots \tau^r$, and $r = 1, \ldots, m$.
(b) $\tilde{\beta}_i^r(j) = 0$ for any $i < \max(a^r, k+1)$ or $k > b^r$.
(c) $\sum_{j=1}^{\tau^r} \tilde{\beta}_i^r(j) \leq \tilde{N}_k^r$, for $i = 1, \ldots T$, $r = 1, \ldots, m$.
(d) $\sum_{i=1}^{T} \tilde{\beta}_i^r(j) = \tilde{N}_k^r$, for any $j = 1, \ldots, \tau^r, r = 1, \ldots, m$.
(e) $\sum_{i=1}^{n+1} \tilde{\beta}_i^r(j+1) \leq \sum_{i=1}^{n} \tilde{\beta}_i^r(j)$, for all $n = 1, \ldots, T-1, j = 1, \ldots, \tau^r$, $r = 1, \ldots, m$.

In the above formulation, $\bar{L}_i$ are known constants for $k = 1, 2, \ldots, T$, as calculated in 0, which indicates the power amount at each time step if the charging schedules for the existing EV loads keep the same. The decision variables are $\tilde{\beta}_i^r(j)$, the number of newly-arrived type-r requests. Note that the optimization is conducted only on the new arrivals. In the situation where traffic of new arrivals is light, this small-size scheduling problem can be conveniently solved with heuristic algorithms such as Largest Energy Consumption First and Longest Process Time First.

In summary, the procedure of partial RH algorithm is listed as Algorithm 4 below.

Algorithm 4 (*Partial Receding Horizon Algorithm*):

Procedures:

Step 1. Set $k = 0$. Solve the original optimization problem P_p to obtain $\beta_i^r(j)$ for $i = 1, \ldots, T$, $j = 1, \ldots, \tau^r$, and $r = 1, \ldots, m$.

Let $\tilde{\beta}_0^r(j) = \beta_0^r(j)$ and

Calculate

$$\bar{L}_i = \sum_{r=1}^{m} \sum_{j=1}^{\tau^r} \beta_i^r(j) p^r(j) \tag{2.23}$$

Step 2. At time step k:

a) Issue charging services $\beta_k^r(j)$ for all j and r.
b) Obtain $D_k^r(j)$ and F_k from the aggregator. Update:

$$\bar{L}_i = \bar{L}_i - \sum_{r=1}^{m} \sum_{j=1}^{\tau^r} D_i^r(j) p^r(j) \tag{2.24}$$

c) Solve the updated receding horizon optimization problem P_p^{PRH}.

Step 3. Repeat *Step 2* procedure with $k = k + 1$ until $k = T$.

The goal of optimization module is to minimize the distribution network losses and to maximize the energy transferred from the grid to PEVs. Hence the optimization problem is a multi objective optimization problem for which we have used a bacterial foraging optimization algorithm. Bacterial Foraging Optimization (BFO) algorithm has been applied to model the E. coli bacteria foraging behavior for solving optimization problems. It is known that bacteria swim by rotating whip-like flagella driven by a reversible motor embedded in the cell wall. For E. coli have 8–10 flagella placed randomly on a cell body. When all flagella rotate counterclockwise, they form a compact, helically propelling the cell along a helical trajectory, which is called run. When the flagella rotate clockwise, they all pull on the bacterium in different directions, which causes the bacteria to tumble. The cycle of optimization can be divided into three parts: Chemotaxis, Reproduction, Elimination and Dispersal. Interested readers can refer to [44] for more detailed explanation about BFO. The following section explains Multi Objective bacterial Foraging Optimization method.

Since the BFO algorithms could solve single-objective optimization problems, the idea of solving multi-objective optimization problems with BFO algorithms was tested. However, the purpose of multi-objective optimization problems is to find all values which are possibly satisfied to all functions. Since different decision makers have different ideas about objective functions, it is not easy to choose a single solution for a multi-objective optimization problem without interaction with the decision makers. Thus, all we could do is to show the set of Pareto optimal solutions to decision makers. The main goal of multi-objective optimization problems is to obtain a non-dominated front which is close to the true Pareto front. The details of the new optimization algorithm based on BFO are given in the following sections.

In what follows we briefly outline the Multi-objective Bacterial Foraging Optimization (MBFO) step by step:

Algorithm 5 (*Multi – Objective Bacterial Foraging Optimization*

<u>**Algorithm MBFO**</u>

```
Begin
        Initialize all the parameters and positions
        While (a terminate-condition is met)
                For (Elimination-dispersal loop)
                        For (Reproduction loop)
                                For (Chemotaxis loop)
                                        Compute two fitness functions J1 and J2 .
                                        Let J last1 = J1, and J last2 = J2
                                        Update of the positions
                                End For (Chemotaxis)
                                Compute two health values Jhealth1 and J health2
                                Sort bacteria based on health values
                                Copy the best bacteria
                        End For (Reproduction)
                        Eliminate and disperse each bacterium with probability Ped
                End For (Elimination-dispersal)
        End While
End
```

2.8 Case Study Simulation and Results

For verification and validation of the proposed smart coordination infrastructure, two test systems were considered for simulation purposes. The first system is a IEEE 13 node distribution system with PEV charging system (Figs. 2.7 and 2.8).

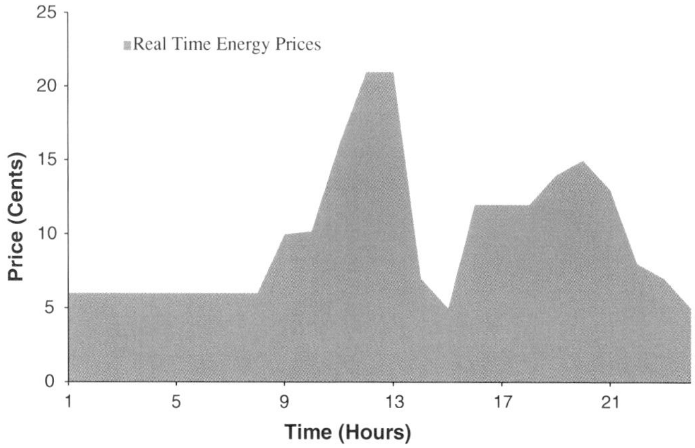

Fig. 2.8 Real time energy prices

Different penetration levels of PEVs have been considered. Reference [45] obtained that the capacity of this network for PEV charging station is equal to 1.2 MW, in case of unity power factor consideration. In this study, we have considered 20, 40 and 60 % penetration level based on the maximum capacity of the charging station, to see the impact of PEVs charging demand's increment on losses in the lines and energy transferred. In addition, the supervision system is considered in parallel to compare the improvement level of voltage profile and losses reduction. The results of supervised charging are labeled as coordinated, whereas unsupervised charging results are marked as uncoordinated.

In order to assess the state of a smart grid subject to PEV charging as well as generation status, voltage profile, and power losses necessary for the objective function and checking of constraints, a modified Newton-based load flow routine is used. All loads are modeled as constant power loads with their real and reactive powers updated through a daily load curve for each time interval the load flow is performed.

IEEE 13 node system The results of simulation are presented in Figs. 2.9 and 2.10. In Fig. 2.9, charging power of arriving vehicles in uncoordinated charging makes a peak demand at 19:00, which is increasing with more penetration level. While in the smart coordinated charging, the same charging power is distributed after the peak-hours until departure time of the vehicles. So the vehicles will be charged at the off-peak hours from 22:00 to 9:00 approximately. In addition, with the help of V2G application, from 13:00 to 20:00, capable vehicles with respect to their arrival SOC, participate in injecting power to the grid, i.e., V2G is enabled.

These results could be considered as global optimum, as a multi-objective bacterial foraging optimization algorithm has been considered to minimize cost of power losses and maximize energy transferred, where at each instant the a priori optimal result is taken into account. Finally, from network losses point of view, it is

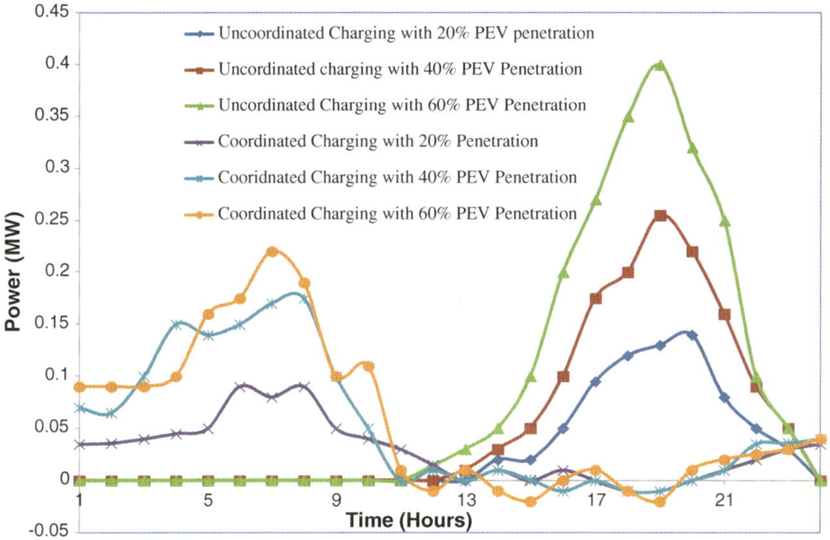

Fig. 2.9 Power drawn by PEVs for charging

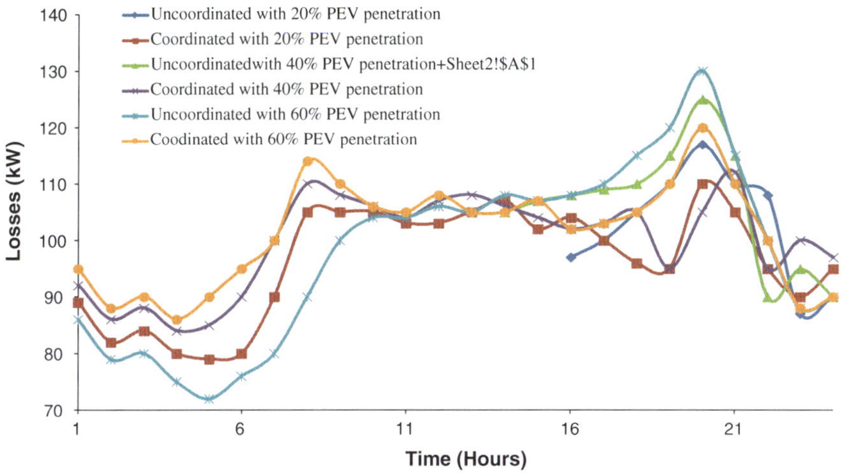

Fig. 2.10 Variation of total power losses in the system

shown in Fig. 2.10 that maximum losses at 20:00 with 50 % penetration is reduced with using coordinated charging strategy. Peak-hours charging avoidance is an important criteria from loss reduction point of view, which in this algorithm is implemented successfully. The multi objective bacterial foraging algorithm has resulted in a minimum loss and also maximum energy transferred to the PEVs. Under uncoordinated operation of the system and, low penetration of PEVs, the

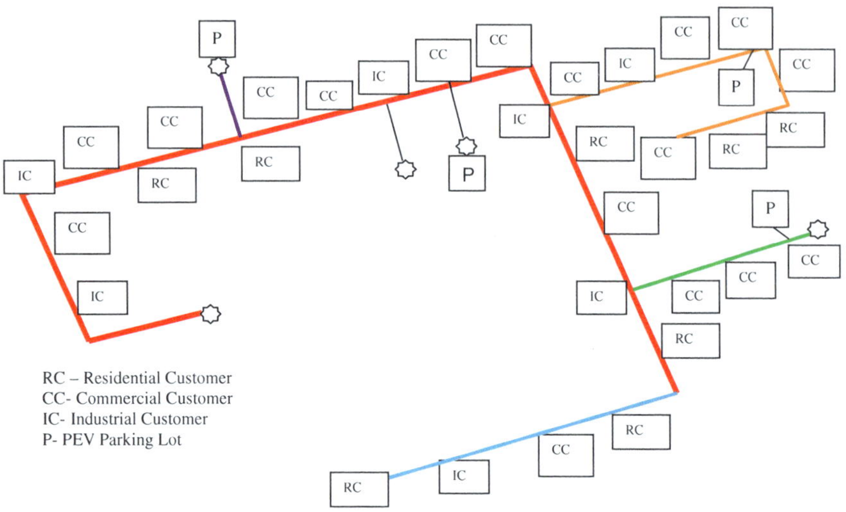

Fig. 2.11 Distribution feeder—38 bus system

system allows all the PEVs to be charged without violating the technical limitations and constraints imposed on the system. On the other hand, the smart coordination method even during the regular load peak, PEV charging is limited due the prediction module and optimization module functioning hand in hand to reduce the power losses and to maximize power transferred so that system reliability is preserved.

Bus Distribution Test Feeder The total system peak load is 4:37 MVA. The system line data, customer type, and load point demand are as given in [46]. The system contains four parking lots as shown in Fig. 2.11. Two cases of PEV penetration levels (20 and 40 %) were considered because above 40 % penetration level, PEV loads resulted in violation of several system constraints under uncoordinated charging. Load profile is given in Fig. 2.12.

Fig. 2.12 Load profile of the 38 bus system with residential, commercial and industrial loads

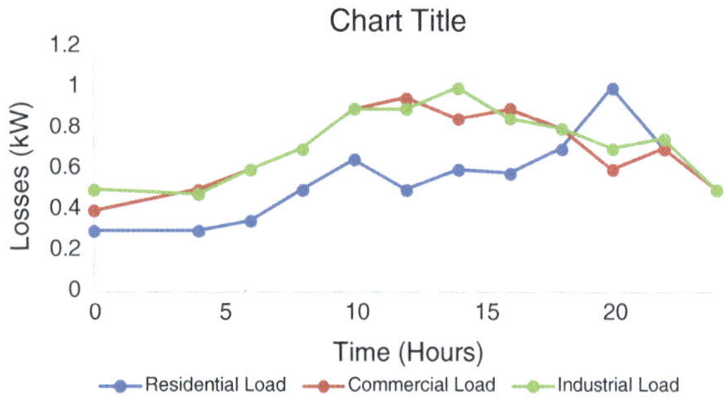

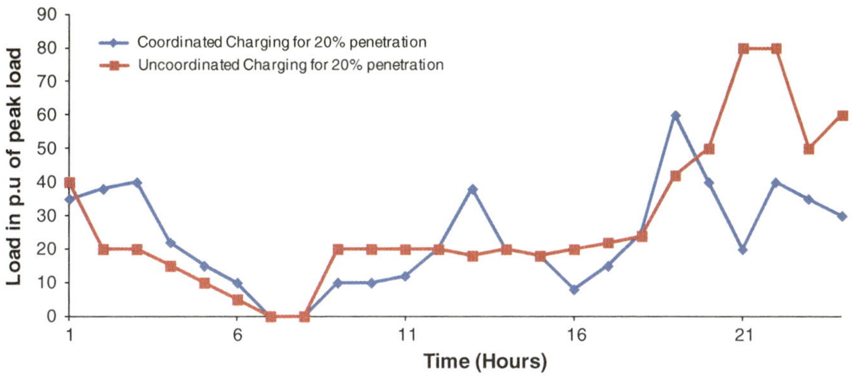

Fig. 2.13 Variation of total power losses in the system for 20 % penetration

The resulting coordinated and uncoordinated charging results for 20 and 40 % penetration of PEV's is shown in Figs. 2.13, 2.14 and 2.15.

Comparing the uncoordinated charging and smart coordination results, it is evident that a significant improvement in performance is achieved with the help of the proposed approach. Most importantly, the system peak demand is reduced which is very advantageous from the standpoint of generation dispatch and preventing overloads. Comparison of results in Figs. 2.9, 2.10, 2.12, 2.13, 2.14 and 2.15 also indicates that energy transferred to the PEV is increased compared to the uncoordinated case. Furthermore, peak power losses have been reduced to a fraction of the uncoordinated case. The computing time required by BFO to arrive at these solutions was in the range of 3 ms for each time period considered.

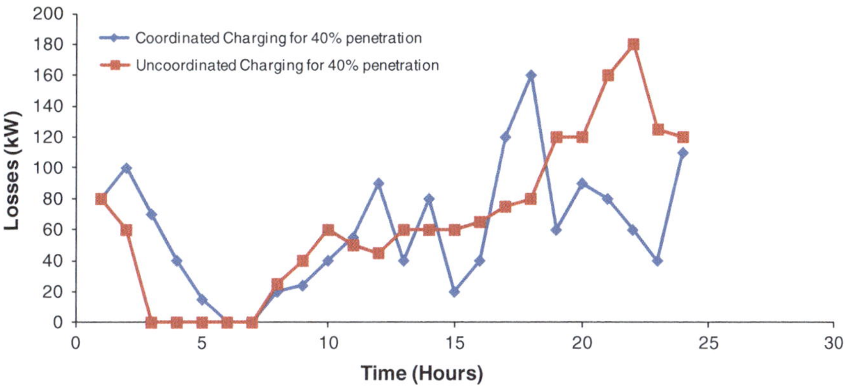

Fig. 2.14 Variation of total power losses in the system for 40 % penetration

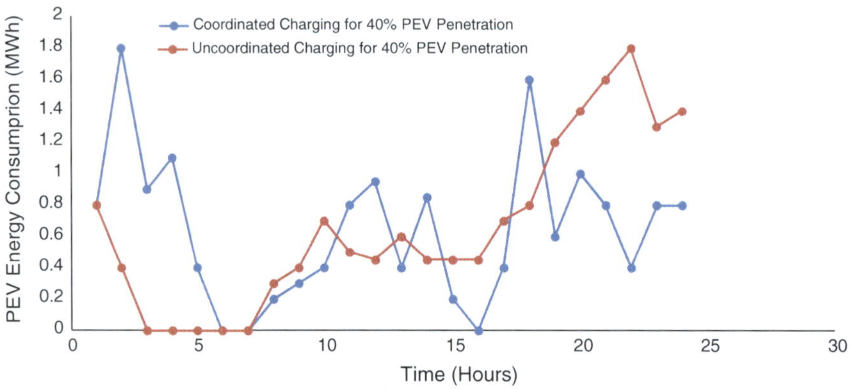

Fig. 2.15 Variation of maximum energy transferred to PEV over a 24 h period in the system for 40 % penetration

2.9 Conclusion

In this chapter, a real time system for managing the dynamics associated with charging/discharging of PEVs in a grid has been proposed. The smart coordination approach for power management incorporates a forecasting module and an optimization module. For a superior coordination of PEV charging/discharging, the forecasting module comprising of a receding time horizon approach provides information about the PEV loads for the next interval. Then the optimization module comprising of BFO based technique guarantees that the power management strategy results in minimum distribution network losses and maximum energy transferred to the PEV. This smart coordination approach has been implemented on two different test systems with varying levels of penetration. The results from simulation demonstrate the versatile performance of the smart coordination approach. This implementation is very appropriate for practical implementation as the computation time required by receding time horizon method and BFO based approach are of the order of milliseconds.

References

1. Juelsgaard M, Andersen P, Wisniewski R (2014) Distribution loss reduction by household consumption coordination in smart grids. IEEE Trans Smart Grid 5(4):2133–2144
2. Boulanger A, Chu A, Maxx S, Waltz D (2011) Vehicle electrification: status and issues. Proc IEEE 99(6):1116–1138
3. Heydt G (1983) The impact of electric vehicle deployment on load management strategies. IEEE Trans Power App Syst 102(5):1253–1259
4. Meliopoulos S, Meisel J, Cokkinides G, Overbye T (2009) Power system level impacts of plug-in hybrid vehicles. Power Systems Engineering Research Center (PSERC), Technical report

5. Clement-Nyns K, Haesen E, Driesen J (2010) The impact of charging plug-in hybrid electric vehicles on a residential distribution grid. IEEE Trans Power Syst 25(1):371–380
6. Fernandez L, Roman T, Cossent R, Domingo C, Frias P (2011) Assessment of the impact of plug-in electric vehicles on distribution networks. IEEE Trans Power Syst 26(1):206–213
7. Lopes J, Soares F, Almeida P (2011) Integration of electric vehicles in the electric power system. Proc IEEE 99(1):168–183
8. Dyke K, Schofield N, Barnes M (2010) The impact of transport electrification on electrical networks. IEEE Trans Ind Electron 57(12):3917–3926
9. Schneider K, Gerkensmeyer C, Kintner-Meyer M, Fletcher R (2008) Impact assessment of plug-in hybrid vehicles on pacific northwest distribution systems. In: Proceedings of power energy society general meeting, pp 1–6, Jul 2008
10. Wu D, Aliprantis D, Gkritza K (2011) Electric energy and power consumption by light-duty plug-in electric vehicles. IEEE Trans Power Syst 26(2):738–746
11. Masoum AS, Deilami S, Moses P, Masoum MAS, Abu-Siada A (2011) Smart load management of plug-in electric vehicles in distribution and residential networks with charging stations for peak shaving and loss minimization considering voltage regulation. IET Proc Gener Transm Distrib 5(8):877–888
12. Deilami S, Masoum AS, Moses P, Masoum MAS (2011) Real-time coordination of plug-in electric vehicle charging in smart grids to minimize power losses and improve voltage profile. IEEE Trans Smart Grid 2(3):456–467
13. Ashtari A, Bibeau E, Shahidinejad S, Molinski T (2012) PEV charging profile prediction and analysis based on vehicle usage data. IEEE Trans Smart Grid 3(1):341–350
14. Wu D, Aliprantis DC, Ying L (2012) Load scheduling and dispatch for aggregators of plug-in electric vehicle. IEEE Trans Smart Grid 3(1):368–376
15. Bashash S, Fathy HK (2012) Transport-based load modeling and sliding mode control of plug-in electric vehicles for robust renewable power tracking. IEEE Trans Smart Grid 3(1):526–534
16. Han S, Han S, Sezaki K (2012) Development of an optimal vehicle-to-grid aggregator for frequency regulation. IEEE Trans Smart Grid 1(1):65–72
17. Sortomme E, El-Sharkawi MA (2011) Optimal charging strategies for unidirectional vehicle-to-grid. IEEE Trans Smart Grid 2(1):131–138
18. Dam QB, Mohagheghi S, Stoupis J (2008) Intelligent demand response scheme for customer side load management. In: Proceedings of IEEE energy 2030 conference, pp 1–7
19. Mauri G, Moneta D, Bettoni C (2009) Energy conservation and smartgrids: new challenge for multimetering infrastructures. In: Proceedings of IEEE Bucharest PowerTech, pp 1–7
20. Chuang A, Gellings C (2009) Demand-side integration for customer choice through variable service subscription. In: Proceedings of IEEE power energy society general meeting, pp 1–7
21. Bruno S, Lamonaca S, Scala ML, Rotondo G, Stecchi U (2009) Load control through smart-metering on distribution networks. In: Proceedings of IEEE Bucharest PowerTech, pp 1–8
22. Rahimi F, Ipakchi A (2010) Demand response as a market resource under the smart grid paradigm. IEEE Trans Smart Grid 1(1):82–88
23. Pillai JR, Bak-Jensen B (2011) Integration of vehicle-to-grid in the Western Danish power system. IEEE Trans Sustain Energ 2(1):12–19
24. Moslehi K, Kumar R (2010) A reliability perspective of the smart grid. IEEE Trans Smart Grid 1(1):57–64
25. Russell BD, Benner CL (2010) Intelligent systems for improved reliability and failure diagnosis in distribution systems. IEEE Trans Smart Grid 1(1):48–56
26. Metke AR, Ekl RL (2010) Security technology for smart grid networks. IEEE Trans Smart Grid 1(1):99–107
27. Jeongje P, Jaeseok C, Shahidehpour M, Lee KY, Billinton R (2010) New efficient reserve rate index of power system including renewable energy generators. In: Proceedings of international conference innovative smart grid technologies (ISGT), pp 1–6, 19–21 Jan 2010
28. Jaeseok C, Jeongje P, Shahidehpour M, Billinton R (2010) Assessment of CO_2 reduction by renewable energy generators. In: Proceedings of international conference innovative smart grid technologies (ISGT), pp 1–5, 19–21 Jan 2010

29. Civanlar S, Grainger JJ (1985) Volt/var control on distribution systems with lateral branches using shunt capacitors and voltage regulators part III: the numerical results. IEEE Trans Power App Syst 104(11):3291–3297
30. Duvall M, Knipping E, Alexander M (2007) Environmental assessment of plug-in hybrid electric vehicles EPRI. Nationwide greenhouse gas emissions, vol 1
31. Turitsyn K, Sinitsyn N, Backhaus S, Chertkov M (2010) Robust broadcast-communication control of electric vehicle charging. In: Proceedings of IEEE smart grid Communications, pp 203–207
32. Pang C, Dutta P, Kezunovic M (2012) BEVs/PHEVs as dispersed energy storage for V2B uses in the smart grid. IEEE Trans Smart Grid 3(1):473–482
33. Cao Y, Tang S, Li C, Zhang P, Tan Y, Zhang Z, Li J (2012) An optimized EV charging model considering TOU price and SOC curve. IEEE Trans Smart Grid 3(1):388–393
34. He Y, Venkatesh B, Guan L (2012) Optimal scheduling for charging and discharging of electric vehicles. IEEE Trans Smart Grid 2(3):1095–1105
35. Bessa RJ, Matos MA, Soares FJ, Lopes JAP (2012) Optimized bidding of a EV aggregation agent in the electricity market. IEEE Trans Smart Grid 3(1):443–452
36. Faruqui A, Hledik R, Levy A, Madian A (2011) Will smart prices induce smart charging of electric vehicles? The Brattle group, discussion paper, July 2011
37. Paracha Z, Doulai P (1998) Load management: techniques and methods in electric power system. Proc Energ Manage Power Deliv 1:213–217
38. Palensky P, Dietrich D (2011) Demand side management: demand response, intelligent energy systems, and smart loads. IEEE Trans Ind Inf 7(3):381–388
39. Mohamed A, Khan MT (2009) A review of electrical energy management techniques: supply and consumer side (industries). J Energ South Afr 20(3):14–21
40. Backer DL (1986) Load management direct control fact or simulation. IEEE Trans Power syst 1(1):82–88
41. Bhattacharya K, Bollen MHJ, Daalder JE (2001) Operation of restructured power systems. In: Pai MA (ed) Springer, US
42. Geng X, Khargonekar P (2012) Electric vehicles as flexible loads: algorithms to optimize aggregate behavior. In: 2012 IEEE third international conference on smart grid communication, pp 430–435
43. Geng X, Khargonekar P, Ramachandran B (2014) Receding horizon power management for electrical vehicle charging. In: Power systems conference, 2014 Clemson University, Clemson, pp 1–7, 11–14 Mar 2014
44. Das S, Biswas A, Dasgupta S, Abraham A (2009) Bacterial foraging optimization algorithm: theoretical foundations, analysis, and applications. Stud Comput Intell 203:23–55
45. Sarabi S, Kefsi L, Asma M, Robyns B (2013) Supervision of plug-in electric vehicles connected to the electric distribution grids. Int J Electr Energ 1(4):256–263
46. Singh D, Misra R, Singh D (2007) Effect of load models in distributed generation planning. IEEE Trans Power Syst 22(4):2204–2212

Chapter 3
Plug-in Electric Vehicles Management in Smart Distribution Systems

Antonio Carlos Zambroni de Souza and Denisson Queiroz Oliveira

Abstract The advent of smart grids brings a set of new concepts not usually employed in current power systems. Plug-in electric vehicles fit this concept, since it is a low carbon emission device. However, an important characteristic of plug-in vehicles lies on the fact that it may become a source of energy during emergency conditions. Another aspect that may not be overlooked is the fact that the advantage of low carbon emission may be faded by the fact that charging these vehicles may deteriorate the network operating conditions. In this sense, a recharging policy must be addressed, so the system losses and voltage profile are adequately managed. This chapter deals with these topics, so the advent of plug-in electric vehicles may be understood as an important component of future smart grids.

Keywords Plug-in electric vehicles · Energy management system · Intelligent control

3.1 Introduction

The Plug-in Electric Vehicles (PEVs) are the result of efforts made by automotive industry to develop a product that reflects the worry about the environment and the future. These vehicles have energy storage systems (ESSs) to supply power to an electric motor and mechanical shaft. They generally have energy recovery systems to help to charge the ESS. Each manufacturer develops its power unit architecture and mechanical transmission by choosing the technologies that best meet the individual requirements. Although many manufacturers are already trading their PEV models, their expensive prices, market uncertainties, and some other points still turn them into the last choice for the most of people.

A.C. Zambroni de Souza (✉) · D.Q. Oliveira
Federal University of Itajuba, Itajuba, Brazil
e-mail: zambroni@unifei.edu.br

S. Rajakaruna et al. (eds.), *Plug In Electric Vehicles in Smart Grids*,
Power Systems, DOI 10.1007/978-981-287-302-6_3

But, as the PEVs become more popular, they will bring some problems to the electrical power distribution systems. As the vehicles are intended to be recharged mainly at home in residential outlets, a sudden load increase due to thousands of PEVs recharging simultaneously in the evening may harm the system. The PEVs charging process will contribute to increase load uncertainty, peak load, power losses, harmonic distortion, voltage deviations, conductors and power transformers overload and early aging. In addition, it may change the philosophy of power systems operation, since the load may become a supplier in emergency conditions. This important issue is about to be discussed and will certainly play a crucial role on the definition of policies and operating modes of power systems in the future.

Beyond, this daily peak load will require reinforcements over the grid and new assets, though this new capacity will remain idle during almost all the time. Furthermore, the PEVs will require the development of a new recharging infrastructure on the streets and public buildings. These recharging stations may belong to the electricity utility or to private agents. However, the domestic recharging is expected to be more ordinary than those options. As depicted above, the PEVs integration represents a great challenge for the electrical power systems, which has to be faced by engineers, power utilities and all electric power industry.

The emerging smart grids concept presents a novel structure for electricity industry with many new features which adds many possibilities for electrical power grid control, operation and management in all stages. The communication infrastructure advanced smart metering and demand management systems are some of these new features, providing real-time information on all system variables, as voltage levels, currents, active and reactive power flows and power losses.

Regarding to PEVs, the smart grids present several solutions to manage their integration to electrical power systems. Real-time communication, demand management systems, demand response and intelligent control systems are some of these features. This integration process comprises the recharging process, the communication between the Distribution System Operator (DSO) and the vehicles and between PEVs and aggregating agents, and finally the vehicle-to-grid capability.

In the literature, several papers suggested many solutions to cope with this problem in smart grid environment. The main solution is the adoption of a recharge policy based on a Demand Management System (DMS) to keep the voltage on acceptable levels, avoid peak loads and transformers and conductors overload, and decreasing losses by using on-load tap change transformers (LTCs) and shunt capacitors.

On the other side, PEV popularization may bring benefits to power systems, like load leveling, increased generation capacity to the grid during high load periods, regulation capacity, emergency reserve and integration of renewable power generation.

This chapter introduces the PEV management problem and the solutions on smart grids environment. For this purpose, some technical issues are first introduced. Then, the issue of management of recharging process is presented. For this sake, a recharging policy that aims to keep the operating conditions within the specified standards is proposed. This is based on a recharging management along an overnight, so the vehicles are charged along the night with no operating violation.

3.2 PEV Management in Smart Distribution Systems

As different PEV models from several manufacturers are presented on specialized events and are available for trade, the consumers become more familiar with this new vehicle concept. It is possible to perceive good feelings about them, as they represent a green option for the automotive market with an environmental-friendly approach based on zero greenhouse gases emissions, free or low fossil fuel usage and low noise.

However it is not expected this type of vehicles become dominant soon, there is a concern about their integration to power systems. As the modern society development is also based on large electricity usage and the population is still growing, a higher pressure on electrical energy generation is expected. The PEVs will have an important role in this scenario. As the prices become cheaper and the PEV turns into a real option for population, a higher load is expected, especially in the evening.

The integration of PEVs in distribution power systems will create a new load to be connected to the grid mainly in the evening, possibly causing low voltage levels on further nodes of feeders, high currents which can cause transformers overloading and exceed thermal limits, power losses and harmonics distortion increase.

In addition to it, this new load claims for investments on distribution grid expansion and assets replacement (substation and lines transformers, conductors, fuses and switch breakers) to be supplied properly. The current system infrastructure seems to be insufficient to the task, but the expensive investment on system expansion seems to not be the better option, as the new load is highly dependent of the time of the day. This means that the entirely system will operate on a lower efficiency point, remaining idle during almost the day.

As the PEV integration to power systems is expected to create so many problems, it is necessary to find a way to overtake them and also take advantage of it. The smart grids have suitable features to solve this problem, e.g. real-time communication, Advanced Metering Infrastructure (AMI) and Demand Management Systems (DMS). For this sake, many researchers have presented the implementation of a vehicle policy recharge controlled by an Intelligent DMS as the possible solution for PEV integration.

This DMS is responsible for managing the controllable loads, i.e. the PEVs. This is done by shifting the PEV recharge process for late night, during a time when the system have low load and it is possible to supply them with sufficient power to reach a suitable level for transportation purposes and control the system operational variables. This approach also avoids large investments for system expansion to supply the new load, making possible to supply a PEV fleet with the current network infrastructure.

As the reader can imagine, the development of smart grids is a fundamental step to help on PEV integration. However some aspects like real-time communication systems and DMS, though not mandatory, tend to facilitate the whole process. Hence, the charging control and management algorithm must receive all the information from the system and assesses the charging schedule for each PEV and the network operational state. By the communication infrastructure, the DMS send signals for all customers with determined schedules.

The DMS may perform its task in a centralized or distributed way. These two approaches are discussed on next sessions.

3.2.1 Centralized Demand Management System Approach

As described before, the DMS role is to manage the controllable loads aiming to shave load on peak times and allowing the PEV recharge only in low demand time. This is equivalent to shift the recharge process to late night, when the system's load is low.

By the time the PEV owner arrives home, he/she plugs the vehicle on the residential outlet. Generally, the vehicles are used for commuting during the weekdays, i.e., a round-trip travel from home to work. In this case, it is expected that the vehicle will be used again only in the next morning. Sometimes the vehicle can be used for unscheduled travels during the night. On the weekends, there will be longer travels that should require more power from the ESS.

As one could see, the vehicle usage has an ordinary schedule represented by the daily commuting and some deviations represented by the unscheduled travels and weekends. Considering that all PEV owners have an individual schedule, this stochastic behavior should be represented by a probabilistic distribution. Although these are random variables, they have a predictable behavior if enough data is stored in order to assess them.

These comments are important to show that there is an optimal ESS State of Charge (SoC) to be reached in every case to assure the transportation daily purposes, which is not necessarily the 100 % of SoC. So, if exists some smart control on the PEV charger, this feature is very welcome. This intelligent control will learn about the owner's driving patterns, store data as standard commuting distance and assess this optimal SoC.

After connecting the vehicle to the outlet, the consumer may choose between a standard charge process, if the vehicle will only be used in the next day and a fast charge process if he wishes to use the vehicle soon. This information is sent by the vehicle charger to the DSO operator through the real-time communication network. At the DSO operational center, it is also expected that there is some information about the customers' charging schedule. Some data like PEV connection probability during the day for each feeder or geographical region is expected to exist and it will help the DSO to assess some important variables like load forecasting.

Besides this information, several data is sent to DSO operational center through Supervisory Control and Data Acquisition Systems (SCADA), e.g. nodal voltages, currents and power measurements, Load-Tap Transformers (LTC), Line Step-Voltage Regulators (SVR) and Shunt Capacitors (SC) status. These measurements have an important feature: they are not acquired at the same hour, i.e. they are not synchronized.

Hence, after receiving all data from SCADA and from real-time communication systems, the DSO operational center has all necessary input for its assessments. The

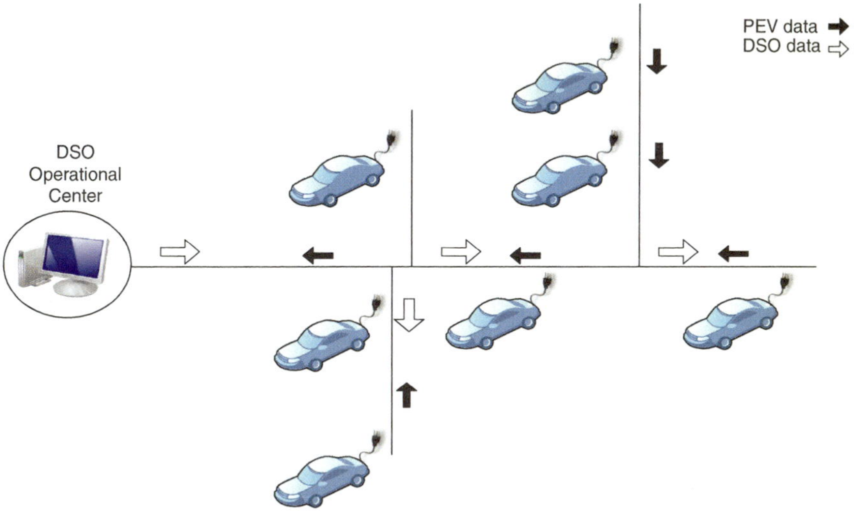

Fig. 3.1 A centralized DMS control for PEV management

DSO is now able to run the optimization and operational algorithms in order to assess the best control actions for proper system operation and also to determine the PEVs charging schedules in order to reach their optimal SoC in expected time and to maintain the systems' variables within limits. The expected outputs of this process are the individual PEVs charging schedules and tape positions for voltage control devices as LTCs, SVRs and SCs. This data is sent back through real-time communication system and voltage control devices can be controlled remotely from the substation.

This approach requires a fast, robust and reliable communication system. The amount of data is high even for a small system. But, by the other side, the centralization of control makes possible to assess the optimal schedule and control actions each time.

Figure 3.1 depicts the main players involved in a centralized DMS structure. There are several PEV units connected to residential outlets which send their recharging schedules directly to the DSO operational center through the real-time communication channel. The flow of the PEVs' data is represented by the dark arrow and as described before, it contains the desired schedule by the PEV owner and other important information as the expected plug-out time. All information is processed by the DSO operational center and the optimal schedule is sent for all connected PEVs. The DSO's data flow is represented by the light arrow.

3.2.2 Distributed Demand Management System

Since the DMS role is to manage the controllable loads aiming to shift the PEVs' recharge process to a low demand time, this also could be done using a distributed or decentralized approach. The decentralized approach is based on the existence of local controllers spread over the entire system. These local controllers are able to receive data from different devices under their control and assess some optimization and control algorithms in order to send back setup configurations to the controlled devices.

The role of these local controllers is played with the help of computational agents. These agents are computational intelligent players that are designed to execute specific tasks. A set of agents working on a system creates a multiagent system. These agents have some interesting features that allow them to perform complex tasks:

- Intelligence.
- Learning capacity.
- Sociability.
- Communication capacity.
- Hierarchy.

During the normal operation of a multiagent system, each agent looks for its own objectives, i.e. they compete for resources considering the system limits and commands from superior agents. In a case of an emergence, they are able to work together under superior agents commands to overtake the critical situation, i.e. they collaborate with each other during the emergence. These agents' features make them a good choice for application on smart grids distributed control.

In the decentralized approach, by the time the PEV owner arrives home, he/she plugs the vehicle on the residential outlet and chooses for a recharging mode, which could be fast or standard. At this moment, the residential agent is responsible to control and assess the PEV schedules according to the information received from the vehicle charger.

The residential agent sends to the upper level agent, called here feeder agent, a request for an amount of power. The feeder agent receives requests from several residential agents. It may assess simple calculations regarding to the available feeder's power and current capacity and should send a requesting message for an upper agent, called here as Substation Agent. The Substation agent, by its turn, communicates with Feeder Agents and may also assess some other variables and control devices operations and control all agents in lower hierarchy. Substation agents can communicate to DSO Agent, which is the highest level. The DSO Agent is on DSO operational center and is able to control and to communicate with all agents in different substations. As the different agents assess their solutions, they send the solutions to other agents in hierarchy, which updates their controls and PEV schedules according to the solution presented.

As this scenario is just an example, in real situations could exist more or less agents than in the situation described above. But, this is a good example to describe the operation of multiagents systems in a smart grid environment.

Figure 3.2 illustrates the structure of a decentralized PEV management DMS control and the different participants. The different agents' data flows are represented using color arrows. Figure 3.2 aims to show the data flow of each agent.

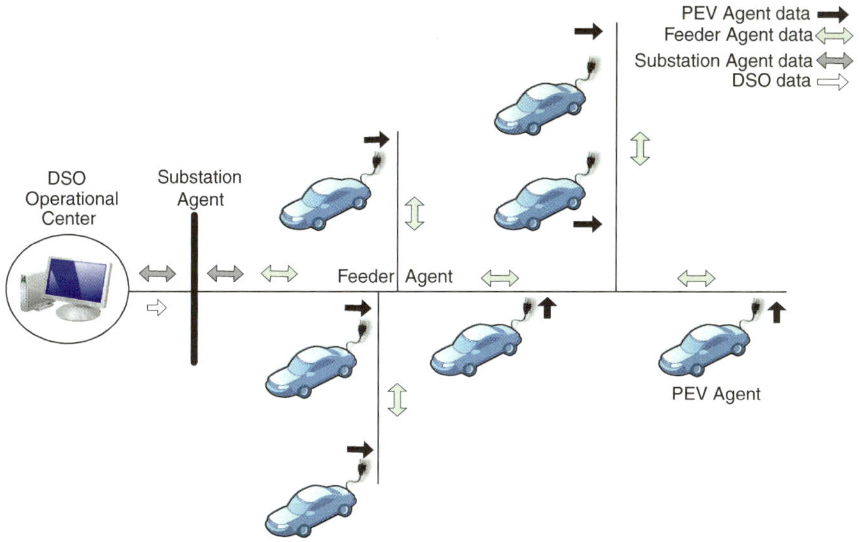

Fig. 3.2 A decentralized DMS control for PEV management

There are some bidirectional arrows, which mean that the respective agent can request/send information and also receive data from other agents.

3.2.3 Technical Issues

The problem of PEV management in smart distribution grids is a complex task. It depends on the existence of some requirements which, until now, are not properly determined. Although the major parts of the necessary technologies already exist, there are some questions that are not answered yet. These questions refer mainly to regulatory aspects and to the definition of standard practices.

The communication issue is one of the most important. As seen before, the real-time communication system has a big role on all the system operation. It is responsible for bringing data from all over the system and for taking them back to every smart device in the network. So, some requirements are mandatory for these communication systems:

- Reliability.
- Robustness.
- Data speed.
- Data capacity.
- Signal range.

Currently, there are a set of technologies which could be employed in smart grids for different purposes. The Power Line Carrier (PLC), the Internet (TCP/IP), ZigBee, WiMax, radio waves, mobile communication lines, optical cables and so are existent and emerging technologies with potential applications. As they have different features regarding to speed, reliability, capacity and signal range it is expected that not only one technology become dominant, but a merge of them is applied in each case.

PLC and wireless communication are promising technologies to implement in distribution level. The PLC carries data on a conductor that is also used simultaneously for electricity supply. The main issues for this technology are the power wires limited ability to carry high frequency signals and interferences with radio services. At its turn, the WiMax is a wireless communication standard designed to provide higher data rates and it refers to interoperable implementations of the IEEE 802.16 family of wireless networks standards. Due to its higher bandwidth and range it is suitable for potential smart grids and metering applications.

Wi-Fi covers almost 100 meters and can communicate with 3G/4G technologies. It is a low cost solution to control and monitor appliances in-house, most used in Home Area Networks (HANs). As for ZigBee, it is a promising technology to be employed indoors (HAN), i.e., there is no need of infrastructure investment, good data rates and high QoS (Quality of Service), but it faces some disadvantages, high monthly fees, very susceptible to interferences and it does not cover some rural areas. Radio Frequency (RF) is a technological resource used to provide data communication via electromagnetic signals propagating through space. Radios are arranged in series to a pre-determined distance so there is no signal loss and good propagation. Cognitive Radio (CR) is a new concept which takes advantage of white space spectrum in the VHF and UHF/TV ranges. It is composed by a transmitter that analyses the environment and adjusts its own operating frequency from a spare one, which has the fewest possible interference.

As depicted before, the communication requirements for PEV integration on smart grids are high, especially for centralized control approach. The amount of data that should be carried is high even for a small system. The required channel should be robust, fast and reliable.

Still on communication requirements, another important aspect is the geographical area. As bigger the geographical area and/or the demographical density, higher is the data amount because there is a high concentration of PEV over these areas while there is low density areas with low number of PEVs. On these areas, faster and more robust systems are required than on low density neighborhoods.

Regarding to decentralized approach, the same concerns described above also apply, but in this case the data flow is more restrict, have less distance and the amount is lower than in the other case. Even the decentralized approach seems to require less resources, the previous requirements are as important as for the centralized approach.

In smart grids environment, there is a big set of data available for control purposes, e.g. voltage levels, currents, power flows, PEVs schedules, real-time load measurements and so. Hence, there is a possibility to adopt a smart tariff program.

This program adopts a variable electricity tariff for different hours of day. In a smart tariff schedule, the electricity price may be higher in the evening to discourage the electricity usage. At late night, this tariff may be lower to stimulate the PEV recharging process at this time.

For PEV integration, the existence of smart tariff programs stimulates the shift of recharge process to low demand times. The choice for fast charge in the PEV schedule can also be charged with a higher tariff by the DSO. Doing so, the DSO can stimulate the choice for standard recharge, and the fast charging will only be used when it is really necessary.

3.2.4 Other Issues

The PEV integration to power systems does not depend only on technical aspects. In addition to the technical considerations, there are some cultural important considerations that will influence all the integration process.

One of the concerning points is related to how a customer will react when his/her vehicle is unavailable due to low charging state, even after connected to the outlet. This can be a common occurrence considering that the DMS will manage the PEVs schedule according to many variables, including the hour of connection and the available load margin.

To cope with these situations, there is the possibility to propose customer-oriented approaches for DMS control. These customer-oriented programs are based on a control strategy that assures the customer settled schedule is matched and the customer convenience is assured. The customer-oriented approach should be based on customers programs with different features and even with different electricity tariffs. If different tariffs cannot be applied, a reward for customers who signed the program should exist. The customers programs should also include higher tariffs for fast charging schedules.

As occurred with other technologies, the development and popularization of PEVs depend on the creation of a public infrastructure to provide a new set of services for those customers. As a clear example, the development and popularization of the combustion engine vehicles was helped by the development of the oil industry, which could supply the market with cheaper gasoline. The infrastructure was improved with more gas stations, more streets and roads, cheaper components and so.

The infrastructure development, as for fossil fueled vehicles, will help the transportation electrification, not only private PEVs, but also electric buses, trucks and other heavy duty vehicles. Charging points on the streets and on parking lots, fast recharge points, cheaper components and new services can help on integration process.

These new services have a huge potential to overcome the resistance to electric vehicles. The combination of radically different technologies and a highly complex multi-agency operating environment theoretically provide the conditions and requirements for such an emergent business models. As discussed deeper in [1], the understanding of the interplay between place, innovation and sustainability suggests that diverse solutions are likely to be characteristic solution rather than ubiquity and standardization.

Some of these innovative business models include renewable energy integration, battery second use, mobility services and so. The battery second use refers to the reuse of batteries in stationary applications after their automotive retirement thus could contribute to lower the costs. The aim of stationary battery usage lies in storing energy and releasing it at more beneficial time periods. The reason can be found in an added disposal value of the used battery reducing initial battery prices of PEVs. The disposal value depends on the battery lifetime after retirement and its application purpose [1, 2].

Regarding the communication technologies discussed previously, the electromagnetic radiation from wireless technologies should also be considered and it should fits on existent standards, even if the consequences for human health are not completely understood.

3.3 Case Study

In this section, let consider a case study applying a centralized DMS approach to investigate some practical aspects of the PEVs management in smart grids aiming their smooth and complete integration to the grid.

3.3.1 Initial Assumptions

First, let choose one distribution power system for the example. The IEEE 34 node test system is an interesting example to be considered [3]. It is a three-phase feeder with two step-voltage regulators (SVR) and two shunt capacitors (SCs), single-phase laterals and unbalanced load. The IEEE 34 node network diagram is depicted on Fig. 3.3.

In order to create a more realistic example to be investigated, let consider a heterogeneous geographic location of PEVs with some low density areas and other higher PEV density areas. This heterogeneous feature is also found in real distribution systems due to demographic aspects. The PEV localization used in this example is shown in Fig. 3.4. It shows the PEV percentage distribution for each hundred vehicles. In the x-axis it is depicted the node number and in the y-axis the respective percentage of connected PEVs. The percentages shown here are kept for all simulations, independently from the quantity of PEVs.

Some different real PEV models are also considered with different ESS capacity and features. These different PEV models are presented on Table 3.1. The first column shows the PEV model, the second column presents the ESS nominal capacity, in kWh, and the third column shows the maximum Depth of Discharge (DoD) for each model. The DoD is the maximum allowable discharge for the ESS. If the ESS is discharged below this point, it will suffer early aging and decrease the

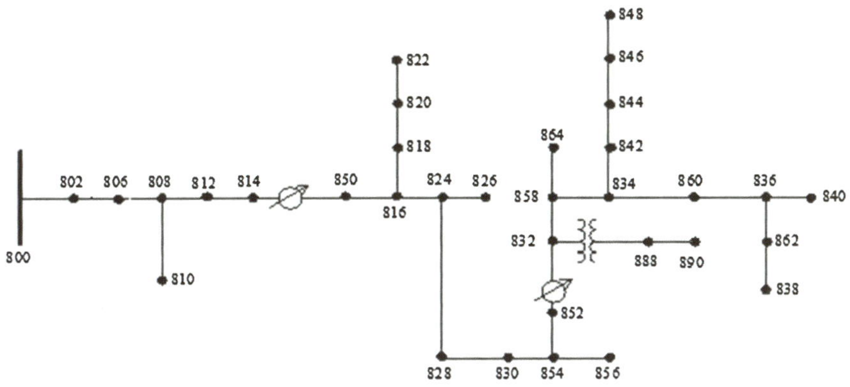

Fig. 3.3 Network diagram of IEEE 34 node distribution system [3]

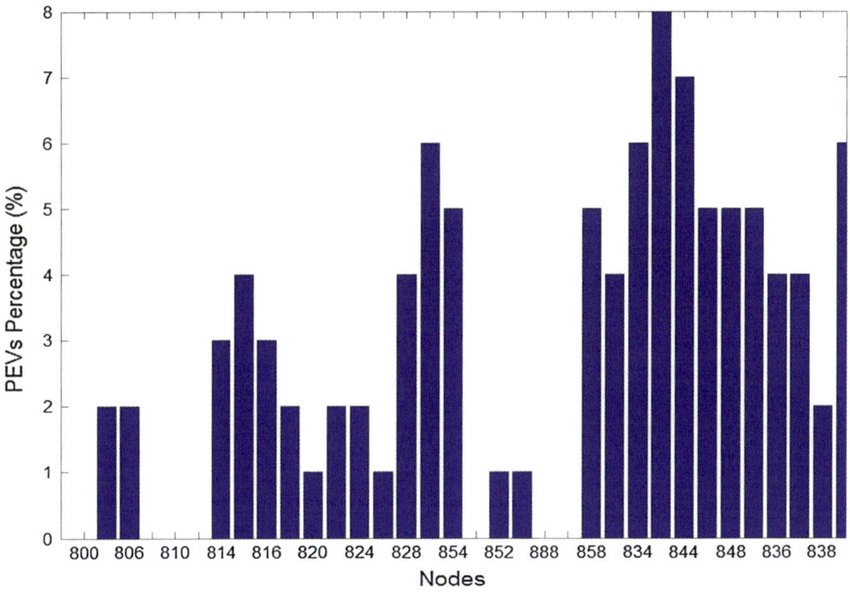

Fig. 3.4 PEV distribution on IEEE 34 node system

Table 3.1 PEV models

PEV models	ESS nominal capacity (kWh)	Maximum DoD (%)	Consumption (km/kWh)
Model 1	42	80	4.9
Model 2	24	65	6.15
Model 3	16	65	5.86

batteries life cycle. In the fourth column, the energy consumption in kilometer per kilowatt-hour of each model is depicted.

Important information for simulations regards to daily commuting distance. It is obvious that all PEVs do not run the same distance and one PEV does not run the same distance every day. So, the PEV's commuting distance can be represented by a normal distribution with a mean and a standard deviation. For the case study, all PEV fleet commuting distance is determined following the feature described in Fig. 3.5.

In the same way the commuting distance is not equal for all PEVs, the connection hour is not the same too. Hence, it is also considered a connection probability during the simulations, as depicted on Fig. 3.6. Looking at the situation pictured here, the connection of a PEV has more probability to occur during these hours of the day. For the case presented here, a PEV can be connected from 4 to 10 p.m.

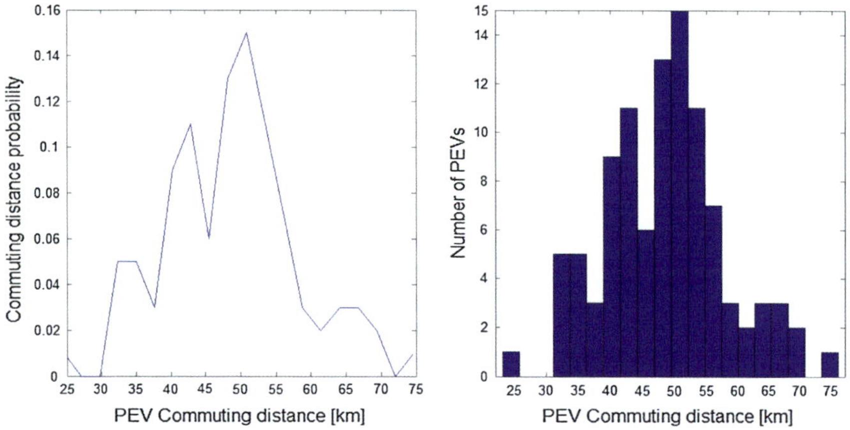

Fig. 3.5 Commuting distance for PEVs

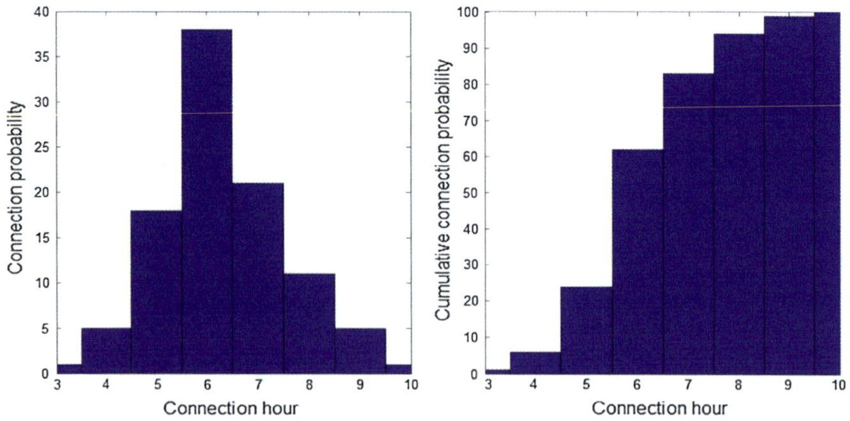

Fig. 3.6 PEV connection probability

When a PEV is connected to the outlet, the customer is supposed to inform desired the recharge schedule. This schedule configuration should include the type of recharge, which can be standard or fast. Beyond this, other information may be included and new data assessed by the intelligent charger, as the expected time of disconnection and the optimal SoC for the PEV.

The DMS is responsible for managing the PEV recharge process in the distribution system, avoiding the problems described before and assuring the optimal SoC for each PEV is reach by the time the vehicle is disconnected the next morning. To accomplish this task, the DMS manages the charging power for each vehicle, controlling the load while shift the demand to late night.

It is possible to describe better the DMS role looking for the power flow equations in (3.1)–(3.2).

$$P_n = V_n \sum_{x=1}^{N} V_m \cdot (G_{nm} \cos \theta_{nm} + B_{nm} \sin \theta_{nm}) \tag{3.1}$$

$$Q_n = V_n \sum_{x=1}^{N} V_m (G_{nm} \sin \theta_{nm} - B_{nm} \cos \theta_{nm}) \tag{3.2}$$

Regarding to active and reactive power for each node, assessed as (3.1)–(3.2), the load power is assessed as (3.3). The DMS should manage the P_{PEV} demand, since this is the controllable load in the system. Regarding to P_L, it refers to non-controllable loads connected to the grid and it is not a variable to be considered in the problem.

$$P_n = P_G - \left(P_L + \sum P_{PEV}\right) \tag{3.3}$$

where

P_n Node active power
P_G Node active power generation
P_L Node active power load
P_{PEV} Node active power PEV load

While managing the PEV recharge, the DMS also looks for improving some system operational conditions, as power losses, load unbalance, and cost of recharging process considering some restrictions, e.g. voltage level, ampacity, and power transformers loading. Here, let consider the DMS searches for power losses minimization through capacitor placement and switching and load management. The objective function for this problem is defined on (3.4).

$$\text{Min} \sum_{r=1}^{R} \text{Re}\left(Z_r \cdot I_r^2\right) \tag{3.4}$$

St.:

- $$Vn(a,b,c)_{min} < Vn(a,b,c) < Vn(a,b,c)_{max} \quad (3.4.1)$$

- $$Ir < Ir_{max} \quad (3.4.2)$$

where:

r	Branch
R	Total number of branches
n	node (phases a, b, and c)
Vmin	0.90 p.u
Vmax	1.05 p.u
Ir	Branch current
Ir_{max}	Maximum conductor ampacity

3.3.2 The Artificial Immune Systems

As optimization tool, the Artificial Immune System (AIS) is applied during the investigation. The AIS is an evolutionary technique which mimics the natural immune systems from the animals. This system works in a decentralized, parallel and adaptive way, desirable features to solve complex problems.

The immune system of the vertebrate animals is a complex of cells, molecules and organs that represent an identification mechanism capable of perceiving and combating dysfunction from our own cells and the action of exogenous infectious agents. It recognizes an almost limitless variety of infectious foreign cells and substances, known as nonself elements, distinguishing them from those native cells. When an infectious foreign agent enters the body, it is detected and mobilized for elimination.

The tissues and organs that compose the immune system are distributed throughout the body. They are known as lymphoid organs, once they are related to the production, growing and development of lymphocytes, the leukocytes that compose the main operative part of the immune system. In the lymphoid organs, the lymphocytes interact with important non-lymphoid cells, either during their maturation process or during the start of the immune response. There are two interrelated systems which the body identifies pathogens: the innate and the adaptive immune systems.

The adaptive immune system uses somatically generated antigen receptors which are clonally distributed on the two types of lymphocytes: B cells and T cells. These antigen receptors are generated by random processes and, as a consequence, the general design of the adaptive immune response is based upon the clonal selection of lymphocytes expressing receptors with particular specificities.

The antibody molecules play a leading role in the adaptive immune system. The receptors used in the adaptive immune response are formed by piecing together gene segments. Each cell uses the available pieces differently to make a unique receptor, enabling the cells to collectively recognize the infectious organisms confronted during a lifetime. Adaptive immunity enables the body to recognize and respond to any agent, even if it has never faced the invader before.

When an animal is exposed to an antigen, some subpopulation of its bone marrow derived cells (B lymphocytes) respond by producing antibodies. Each cell secretes a single type of antibody, which is relatively specific for the antigen. By binding to these antibodies (cell receptors), and with a second signal from accessory cells, such as the T-helper cell, the antigen stimulates the B cell to proliferate (divide) and mature into terminal (non-dividing) antibody secreting cells, called plasma cells. The process of cell division (mitosis) generates a clone, i.e., a cell or set of cells that are the progenies of a single cell.

While plasma cells are the most active antibody secretors, large B lymphocytes, which divide rapidly, also secrete antibodies, albeit at a lower rate. On the other hand, T-cells play a central role in the regulation of the B cell response and are preeminent in cell mediated immune responses, but will not be explicitly accounted for the development of the model. Lymphocytes, in addition to proliferating and/or differentiating into plasma cells, can differentiate into long-lived B memory cells. Memory cells circulate through the blood, lymph and tissues, and when exposed to a second antigenic stimulus commence to differentiate into large lymphocytes capable of producing high affinity antibodies, pre-selected for the specific antigen that had stimulated the primary response [4, 5].

The AISs try to reproduce the immune system features in a computational way by reproducing some interesting abilities described below [4, 5]:

- Pattern recognition: the ability to recognize different patterns from known and unknown cells.
- Reinforced learning: AIS is able to learn when solving one problem in such a way that in the next time the same problem is faced, the solution is faster and more robust.
- Memory: as the natural immune system, AIS also has a memory for previous known problems.
- Imperfect detection: an absolute recognition of the pathogens is not required, hence the system is flexible.

The basic steps for AISs evolutionary algorithms is: (i) reproduction; (ii) maturation; (iii) selection, and; (iv) receptor editing. These four steps are responsible for the creation of the random initial population, reproduction of the individuals, and for the global and local search steps through random mutations in the individuals according to their viability/fitness index.

As other evolutionary techniques inspired on biological processes and animal behavior, the AISs also apply some terms from biology as metaphors for computational problems. In order to help the reader to understand better these metaphors, some terms are briefly explained. For more information, please refer to glossary on [2].

- Antibody: a soluble protein molecule produced and secreted by B lymphocytes in response to an antigen. Translating to an optimization problem, the antibody is the solution for the objective function. Its dimensions and length depend on the variables of the problem.
- Antigen: any substance that, when introduced into the body, is recognized as an enemy by the immune system. For optimization problems, the antigen is the faced problem or any objective function to be solved.
- Clone: a group of genetically identical cells or organisms from a single common ancestor. In evolutionary techniques, a clone is a copy of a candidate solution.
- Hypermutation: mutations somatically introduced into an antibody gene at a high rate. For AISs, the hypermutation is the name of maturation process and corresponds to the local search of the evolutionary algorithm.

For further discussion on immune systems and on AIS, please refer to [4–7].

3.3.3 Simulations and Results

For this case, let consider that one hundred PEVs are expected to be connected to the power system according to the described features. The PEV models are chosen from those presented on Table 3.1 and are divided in 20 % of Model 1, 45 % of Model 2 and 35 % of Model 3. Their individual commuting distance is randomly chosen from Fig. 3.5 and their connection hour is in accordance to Fig. 3.6, which depicts the PEV connection probability according to the hour of the day.

Now, to simplify the study case, let consider that when disconnected in the next morning, all PEVs should have a 100 % SoC. So, in this case it is not considered the customers programs described before. As the evening time is the most interesting situation, let consider a simulation from 4 p.m. to 7 a.m. of the next day, resulting a 15-hour simulation divided into steps of 15 min.

First, it is interesting to see what happens when there is not any DMS control to manage the PEV load during the recharging process. As depicted on Fig. 3.7, the low voltage level in all three phases during some hours is evident as the need for a managing control application. The minimum voltage level is represented by continuous line in phase ‘a’, by dashed line in phase ‘b’ and by dotted line in phase ‘c’. When connected, the PEV starts to be recharged with the nominal charger power. Here, this maximum power is 4 kW for a residential outlet, even if the nominal charger power is higher. When more PEVs are considered, the scenario is worse than this one.

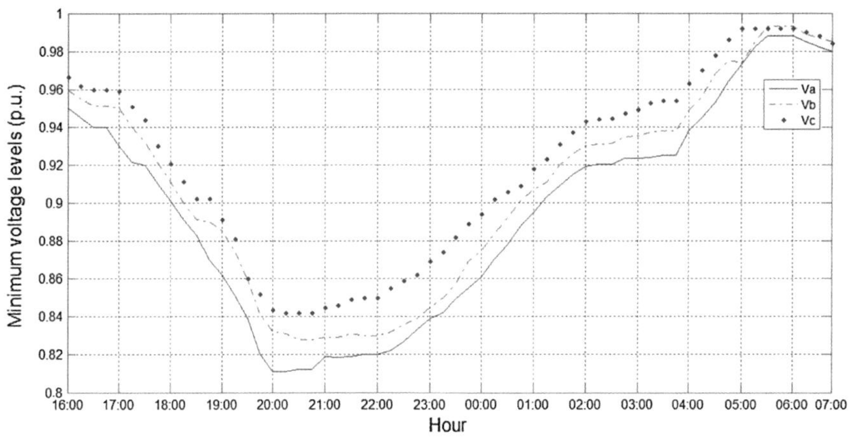

Fig. 3.7 Minimum voltage levels without a DMS control

Repeating the simulation using the same set of vehicles and the same initial conditions, the recharging process applying the DMS control presents better results regarding to voltage levels and power losses, as depicted on Fig. 3.8. As the previous example, the figure shows the minimum voltage levels in phase 'a' using continuous lines, in phase 'b' using dashed lines and in phase 'c' using dotted line. The DMS manages the individual PEV schedule according to the load margin of the system. Each PEV recharge power is assessed using the load margin of the system, in such a way that the minimum voltage level is matched in every hour.

Regarding the power losses, the application of a DMS control also decreases the losses from 0.0473 to 0.0312 p.u. This result is obtained using the antibody [840 852 864 828]. This solution is found after the AIS optimization and means

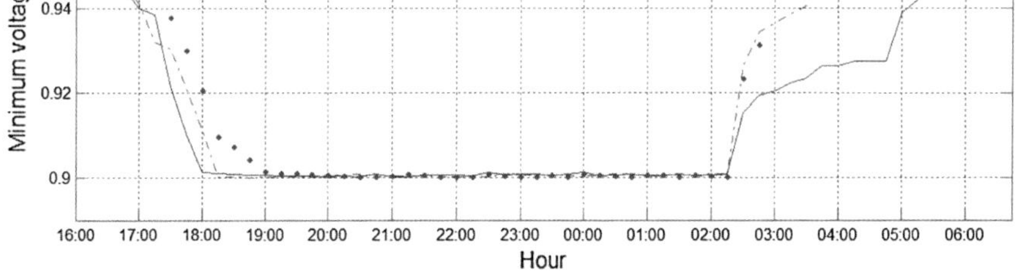

Fig. 3.8 Minimum voltage levels with DMS control

that those nodes are potential candidates to capacitor placement on planning stage or switching during operating stage aiming for power losses minimization, according to (3.4) and the restrictions (3.4.1)–(3.4.2). As the objective here is not developing the optimization technique itself, more information and examples may be found on [8].

By observing Fig. 3.8, it is possible to see that the recharging process is slower than on the uncontrolled previous case due to the power limitation imposed by the DMS after assessing each PEV schedule. This schedule is determined using a previous assessment of the PV curve, determining the maximum load in order to maintain the voltage levels at each node. The available power is shared between the PEVs according to their schedule and setup recharge configurations. Although the recharging process is slower, the minimum voltage level requirement is matched for all nodes.

During the process, there is a possibility of communication failure due to any cause. In this case, it is impossible to inform the assessed schedule for a set of PEVs. As this is a possible event in smart grids, many researchers proposed to use local measurements as reference signals for control functions to keep the frequency and voltage control [9, 10]. Even if this solution is proposed for parallel inverters from renewable distributed generation connected to power distribution system, it is possible to apply it for PEV chargers as a secondary control reference signal when the DSO signal is not available.

3.4 Discussions on the Results

The PEV is one of the most promising products of the automobile industry. It has a great potential to change the way people see the transportation concept. But, it is not expected that the fossil-fuelled vehicle will disappear as it can be presented as a different option best suited for other applications than daily commuting in urban environment.

As the PEV will become a real option as the prices decrease and a younger generation more used to take advantage of new technologies will be potential PEV owners, the integration of these vehicles to power system is a task to be planned now. Assuming this planning is performed now, any of those problems described before will take place.

The smart grids features will help in this process in a definitive way. A DMS control will not be responsible only for PEV loads, but for a wider set of smart appliances and applications which will be developed in the future. But, even with all this expectations and possibilities, the core of a DMS will not change. It is a system responsible for managing the controllable loads according to predefined rules in order to assure some operational restrictions for quality and safety purposes.

The assessment is based on the available system's load margin and the PEVs' schedules depend on this solution and also on some other combinatorial situations represented by probabilities, as the daily commuting distance, the hour of connections and the predefined charge setup chosen by the customer.

Regarding to the best approach to be applied, the centralized and decentralized ones are described here. As their features are different but they perform the same task, the choice for one model depends more on other variables as costs, current infrastructure, technology availability and revenue expectations.

References

1. Christensen TB, Wells P, Cipcigan L (2012) Can innovative business models overcome resistance to electric vehicles? Better place and battery electric cars in Denmark. Energy Policy 48:498–505
2. Rezzania R, Pruggler W (2012) Business models for the integration of electric vehicles into the Austrian energy system. In: 9th international conference on the European energy market, pp 1–8
3. IEEE PES Distribution System Analysis Subcommittee's. Distribution Test Feeder Working Group. Available on: http://ewh.ieee.org/soc/pes/dsacom/testfeeders/index.html
4. de Castro LN, Von Zuber FJ (1999) Artificial immune systems: part I—basic theory and applications. Technical report DCA-TR 01/99, UNICAMP, Brazil
5. de Castro LN, Von Zuben FJ (2002) Learning and optimization using the clonal selection principle. IEEE Trans Evol Comput 6(3):239–251
6. de Castro LN, Timmis J (2002) Artificial immune systems: a novel paradigm to pattern recognition. In: Corchado JM, Alonso L, Fyfe C (eds) Artificial neural networks in pattern recognition. University of Paisley, Paisley, pp 67–84
7. Dasgupta D, Yu S, Nino F (2011) Recent advances in artificial immune systems: models and applications. Appl Soft Comput 11(2):1650–1658
8. Oliveira DQ, Zambroni de Souza AC, Delboni LFN (2013) Optimal plug-in hybrid electric vehicles recharge in distribution power systems. Electr Power Syst Res 98:77–85
9. Simpson-Porco JW, Dörfler F, Bullo F (2013) Synchronization and power sharing of droop-controlled inverters in islanded microgrids. Automatica 49:2603–2611
10. Chandorkar MC, Divan DM, Adapa R (1993) Control of parallel connected inverters in standalone AC supply systems. IEEE Trans Ind Appl 29(1):136–143

Chapter 4
An Optimal and Distributed Control Strategy for Charging Plug-in Electrical Vehicles in the Future Smart Grid

Zhao Tan, Peng Yang and Arye Nehorai

Abstract In this chapter, we propose an optimal and distributed control strategy for plug-in electric vehicles' (PEVs) charging as part of demand response in the smart grid. We consider an electricity market where users have the flexibility to sell back the energy stored in their PEVs or the energy generated from their distributed generators. The smart grid model in this chapter integrates a two-way communication system between the utility company and consumers. A price scheme considering fluctuation cost is developed to encourage consumers to lower the fluctuation in the demand response by charging and discharging their PEVs reasonably. A distributed optimization algorithm based on the alternating direction method of multipliers is applied to solve the optimization problem, in which consumers need to report their aggregated loads only to the utility company, thus ensuring their privacy. Consumers update the scheduling of their loads simultaneously and locally to speed up the optimization computing. We also extend the distributed algorithm to the asynchronous case, where communication loss exists in the smart grid. Using numerical examples, we show that the demand curve is flattened after the optimal PEV charging and load scheduling. We also show the robustness of the proposed method by considering estimation uncertainty on the overall next day load, and also the renewable energy. The distributed algorithms are shown to reduce the users' daily bills with respect to different scenarios, thus motivating consumers to participate in the proposed framework.

Keywords Plug-in electric vehicles · Charging and discharging · Demand response · Distributed optimization · Fluctuation cost · Smart grid · Alternating direction method of multipliers

---

Z. Tan · P. Yang · A. Nehorai (✉)
The Preston M. Green Department of Electrical and System Engineering, Washington University in St. Louis, St. Louis, USA
e-mail: nehorai@ese.wustl.edu

Z. Tan
e-mail: tanz@ese.wustl.edu

P. Yang
e-mail: yangp@ese.wustl.edu

S. Rajakaruna et al. (eds.), *Plug In Electric Vehicles in Smart Grids*,
Power Systems, DOI 10.1007/978-981-287-302-6_4

4.1 Introduction

In the electricity market, demand response [1] is a mechanism to manage users' consumption behavior under specific supply conditions. The goal of demand response is to benefit both consumers and utilities via a more intelligent resource scheduling method. While the classical rule for operating a power system is to supply all the demand whenever it occurs, the new philosophy focuses on the concept that the system will be more efficient when the fluctuations in demand are kept as small as possible [2]. With a fixed amount of electricity generation, demand fluctuation can add significant ancillary cost to the suppliers due to the inefficient usage of thermal plants in the power grid. Therefore the goal of demand response is to flatten the demand curve by shifting the peak hour loads to off-peak hours. Traditionally it is achieved by setting a time of use (TOU) price scheme [3, 4], which normally assigns high prices to the peak hours and low prices to the off-peak hours. Thus consumers will try to move some of their schedulable power usage to the off-peak hours in order to reduce their electricity bill. Overall, this behavior will reduce the fluctuation in the power consumption level. The TOU price scheme works well when the schedulable power usage is not dominant.

With the incorporation of plug-in electric vehicles (PEVs) into the power grid [5–7] consumers have more flexibility to schedule their loads and tend to charge their PEVs when the electricity price is low. Therefore with high PEV penetration the effect of the traditional TOU price scheme is simply to move the peak demand from previous peak hours into previous off-peak hours. The cost arising from load variation still remains high in this situation.

In a smart grid, an advanced metering infrastructure (AMI) [8] and an energy-management controller (EMC) [9, 10] are widely employed, and they enable communications between the users and the utility company, which makes it possible to implement more effective PEV scheduling strategies. The AMI device collects data on the electricity usage and communicates with other AMI devices and the system controller. The EMC device helps the users to manage and schedule their consumption of electricity in order to minimize their cost and inconvenience/dissatisfaction. With the introduction of the AMI and EMC more effective PEV charging strategies have been proposed, and they fall into two categories. One is based on the centralized control, in which every consumer transmits his or her PEV's information to the centralized controller. Then the centralized controller determines the charging portfolio of each consumer by minimizing several objective functions, such as power loss, load variance, and so on [11, 12]. The other category aims to solve the peak shifting and electric vehicle charging problems [10, 13, 14] in a distributed fashion. In [13], an optimal distributed charging algorithm is shown, but it is limited to the case where all PEVs have the same behavior, i.e., all PEVs have the same starting time and also the same deadline. In [10], the authors apply a game-theoretical approach for consumers to schedule their loads. A distributed algorithm is proposed and guaranteed to find the Nash equilibrium of the game. But their distributed algorithm can be applied only sequentially among the consumers, and the communication time and cost reduce the effectiveness of the method. Moreover, consumers have to report their own usage curves to all the other consumers, so the privacy of each consumer is not protected. In [14], the authors propose a parallel distributed optimization algorithm for the PEV charging problem and present a convergence proof. A similar distributed approach is proposed for the demand response for smart grid with penetration of distributed renewable generators [15]. In our numerical example at the end of this chapter, we will show that the convergence behavior of this method is sensitive to the choice of parameters in the computing algorithm.

In this work (see also [16, 17]) we consider a smart grid with a certain penetration level of PEVs and also with some on-site renewable distributed generators [18–20], such as solar panels and wind turbines. The smart grid model is provided in Fig. 4.1, in which we can see that consumers exchange information with the utility company. We consider the case where users can sell back the energy they generate to the grid. The PEVs can also be used as batteries to store electricity, which can be either consumed or sold back to the grid whenever is more advantageous. The price model in this chapter consists of two parts. The first part considers the base price, and the second part takes the fluctuation cost into account. The fluctuation price encourages consumers to cooperate with each other when they calculate their charging portfolio. We implement the alternating direction method of multipliers (ADMM) to solve the optimization [21]. Unlike in [10], our algorithm is computed in parallel. Each user needs to report their usage curve only to the utility company, thus privacy can be guaranteed. The convergence of the ADMM requires only convexity and the saddle point condition [21]; therefore the convergence of the distributed algorithm in this paper can be easily obtained. We extend the traditional ADMM approach to consider the case when AMI messages are lost during the communication between the utility company and consumers, and an asynchronous ADMM is formulated in this chapter to deal with loss of data.

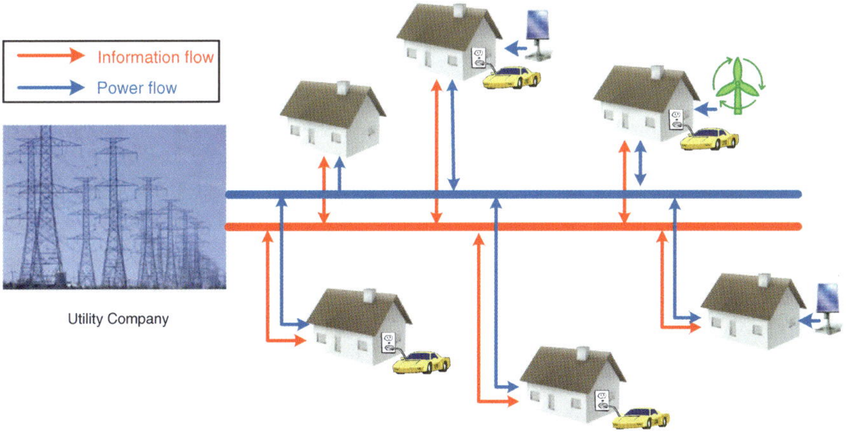

Fig. 4.1 Schematic diagram of a smart grid model with PEVs and renewable generators

The rest of the chapter is organized as follows. In Sect. 4.2, we build a mathematical model for PEVs and other loads. We then set the pricing policy for the utility company. We also formulate the model with random prediction error in the base load and distributed generations. In Sect. 4.3, using the alternating direction method of multipliers, we reformulate the optimization problem into a distributed optimization problem. We provide the convergence analysis and also extend the traditional distributed approach to take communication loss into account. In Sect. 4.4, we show numerical examples to demonstrate the performance of the proposed method. In Sect. 4.5, we conclude the chapter and point out directions for future research.

4.2 Smart Grid Model with PEVs

We consider a smart grid model with certain number of residences provided with electricity from the same utility company. Each consumer has an EMC that controls and communicates with different appliances within the household and also has an AMI to perform two-way communication with the utility company. We also assume that there are a certain number of user-owned distributed generators and PEVs in the grid. Users can sell back the energy generated from their own distributed generators or store this energy in the batteries of their PEVs. Users can also sell the energy left in their PEV batteries back to the grid whenever it is profitable. The price contains two parts: the first part is based on the non-schedulable load at each time, which we will refer to as the base price; the second part is based on the fluctuation of the load throughout a day.

We divide a day into T time periods, i.e. we have $t = 1,2\ldots T$. We assume there are four types of loads, both positive and negative, in our model: the base load, schedulable load, PEV load, and the distributed generation. The base load supplies users' basic needs, such as lighting, which cannot be scheduled. The schedulable load can be scheduled, but needs to maintain a certain quantity during a day, such as refrigerators, air conditioning, laundry machines, and dishwashers. Please note that the minimum load requirement for the schedulable load can also be included in the base load. The PEV load denotes the electricity usage of PEVs. It can be a negative number at time slots when users sell back the electricity stored in their PEVs. The distributed generation is considered as a negative load generated by solar or wind generators and users can sell this energy back to the grid when they have surplus. We use $l_k^I(t)$ to represent the type I load for user k at time point t. Here, the superscript $I = B, S, P,$ and D, to indicate the base load, schedulable load, PEV load, and distributed generation, respectively. Then the total load of user k at time point t can be expressed as the sum of these four types of loads:

$$l_k(t) = l_k^B(t) + l_k^S(t) + l_k^P(t) + l_k^D(t) \tag{4.1}$$

The above equation is satisfied for all k and t.

$l_k^B(t)$ denotes the base load of user k at time point t. $l_k^S(t)$ is the schedulable load of user k at time t. It can be scheduled during a day, but it must satisfy the following sum constraint to meet the satisfaction of consumers. It should also meet the maximum physical usage rate constraint. The constraints for schedule load of user k are stated as follows:

$$\sum_{t=1}^{T} l_k^S(t) = l_k^S, \; and \, 0 \le l_k^S(t) \le l_k^{max}(t). \tag{4.2}$$

Here l_k^s is the total amount of schedule load for user k.

$l_k^P(t)$ denotes the electric vehicle load of user k. It can be decomposed into two parts, namely the charging energy and discharging energy. The equation can be written as

$$l_k^P(t) = \frac{l_k^{P+}(t)}{\mu_c} + \mu_d l_k^{P-}(t), \tag{4.3}$$

where μ_c and μ_d denote the charging efficiency and discharging efficiency of the electric vehicle in this chapter. The terms $l_k^{\mathrm{P}+}(\mathrm{t})$ and $l_k^{\mathrm{P}-}(\mathrm{t})$ represent the charging and discharging electricity seen by the battery of the PEV. Since a PEV cannot charge and discharge at the same time, $l_k^{\mathrm{P}+}(\mathrm{t})$ and $l_k^{\mathrm{P}-}(\mathrm{t})$ follows a constraint, expressed as

$$l_k^{P+}(t) \cdot l_k^{P-}(t) = 0. \tag{4.4}$$

The charging and discharging profile should also satisfy the sum constraint. In addition, the user can use the electric vehicle as a battery to store energy when the electricity price is low and sell it back when the price is high. The charging and discharging rate also have upper bounds to meet physical constraints. The energy remaining in a battery at every time slot should also be larger than zero and less than the battery size. Let E^k denote the battery size of user k, and let $r_k^{\min}$ and $r_k^{\max}$ denote the maximum discharging rate and maximum charging rate respectively. Let l_k^P denote the energy left in the battery when user k's PEV arrives home, and let l_k^Q denote the energy required for the next trip for user k. Then these constraints can be stated as follows:

$$\sum_{t=t_k^{ar}}^{t_k^{dep}-1} l_k^{P+}(t) + \sum_{t=t_k^{ar}}^{t_k^{dep}-1} l_k^{P-}(t) = l_k^Q - l_k^P, \tag{4.5}$$

$$0 \le l_k^P + \sum_{t=t_k^{ar}}^{t_0} l_k^{P+}(t) + \sum_{t=t_k^{ar}}^{t_0} l_k^{P-}(t) \le E_k, \quad t_0 = t_k^{ar}, t_k^{ar}+1, \ldots, t_k^{dep}-1, \tag{4.6}$$

$$r_k^{\min} \le l_k^{P-}(t) \le 0, \; 0 \le l_k^{P+}(t) \le r_k^{\max}, \tag{4.7}$$

where t_k^{ar} is the arrival time of user k's PEV and t_k^{dep} indicates the time of departure. Equation (4.5) means that the battery of PEV of user k is charged to satisfy the preset energy level before departure. Equation (4.6) expresses that the energy left in one battery is always larger than zero and less than the capacity of that battery. The physical constraint of charging speed is given in Eq. (4.7). If we are given a charging profile with multiple charging requests for the PEV of user k, it can be easily decomposed into several constraints, which are similar to constraints (4.5–4.7) given above.

Due to the fact that charging and discharging cannot happen at the same time, these constraints for PEVs, i.e. (4.3–4.7) are not convex when μ_c and μ_d are less than one. When $\mu_c = \mu_d = 1$, these constraints define a convex set:

$$\sum_{t=t_k^{ar}}^{t_k^{dep}-1} l_k^P(t) = l_k^Q - l_k^P, \tag{4.8}$$

$$0 \le l_k^P + \sum_{t=t_k^{ar}}^{t_0} l_k^{P+}(t) + \sum_{t=t_k^{ar}}^{t_0} l_k^{P-}(t) \le E, \; t_0 = t_k^{ar}, t_k^{ar}+1, \ldots, t_k^{dep}-1, \tag{4.9}$$

$$r_k^{\min} \le l_k^P(t) \le r_k^{\max}. \tag{4.10}$$

In this chapter, we let $l_k^D(\mathrm{t})$ denote the distributed generation of user k at time point t. It is a negative value since it is obtained from an external clean energy source and can be sold back to the grid when the user has a surplus.

4.3 Electricity Pricing Policy

In this section, we first describe the cost model for electricity generation and then propose a corresponding pricing scheme related to this cost model. Let c_t denote the marginal generation cost for one unit of electricity from thermal plants at time t. In an electrical power system, demand fluctuation can result in ancillary cost to the utility company since larger fluctuation will also lead to inefficient usage of the plants and the need for secondary thermal plants during the peak hours. We model this fluctuation cost as a function of the variance of the electricity load [22]. Therefore the total generation cost model for the utility company can be described as

$$\text{Cost} = \sum_{t=1}^{T} c_t \sum_k l_k(t) + \mu \sum_{t=1}^{T} \left(\sum_k l_k(t) - m \right)^2, \tag{4.11}$$

where $l_k(t)$ is the load of user k at time t as defined in Eq. (4.1) and m is the mean usage during a day defined by $\frac{1}{T}\sum_{t=0}^{T}\sum_k l_k(t)$. The parameter μ is chosen to describe the fluctuation cost.

Electricity pricing is a complicated procedure and depends on the entire electricity market. The price discussed in this chapter can be understood as a price indicator. For simplicity we will still call it "price" in the rest of the chapter. The price function contains two parts, namely the base price and the price arising from the demand fluctuation. They are related to the two parts of the generation cost of electricity.

The base price $p_B(t)$ is determined by the sum of the base loads of all individual users at time t. Since the sum of the base loads can be well predicted in a day-ahead market, the base price is well defined. We assume that the base price $p_B(t)$ is proportional to the sum of base loads at time t:

$$p_B(t) = C_1 \left(\sum_k l_k^B(t) \right)^\alpha, \tag{4.12}$$

where C_1 is chosen so that $c_t \le p_B(t)$ for all time points t. This choice will guarantee that revenue can cover the regular electricity generation cost as long as there is more energy demand than the generation from the on-site renewable generators when fluctuation cost is not considered. The parameter α is within the range [0, 1], which can influence changes in the base electricity price between the different time slots and therefore influence the users' usage pattern. We will show this influence in the numerical examples.

If only base price is used, consumers have limited motivation to reschedule their demand response to lower the fluctuation in the demand curve. When the percentage of schedulable load and the penetration level of PEVs are high, another peak will be created in the time periods with low base prices. In order to align the incentives of consumers to lower the fluctuation cost in (4.11), an extra price based on how much they contribute to this demand curve fluctuation needs to be introduced. Let f_0 denote the variance of the aggregated demand load:

$$f_0 = \sum_{t=1}^{T} \left(\sum_k l_k(t) - m \right)^2. \tag{4.13}$$

An extra price term $p_F(t)$ related to this variance is introduced in our price model. This price term is added to the base price only in the time periods when the total load of all users is larger than the mean usage. Let Ω denote the set containing these time points. The fluctuation price $p_F(t)$ can be written as

$$p_F(t) = \begin{cases} C_2 f_0 \frac{\sum_{k'} l_{k'}(t) - m}{\sum_{t' \in \Omega} \left(\sum_{k'} l_{k'}(t') - m\right)} \frac{l_k(t)}{\sum_{k'} l_{k'}(t)}, & if\ t \in \Omega \\ 0 & otherwise \end{cases} \tag{4.14}$$

Therefore the actual price charged on the consumers is $p(t) = p_B(t) + p_F(t)$. The electrical bill to each individual user during a day can be summarized as

$$c_k = \sum_{t=1}^{T} (p_B(t) + p_F(t)) l_k(t). \tag{4.15}$$

By summing c_k over all the consumers, we have the total revenue for the utility company as

$$R = \sum_{t=1}^{T} p_B(t) \sum_k l_k(t) + C_2 \sum_{t=1}^{T} \left(\sum_k l_k(t) - m \right)^2, \tag{4.16}$$

in which C_2 is chosen to be larger than μ from (4.11) to cover the generation cost.

With the introduction of the fluctuation price, consumers will cooperate with each other to reduce the variance of the overall load curve, and therefore lower the generation cost to the utility company and their own bills. The savings result from a more efficient utilization of the generation infrastructures, which can be shared between the utility company and consumers. The saving for consumers is directly reflected in a reduction of their daily bills, which will be shown in the numerical example. The extension to the unbundling situation, i.e., when the generation companies and retailers are separated entities in the electricity market, is an interesting, but more complicated problem. Discussion on the unbundling model is beyond the scope of this preliminary work.

4.4 Prediction and Uncertainty in the Smart Grid

In order to perform the optimization in the next section, we need to predict the base load, the distributed generation, and also parameters of schedulable load and PEV usage. Users can input the parameters related to PEVs and schedulable loads directly through their own EMC device, and there is no need to report them to the utility company if the distributed algorithm is used. The prediction of the base load has been investigated in the literature, using artificial neural networks [23] and pattern analysis [24]. The prediction can be quite precise. Therefore in our model, we assume the predicted base load for each individual user is a Gaussian random variable as follows:

$$l_k^{\mathrm{B}}(t) = l_k^{\mathrm{B0}}(t) + \varepsilon_k^B(t), \tag{4.17}$$

where $l_k^{B0}(t)$ denotes the actual load of user k at time t, and $\varepsilon_k^B(t)$ is a random Gaussian noise with distribution $\mathcal{N}(0, \sigma_B^2)$, which shows the prediction error in the mathematical model. We assume that the noise term $\varepsilon_k^B(t)$ is uncorrelated among all the users.

The renewable energy can be predicated using a short-term prediction method [25]. By assuming the prediction error as an addictive Gaussian noise [26], we will have a linear model as,

$$l_k^D(\mathrm{t}) = l_k^{D0}(t) + \varepsilon_k^D(t), \tag{4.18}$$

where $l_k^{D0}(t)$ denotes the actual distributed generation of user k at time t, and $\varepsilon_k^D(t)$ is the prediction error, which can be regarded as random noise in the model. Unlike the users' base load, we assume that the noise term $\varepsilon_k^D(t)$ is highly correlated among all the users since they are all affected by the same weather condition if they are located in the same geographic area. Then we can write the noise term as

$$\varepsilon_k^D(t) = e^D(t) + e_k^D(t) \tag{4.19}$$

The term $e^D(t)$ follows a Gaussian distribution $\mathcal{N}(0, \sigma_0^2)$, and it shows that the prediction errors of distribution generators in different households are correlated. The term $e_k^D(t)$ follows a Gaussian distribution $\mathcal{N}(0, \sigma_D^2)$. The impact of the uncertainties in the base load and also the distributed generation is analyzed in the numerical example section.

4.5 A Distributed Optimization Algorithm

4.5.1 Centralized Optimization of the Loads

The goal of users is to minimize their bills. The utility company also has the incentive to minimize this total bill. Since minimizing the total bills of all the users will lead to a more flattened load curve, this will lower the fluctuation cost for the utility company. In other words, the utility company should minimize the total electricity bill under the condition that its revenue can cover its own cost, which can be guaranteed with a proper choice of C_1 and C_2. Then, with fixed C_1 and C_2, the goal of the utility company is to minimize the total bill to all the users. Therefore according (4.16) to the centralized optimization problem can be formulated as

$$\begin{aligned} \min_{\{\boldsymbol{l}_k\}} \sum_k \boldsymbol{p}_B^{\mathrm{T}} \boldsymbol{l}_k + C_2 f_0\left(\sum_k \boldsymbol{l}_k\right), \\ \text{subject to } \boldsymbol{l}_k \in F_k, \forall k. \end{aligned} \tag{4.20}$$

Here $\boldsymbol{p}_B = [p_B(1), p_B(2), \ldots, p_B(T)]^{\mathrm{T}}$ which includes information of base prices at all time points in one vector, and $\boldsymbol{l}_k = [l_k(1), l_k(2), \ldots, l_k(T)]^{\mathrm{T}}$ includes the load scheduling of user k at all time points. F_k denotes the feasible set for load l_k, which means that $F_k = \{\boldsymbol{l}_k : \boldsymbol{l}_k$ satisfies conditions (4.1–4.7)$\}$.

Since in real world application, the charging and discharging efficiencies are less than 1, we can set them to be 1 in the optimization problem (4.20) to ensure convexity of the problem. Therefor the set F_k becomes convex and defined by $F_k = \{\boldsymbol{l}_k : \boldsymbol{l}_k$ satisfies conditions (4.1–4.3) and (4.8–4.10)$\}$ (After obtaining the optimal solution, we post-process the load of the electric vehicles to get a suboptimal solution of the distributed optimization algorithm. Solving the optimization problem (4.20) in a centralized way is inefficient due to the huge dimensionality and thousands of constraints. The size of this optimization problem increases with the number of households. In addition consumers need to report their specific load usage to the utility company, which will lead to privacy issues. In order to speedup the calculation, we implement a distributed algorithm in which every household has a small computing unite center to deal with a relative small scale optimization problem and report their results to the utility company. In order to make this chapter self-contained, we briefly introduce the general theory about Alternating Direction Method of Multipliers (ADMM). For more technical details, please check the review paper on this subject [21].

4.5.2 Mathematical Preliminary: Alternating Direction Method of Multipliers

The standard ADMM is intended to solve a separable convex optimization. The objective function is the sum of several convex functions, and the variables are coupled through linear equations. The general form is

$$\begin{aligned} &\min_{\boldsymbol{x},z} F(\boldsymbol{x}) + G(z) \\ &\text{subject to } A\boldsymbol{x} + Bz = \boldsymbol{c}, \end{aligned} \tag{4.21}$$

in which $\boldsymbol{x} \in \mathbb{R}^n$, $z \in \mathbb{R}^m$, $A \in \mathbb{R}^{p \times n}$, $B \in \mathbb{R}^{p \times m}$, and $\boldsymbol{c} \in \mathbb{R}^p$.

We form the augmented Lagrangian by adding a quadratic penalty for feasibility violation to the original Lagrangian function:

$$L(\boldsymbol{x}, z, \boldsymbol{v}) = F(\boldsymbol{x}) + G(z) + \boldsymbol{v}^T(A\boldsymbol{x} + Bz - \boldsymbol{c}) + \frac{\rho}{2}\|A\boldsymbol{x} + Bz - \boldsymbol{c}\|_2^2, \tag{4.22}$$

where $\boldsymbol{v} \in \mathbb{R}^p$ is the dual variable, and $\rho > 0$ is a penalty parameter. The method of multipliers cycles between the following two steps:

$$\left(\boldsymbol{x}^{i+1}, z^{i+1}\right) = \underset{x,z}{\operatorname{argmin}}\ L\left(\boldsymbol{x}, z, \boldsymbol{v}^i\right), \tag{4.23}$$

$$\boldsymbol{v}^{i+1} = \boldsymbol{v}^i + \rho\left(A\boldsymbol{x}^{i+1} + Bz^{i+1} - \boldsymbol{c}\right). \tag{4.24}$$

Instead of solving $\boldsymbol{x}, z$ jointly as we showed in (4.23), ADMM updates $\boldsymbol{x}$, z in an alternating fashion. It implements a single Gauss-Seidel pass [27] over the augmented Lagrangian multiplier, followed by updating the dual variable with a gradient descent. Specifically, the ADMM consists of the following iterations:

$$\boldsymbol{x}^{i+1} = \underset{x}{\operatorname{argmin}}\ L\left(\boldsymbol{x}, z^i, \boldsymbol{v}^i\right), \tag{4.25}$$

$$z^{i+1} = \underset{z}{\operatorname{argmin}}\ L\left(\boldsymbol{x}^{i+1}, z, \boldsymbol{v}^i\right), \tag{4.26}$$

$$\boldsymbol{v}^{i+1} = \boldsymbol{v}^i + \rho\left(A\boldsymbol{x}^{i+1} + Bz^{i+1} - \boldsymbol{c}\right). \tag{4.27}$$

Note that the step-size for the gradient descent is ρ, which is the same penalty parameter as found in the quadratic term in (4.22).

The convergence analysis of ADMM is based on two theoretical assumptions, which are satisfied for most applications.

Assumption 1 The extended real valued functions of $F(\boldsymbol{x})$ and $G(z)$ are closed, proper and convex.

Assumption 2 The unaugmented Lagrangian function of (4.21) has a saddle point, i.e., there exists $(\boldsymbol{x}^*, z^*, \boldsymbol{v}^*)$ such that

$$L_0(\boldsymbol{x}^*, z^*, \boldsymbol{v}) \leq L_0(\boldsymbol{x}^*, z^*, \boldsymbol{v}^*) \leq L_0(\boldsymbol{x}, z, \boldsymbol{v}^*), \tag{4.28}$$

in which

$$L_0(\boldsymbol{x}, z, \boldsymbol{v}) = F(\boldsymbol{x}) + G(z) + \boldsymbol{v}^{\mathrm{T}}(A\boldsymbol{x} + Bz - \boldsymbol{c}). \tag{4.29}$$

Under these assumptions, the existence of a saddle point implies the existence of a primal dual solution pair. In [21], the authors show that with these assumptions, the ADMM, which cycles among (4.25–4.27), has residual convergence, objective convergence, and also dual variable convergence. For details, please see Sect. 3.2.1 from [21].

4.5.3 Distributed Optimization of the Loads

The objective function in problem (4.20) is a sharing problem and therefore can be decentralized into parallel programming using ADMM. The optimization problem (4.20) is reformulated by introducing auxiliary variables $\{z_k\}$ as follows:

$$\min_{\{l_k\},\{z_k\}} \sum_k \boldsymbol{p}_B^{\mathrm{T}} \boldsymbol{l}_k + C_2 f_0\left(\sum_k z_k\right),$$
$$\text{subject to } \boldsymbol{l}_k = z_k,\ \boldsymbol{l}_k \in F_k,\ \forall k. \tag{4.30}$$

Note that feasible set for each consumer is independent. Thus by introducing indicator functions $f_k(F_k)$ for feasible sets, the centralized optimization can be reformulated as

$$\min_{\{l_k\},\{z_k\}} \sum_k \left(\boldsymbol{p}_B^{\mathrm{T}} \boldsymbol{l}_k + f_k(F_k)\right) + C_2 f_0\left(\sum_k z_k\right),$$
$$\text{subject to } \boldsymbol{l}_k = z_k,\ \forall k. \tag{4.31}$$

The indicator function $f_k(F_k)$ is defined as

$$f_k(F_k) = \begin{cases} 0, & \boldsymbol{l}_k \in F_k, \\ +\infty, & \text{otherwise.} \end{cases} \tag{4.32}$$

The optimization (4.31) leads to the same optimal solution as in (4.20), and can be solved by a primal-dual method. However, this approach is very sensitivity to the selection of the step size. In order to increase the robustness of the algorithm, a quadratic term is added to the objective function in (4.31) without changing the optimal solution.

$$\min_{\{l_k\},\{z_k\}} \sum_k \left(\boldsymbol{p}_B^{\mathrm{T}} \boldsymbol{l}_k + \frac{\rho}{2} \|\boldsymbol{l}_k - z_k\|_2^2 + f_k(F_k)\right) + C_2 f_0\left(\sum_k z_k\right),$$
$$\text{subject to } \boldsymbol{l}_k = z_k,\ \forall k. \tag{4.33}$$

The parameter ρ is a regularization parameter for the quadratic term, and the performance of the algorithm is not sensitive to the choice of ρ, as we will show in the numerical examples. Introducing the Lagrange multipliers $\boldsymbol{v}_k$ for each $\boldsymbol{l}_k = z_k$ constraint in the above optimization, we can obtain the augmented Lagrangian function as

$$L(\{\boldsymbol{l}_k\}, \{z_k\}, \{\boldsymbol{v}_k\}) = \sum_k \left(\boldsymbol{p}_B^{\mathrm{T}} \boldsymbol{l}_k + \boldsymbol{v}_k^{\mathrm{T}}(\boldsymbol{l}_k - z_k) + \frac{\rho}{2} \|\boldsymbol{l}_k - z_k\|_2^2 + f_k(F_k)\right) + C_2 f_0\left(\sum_k z_k\right). \tag{4.34}$$

The original optimal problem can be solved using a Gauss-Seidel algorithm on the augmented Lagrangian function $L(\{\boldsymbol{l}_k\}, \{z_k\}, \{\boldsymbol{v}_k\})$ as we shown in the previous section. Note that (4.25) for the smart grid application is given as

$$\{\boldsymbol{l}_k^{i+1}\} = \operatorname{argmin} \sum_k \left(\boldsymbol{p}_B^{\mathrm{T}} \boldsymbol{l}_k + \boldsymbol{v}_k^{i\mathrm{T}} \left(\boldsymbol{l}_k - z_k^i \right) + \frac{\rho}{2} \| \boldsymbol{l}_k - z_k^i \|_2^2 + f_k(F_k) \right). \quad (4.35)$$

Since the objective function is separable among different customers, it can be solved in parallel and locally. Basically, the Gauss-Seidel algorithm for ADMM cycles through the following steps until convergence is reached:

$$\boldsymbol{l}_k^{i+1} = \operatorname{argmin} \left(\boldsymbol{p}_B^{\mathrm{T}} \boldsymbol{l}_k + v_k^{i\mathrm{T}} \left(\boldsymbol{l}_k - z_k^i \right) + \frac{\rho}{2} \| \boldsymbol{l}_k - z_k^i \|_2^2 + f_k(F_k) \right), \quad (4.36)$$

$$\{z_k^{i+1}\} = \underset{\{z_k^{i+1}\}}{\operatorname{argmin}} \sum_k \left(\boldsymbol{v}_k^{i\mathrm{T}} \left(\boldsymbol{l}_k^{i+1} - z_k \right) + \frac{\rho}{2} \| \boldsymbol{l}_k^{i+1} - z_k \|_2^2 \right) + C_2 f_0 \left(\sum_k z_k \right), \quad (4.37)$$

$$\boldsymbol{v}_k^{i+1} = \boldsymbol{v}_k^i + \rho \left(\boldsymbol{l}_k^{i+1} - z_k^{i+1} \right) \quad (4.38)$$

In (4.36), $\boldsymbol{l}_k$ is updated by solving a convex optimization problem while keeping z_k and $\boldsymbol{v}_k$ fixed to the values from the previous iteration. Likewise, we solve for z_k. Equation (4.38) is a gradient descent of the augmented Lagrangian multiplier, with step size ρ. The optimization problems (4.36 and 4.38) can be solved locally and also in parallel. As shown in Fig. 4.2, each user needs to report his/her total usage during each time slot only to the utility company, which then solves the optimization problem (4.37). They also need to provide the utility company with their Lagrangian multipliers. In every iteration of ADMM, the utility company also needs to send the parameter z_k^i to user k. As an intuitive interpretation of the ADMM procedure, we can regard z_k as the load which is suggested by the utility company to minimize the fluctuation in the demand response, and we regard $\boldsymbol{l}_k$ as the load according to the user's own benefit. The whole algorithm is a process of negotiation between each user and the utility company. $\boldsymbol{v}_k$ and $\rho/2$ are the penalty coefficients for the first and second order terms of disagreement. The optimization problem (4.36) can be reformulated as a constrained convex optimization.

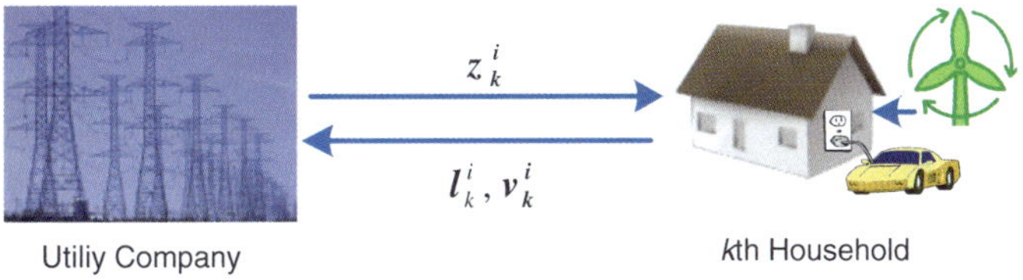

Fig. 4.2 Information exchange between utility company and *k*th household at *i*th iteration

4.5.4 Theoretical Convergence Behavior of ADMM

The conditions for the convergence of the ADMM scheduling method in this chapter are based on one theorem given in [21]. Before introducing the convergence results, a key lemma is introduced.

Lemma 1 *Suppose set D is a closed nonempty convex set and f(D) is an indicator function, then the epigraph of the indicator function f(D) is also a closed nonempty convex set.*

Proof Since set D is nonempty, there exists an element $\boldsymbol{x} \in D$, so that we have $f(\boldsymbol{x}) = 0 \leq 1$. Therefore $(\boldsymbol{x}, 1) \in \text{epi} f$. Thus the epigraph is nonempty. Convexity can be easily obtained by using the convexity of set D. Suppose we have a sequence $(\boldsymbol{x}_k, t_k) \in \text{epi} f$ satisfying $f(\boldsymbol{x}_k) \leq t_k$ that converges to point $(\boldsymbol{x}^*, t^*)$. We know that $\boldsymbol{x}_k \in D$ for all k. Since the sequence $\{\boldsymbol{x}_k\}$ converges to $\boldsymbol{x}^*$, and D is closed, then $\boldsymbol{x}^* \in D$. Therefore we know $f(\boldsymbol{x}^*) = 0 \leq \lim_{k\to\infty} t_k = t^*$. Therefore the point $(\boldsymbol{x}^*, t^*)$ is also in $\mathbf{epi} f$, and the epigraph is closed. □

With the above lemma we can guarantee that the objective function converges to the optimal solution with ADMM iterations as shown in the following theorem.

Theorem 1 *If the charging and discharging efficiency μ_c and μ_d both equal to one and the charging and scheduling constraints are feasible, which means there is at least one $\{\boldsymbol{l}_k\}$ satisfies conditions* (4.1–4.3) *and* (4.8–4.10); *Then by iterating between* (4.36), (4.37) *and* (4.38), *we have the residual convergence behavior. Also, the objective function converges to the optimal value p^*, and the Lagrangian multipliers $\boldsymbol{v}_k$ converge to the optimal solution $\boldsymbol{v}^*$:*

$$\lim_{i\to\infty} \left(\mathbf{l}_k^i - \mathbf{z}_k^i\right) = 0, \quad \forall k, \tag{4.39}$$

$$\lim_{i\to\infty} \left(\sum_k g_k(\boldsymbol{l}_k) + f_0\left(\sum_k z_k^i \right) \right) = p^*, \tag{4.40}$$

$$\lim_{i\to\infty} \boldsymbol{v}_k^i = \boldsymbol{v}_k^*, \forall k, \tag{4.41}$$

where $g_k(\boldsymbol{l}_k) = \boldsymbol{p}_B^T \boldsymbol{l}_k + f(F_k)$.

Proof Function $f_0\left(\sum_k z_k^i\right)$ is a quadratic function in this paper, therefore it is easy to show the epigraph of this function is a nonempty close convex set. When μ_c and μ_d both equal to one, the constraints in the optimization defines a convex set. Combining the solution feasibility and Lemma 1, we can easily show that the epigraph of $g_k(\boldsymbol{l}_k)$ is also a nonempty closed convex set. Since strong duality holds for the original optimization (4.30), the optimal solution is the saddle point for the unaugmented saddle point, satisfying

$$L_0(\{l_k^*\},\{z_k^*\},\{v_k\}) \le L_0(\{l_k^*\},\{z_k^*\},\{v_k^*\}) \le L_0(\{l_k\},\{z_k\},\{v_k\}) \tag{4.42}$$

for $\rho = 0$, where $\{l_k^*\}$, $\{z_k^*\}$ are the optimal solution for the primal problem and $\{v_k^*\}$ are the optimal dual variables. Combining this property with the properties of the epigraphs we get the convergence results [18]. □

4.5.5 Distributed Optimization of the Loads with Lost AMI Messages

The distributed algorithm implemented in the previous section requires the utility company and every household to exchange information in each iteration. In practice, AMI messages get lost if there is a malfunction in the transmitter or due to the random noise in the transmitting line. Therefore we need to extend the ADMM distributed optimization to the case when a certain percent of the transmitted information is lost during the information exchange. We refer to this extended ADMM as an asynchronous ADMM.

The key idea of the asynchronous ADMM is that whenever loss of AMI messages occurs, we use the value from the previous transmission to substitute the missing information. Let Γ_i denote the set of users who successful receive z_k^i from the utility company, and let $\prod_i$ denote the set of users who successful upload their l_k^i, v_k^i to the utility company. We use $\tilde{z}_k^i$ to represent the information received by user k at iteration i. Let $\tilde{l}_k^i$, $\tilde{v}_k^i$ denote the information received by the utility company from user k at iteration i. Then the asynchronous ADMM can be represented as follows:

$$l_k^{i+1} = \underset{l_k \in F_k}{\operatorname{argmin}}\, p_B^T l_k + v_k^{iT}\left(l_k - \tilde{z}_k^i\right) + \frac{\rho}{2}\left\|l_k - \tilde{z}_k^i\right\|_2^2,\ \tilde{z}_k^i = \begin{cases} z_k^i, & k \in \Gamma_{i-1} \\ \tilde{z}_k^i, & k \notin \Gamma_{i-1} \end{cases}, \tag{4.43}$$

$$\begin{gathered}\{z_k^{i+1}\} = \underset{\{z_k^{i+1}\}}{\operatorname{argmin}} \sum_k \left(\tilde{v}_k^{iT}\left(\tilde{l}_k^{i+1} - z_k\right) + \frac{\rho}{2}\left\|\tilde{l}_k^{i+1} - z_k\right\|_2^2\right) + C_2 f_0\left(\sum_k z_k\right), \\ with\, \tilde{l}_k^{i+1} = \begin{cases} l_k^{i+1}, & k \in \Pi_i \\ \tilde{l}_k^i, & k \notin \Pi_i \end{cases}, \tilde{v}_k^i = \begin{cases} v_k^i, & k \in \Pi_i \\ \tilde{v}_k^{i-1}, & k \notin \Pi_i \end{cases},\end{gathered} \tag{4.44}$$

$$v_k^{i+1} = v_k^i + \rho\left(l_k^{i+1} - \tilde{z}_k^{i+1}\right),\ \tilde{z}_k^{i+1} = \begin{cases} z_k^{i+1}, & k \in \Gamma_i \\ \tilde{z}_k^i, & k \notin \Gamma_i \end{cases}. \tag{4.45}$$

The theoretical analysis of the convergence for this asynchronous ADMM is still an open problem. Note that when there is no loss of AMI messages, asynchronous

ADMM is exactly the same as original ADMM. In the section of numerical results, we use this algorithm to show that the ADMM-based algorithm is robust to communication loss, and it can still achieve satisfactory scheduling results for load management when there are PEVs in the smart grid.

4.6 Numerical Results

In this simulation we consider the case where there is only one electricity supplier and 120 households in the smart grid. There are 60 households with both wind-distributed generators and PEVs, 60 households with only distributed generators. The sum of the base load demand and schedulable demand from each household is generated randomly according to the MISO daily report by the U.S. Federal Regulatory Commission (FERC) [28]. The distributed wind generation values are taken from the Ontario Power Authority [29]. We set the battery size of the PEV as either 10 kWh or 20 kWh, to reflect different kinds of vehicles. The maximum charging rate is assumed to be 3.3 kW, and the maximum discharging rate is 1.5 kW. We also assume the charging and discharging rates can change continuously between the maximum discharging rate and maximum charging rate. The statistical mean of arrival time and departure are 18:00 and 8:00, respectively. The specific time slots are generated according to Gaussian distributions.

In our simulation, we first show three different types of distributed scheduling algorithms. The first one uses no optimization, and is widely used when the EMC and AMI devices are not applied. Users randomly select time slots for their schedulable loads, and charge their PEVs as soon as their vehicles are in the garages. The second algorithm is a greedy algorithm, in which everyone tries to lower his or her total electricity bill according to the pre-determined base price. This simulates the scenario when there is no fluctuation fee in the cost (4.15) and users are not cooperating with each other when making decisions. We compare them with the ADMM scheduling method proposed earlier. We show the effect of discharging the PEVs' batteries by comparing the case when discharging is not allowed with the case when discharging is enabled. Then we show the impact of α in pricing (4.12) to reduce the daily bill of each individual user. The effect of the proportion of schedulable load is also discussed in this section. Consumers participate in this algorithm when they see the reduction of their daily bills compared with the greedy method. We also show how prediction errors in base load and distributed generation will affect the performance of the ADMM scheduling method. Additionally, we show the robustness of asynchronous ADMM by introducing lost messages during the communication between the utility company and different customers. Last, we compare our method with another distributed optimization algorithm, proposed in [14]. We show that the convergence behavior of ADMM is less sensitive to the choice of parameter in the algorithm compared with the method in [14].

4.6.1 Valley Filling Properties of the ADMM Scheduling Method

In the first part of the results, we present the experiment when 20 % of the load (except PEV and load supplied by distributed generators) of an individual user is schedulable. Thus the ratio of the base load and schedulable load is 4:1. The α in (4.12) is set to be 1. Both charging efficiency and discharging efficiency are set to be 0.9. ρ in the ADMM iteration is chosen to be 0.006. In the pricing model, we have $C_1 = 7.8 \times 10^{-5}$ and $C_2 = 5 \times 10^{-4}$. We show the valley filling results of the ADMM scheduling algorithm in Fig. 4.3. We can see that the load without optimization creates a peak when most PEVs arrive home. The greedy algorithm simply moves the peak to another time period, which has the lowest base electricity price. The ADMM-based distributed optimization model proposed in this paper can fill the valley of the original base load and thus lead to a lower fluctuation of the load curve.

In Fig. 4.4, we show the specific load arrangement with our proposed method, the method without optimization and also the greedy algorithm of one user with distributed generator and PEV. We can see that the greedy user moves his entire schedulable load into the periods with cheaper electricity prices. She/he also tends to charge his/her PEV at times when the electricity prices are low and discharge his/her PEV at times when electricity prices are high. Comparing part (b) with Fig. 4.3, we can see that actually this decision pattern will lead to a high fluctuation price for this user. Therefore when the proportion of the fluctuation price is high, it is not worth doing this greedy algorithm. In part (c) of this figure, we can see that the user under ADMM strategy is more rational and cooperative. With this kind of cooperation

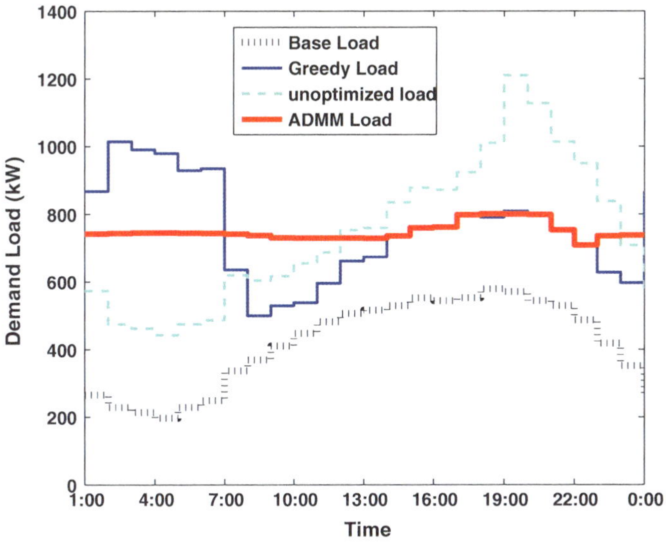

Fig. 4.3 Total loads of three scheduling methods as a function of time

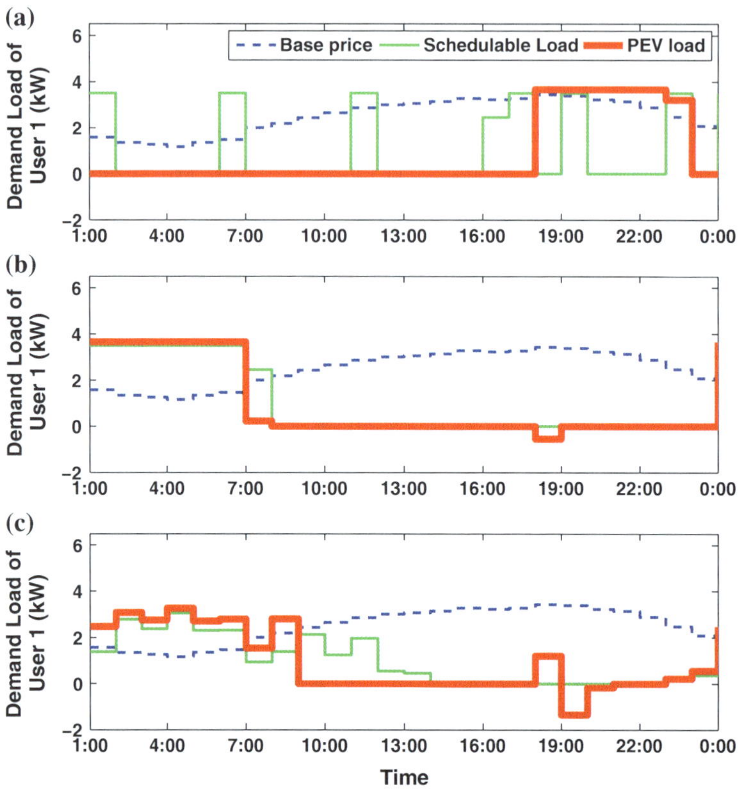

Fig. 4.4 Comparison of user's load scheduling strategy under three different scheduling methods

among different users, the total fluctuation cost will be lowered and therefore reduce the bill every user pays for their utility usage.

The total daily bill of all users is also compared among these three methods. The unoptimized method has a total bill as high as \$725, and the greedy method has a bill amount as \$550. While the ADMM scheduling method leads to a total bill as low as \$455.

4.6.2 Effect of Battery Discharging in the ADMM

In this section, we compare the case when discharging is not allowed with the case when both charging and discharging are implemented in battery management. The system parameters are the same as in the previous section. The results of total load scheduling are shown in Fig. 4.5. We can see that allowing discharging leads to a smaller variation in the demand response. The load schedule of one specific user is shown in Fig. 4.6.

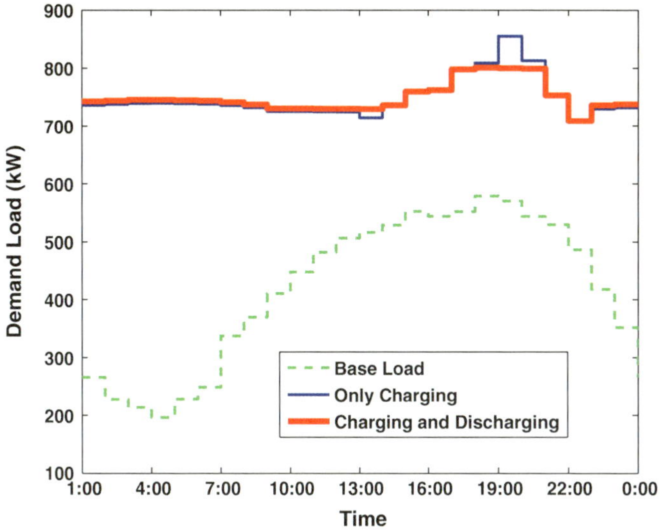

Fig. 4.5 Total loads of two strategies, one with only charging, the other with charging and discharging

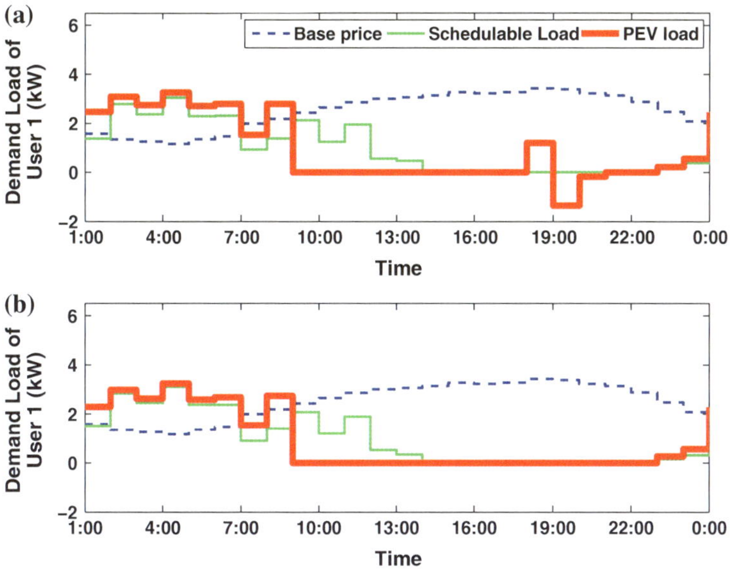

Fig. 4.6 Comparison of one user's load scheduling strategy under two different scheduling methods

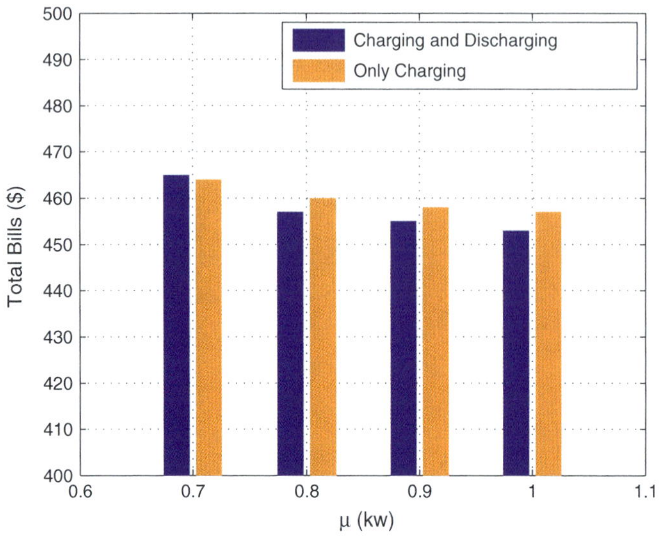

Fig. 4.7 Comparison between two load scheduling strategies with varying charging/discharging efficiency

The total bill for ADMM with both charging and discharging for batteries of PEVs is $455, while the total bill for only charging ADMM is $458. The saving is due to the lower variance when discharging is enabled, as we can observe from time 13:00 to 22:00 in Fig. 4.6. Next we vary the charging and discharging efficiency of PEVs' batteries and show the total bills for different scenarios in the Fig. 4.7. One interesting fact is that when charging and discharging efficiency is reduced to 0.7, the strategy with only charging achieves a better scheduling result. This is because when the efficiency is less than 1, the ADMM strategy for simultaneous charging and discharging obtains a suboptimal solution for the load scheduling, while the ADMM strategy for only charging achieves an optimal solution.

People can decide whether to turn on their PEVs' discharging ability by comparing the savings with the money that needed to replace their battery when battery life is shortened due to frequent discharging. This simulation is a preliminary result, and we can hope for a better bill reduction performance using discharging with more diversified PEV behaviors.

4.6.3 Load Scheduling with Loss of AMI Messages

In this section we show the robustness of the proposed asynchronous ADMM strategy in Sect. 4.5.5 with respect to loss of AMI messages during communication between the utility company and consumers. The efficiency of PEVs' batteries is assumed to be 0.7 in this simulation. We assume that the information loss occurs

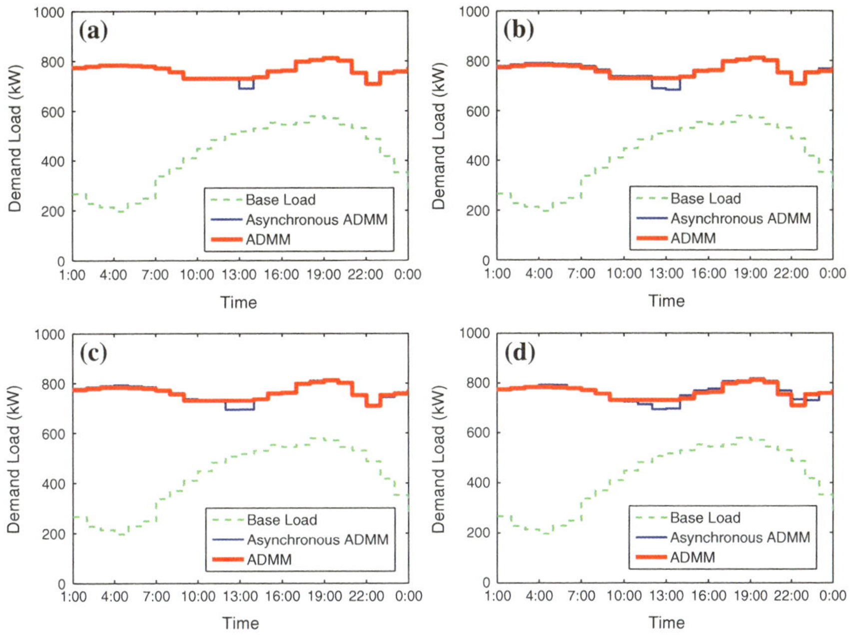

Fig. 4.8 Robustness of asynchronous ADMM with loss of AMI messages

randomly. In this simulation, we assume that different percentages of the information are lost due to transmitting errors. As we show in Fig. 4.8, we can observe that asynchronous ADMM strategy still achieves a desirable load scheduling result, comparable to the case where communication is 100 % successful. The total bill for the asynchronous ADMM with 10 % loss of data is \$466; it is \$468 for both 20 and 30 % loss of data; it is \$470 for 40 % loss of data. By comparison, the total bill for ADMM with no loss of information is \$465.

4.6.4 Impact of Pricing Parameter α

From now on, we consider the case where there are 30 households with both wind distributed generators and PEVs, 20 households with only distributed generators, 30 households with only PEVs, and 40 households with none of these. The efficiency of PEVs' battery is 0.8 for both charging and discharging. As we mentioned earlier in the pricing policy section, the parameter α in (4.12) has the ability to affect users' behavior. When α is set to be 1, the base price is proportional to the base load; therefore, the difference between prices from different time slots is relatively large, and then user will become more sensitive to this base price. When α is set to be 1/2, the base price is proportional to the square root of the base price, and the

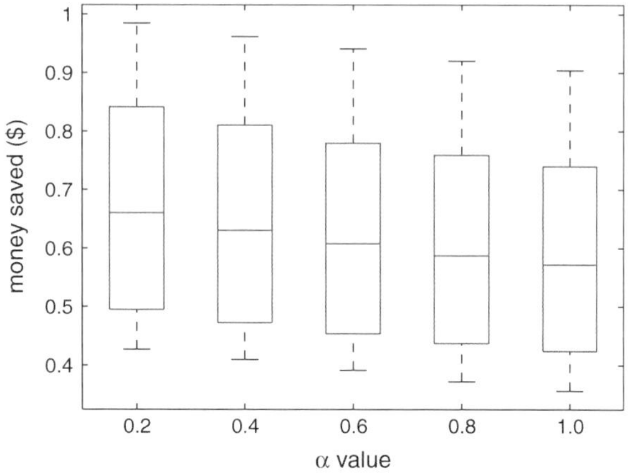

Fig. 4.9 Money saved with changing α (Percentage of schedulable load is 20 %)

user will become less sensitive to which time slot is the cheapest for their schedulable energy. When α is set to be zero, the user will not care about which time slots to allocate their energy and their goal becomes lowering the load variation as much as possible.

In Fig. 4.9, we range the value of α from 0.2 to 1, and plot the money saved for each individual user when consumers use the ADMM method compared with the greedy scheduling method. In each box in Fig. 4.9, the central mark is the median, the edges of the box are the 25th and 75th percentiles, and the whiskers extend to the most extreme data. The portion of schedulable load is 20 %, and ρ is chosen to be 0.006. We can see that even when α is high, which makes the greedy algorithm perform well, the consumer will still save money using the ADMM method. Overall every consumer will save his or her money every day under the ADMM scheduling method.

4.6.4.1 Results with Different Proportions of Schedulable Energy

In Fig. 4.10, we show the daily bill reduction achieved by using the proposed distributed ADMM optimization algorithm, compared with the greedy algorithm. The percentage of schedulable load of each individual user changes from 10 to 50 %. α is set to be 1, and ρ equals 0.006 in the algorithm. Every user will gain under the proposed distributed algorithm, and thus everyone has the incentive to be cooperative. In fact, if some users have perfect information about others' usage, they can gain even more by deviating from the ADMM solution for themselves, but it is impossible to get such information in real applications. In Fig. 4.10 we notice that when percentage of schedulable load is 20 and 30 %, the difference between ADMM scheduling method and greedy method is less than other cases. This is due to the fact that with this percentage of schedulable energy, the valley in the base

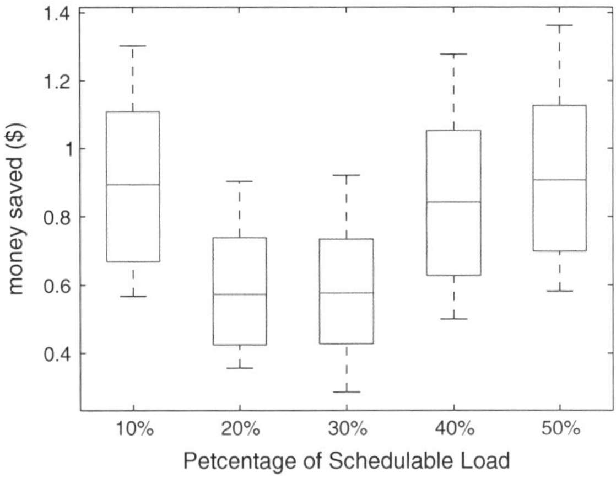

Fig. 4.10 Money saved with changing proportion of schedulable loads ($\alpha = 1$)

load can be filled without creating a new peak at the same time therefore greedy method is more efficient than greedy algorithm in other cases.

4.6.5 Results with Uncertainty in the Smart Grid Model

In order to set the base price for the next day, the utility company needs to predict the sum of the base load for all the households. The prediction will suffer from some random noise. We will see how much the prediction error will deteriorate the performance of ADMM. We will also show the impact of randomness in the distributed generation on the performance of the ADMM scheduling method. The models describing this randomness are shown in Sect. 4.4. The percentage of schedulable energy is 20 %. We let $\alpha = 1$ and $\rho = 0.006$ in this simulation. We range the standard deviation σ_b for the base load prediction error from 0.5 to 2.5 kW. The scale for the base load for a household at a certain time point ranges from 0 to 7 kW. We set σ_d to be 0.01 kW, and change σ_0 from 0.2 to 1 kW to see the impact of randomness in the distributed generation.

From Fig. 4.11, we can see that the ADMM scheduling method still outperform the greedy method and also the method without optimization. The total bill will increase with more uncertainties in the model. The ADMM method is more sensitive to the randomness in the distributed generation than that in the base load since the prediction error terms in the distributed generation are correlated among different households.

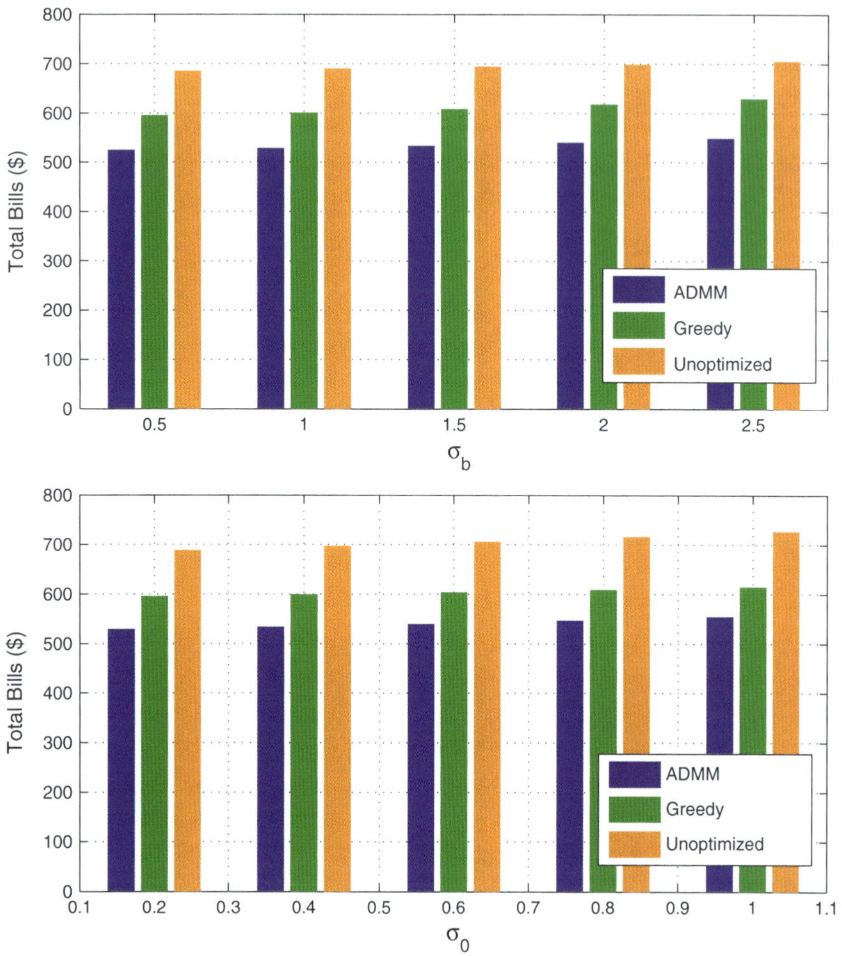

Fig. 4.11 Total bills for three scheduling methods with uncertainty in the model. The *upper figure* shows the impact of uncertainty in the base load, and the *lower figure* shows the impact of uncertainty in the distributed generation

4.6.6 Comparison with Another Distributed PEV Charging Method

In this section, we compare our distributed scheduling method with the optimal decentralized charging method (ODC) proposed in [14]. The ODC algorithm basically cycles between the following two operations until some kind of convergence criteria is satisfied:

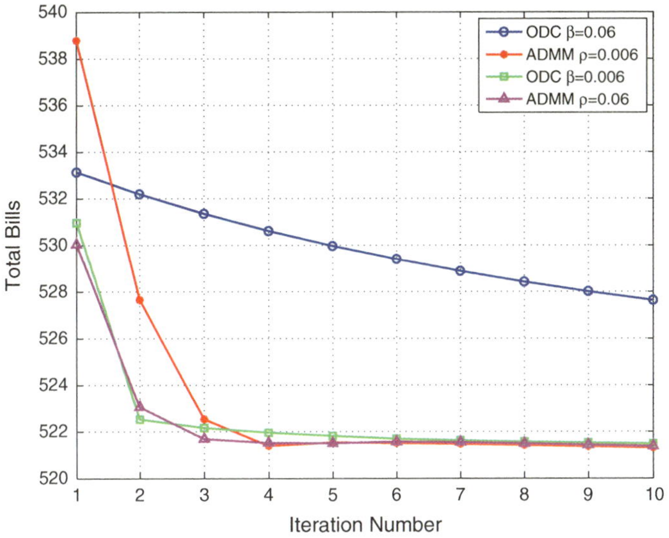

Fig. 4.12 Total bills with respect to number of iterations for ADMM and ODC

$$\boldsymbol{l}_k^{i+1} = \underset{l_k}{\operatorname{argmin}}\ \boldsymbol{p}^{iT}\boldsymbol{l}_k + \beta\left\|\boldsymbol{l}_k - \boldsymbol{l}_k^i\right\|_2^2,\ \boldsymbol{l}_k \in F_k,\ \forall k \tag{4.46}$$

$$\boldsymbol{p}^{i+1} = \gamma \sum_k \boldsymbol{l}_k^{i+1}. \tag{4.47}$$

We can see from optimization problem (4.46) that the ODC algorithm is similar to the optimization problem (4.36) in the ADMM scheduling method. Rather than being given a reference usage pattern by the utility company, the users in ODC are required to choose their usage pattern only within the neighborhood of the usage pattern in previous iteration. The pricing parameter γ is chosen to be 10^{-4}. In the pricing model of ADMM, we let $C_1 = 7.8 \times 10^{-5}$, and $C_2 = 5 \times 10^{-4}$. We test the performance of ADMM and ODC algorithms when β and ρ are set to be 0.0006, 0.006, and 0.06 respectively. In this simulation, the percentage of schedulable load is 20 %, and α is set to 1 in the pricing model.

From Fig. 4.12, we can see that when both β and ρ are set to be 0.006, the ADMM and ODC will have almost the same optimal convergence behavior after three iterations. When we increase the value of the parameters to 0.06, the ODC converges much slower than the ADMM scheduling method. When we decrease the value of these two parameter to 0.0006, which is not shown in the figure, the ODC algorithm will diverge, while the ADMM gives a total bill of \$537 after ten iterations. From this numerical example, we can see that the convergence of the ADMM scheduling method is less sensitive than ODC method to the parameter used.

4.7 Conclusions

In this chapter, we first built an electricity usage model for the PEV charging problem and also provided models for other loads in the smart grid, namely the base load, schedulable load, and distributed generation. A price scheme considering both base price and demand fluctuation in the demand response was proposed to encourage different consumers to corporate with each other to lower the fluctuations in the total loads. By applying the alternating direction method of multipliers, we decomposed the centralized optimization problem into distributed and parallel optimization problems. The decentralized scheme reduces the workload of the centralized controller while protecting the privacy of each consumer. Theoretical convergence proof was given for the ADMM case. We also extended the traditional ADMM to asynchronous ADMM to consider the case where AMI messages are lost during communication.

By showing numerical examples, we demonstrated that the demand response was flattened by using the ADMM based distributed scheduling method. We also compared the case where discharging is not allowed with the case where both charging and discharging are allowed. The results provide guidance for consumers to decide whether they want their PEVs to discharge or not during the load scheduling process. We tested the robustness of the asynchronous ADMM by considering different levels of AMI message loss in the communication system, and we showed that the proposed approach worked well. We also showed the robustness of the proposed method by considering estimation uncertainties for the overall next day load and the renewable energy. We demonstrated that all consumers would reduce their bills under several circumstances, which gives them incentives to participate in the distributed optimization program to charge their PEVs and schedule their appliances.

In our future work, we will employ a more detailed cost function for the utility company. We will also develop more advanced machine learning models to predict the users' future electricity usage behavior and will propose detailed strategies for choosing pricing parameters in real-world applications. The theoretical proof of asynchronous ADMM will also be of interest in the area of distributed computing. Extension of the price scheme proposed in this paper to an unbundling market situation, i.e., when the generation companies and retailers are separated entities in the electricity market and they have different interests, is another interesting open problem, and will also be included in our future work.

Acknowledgments This work was supported in part by the International Center for Advanced Renewable Energy and Sustainability (I-CARES) in Washington University in St. Louis, MO, USA.

References

1. Gellings C (1985) The concept of demand-side management for electric utilities. In: Proceedings of the IEEE, vol 70, no 10, pp 1468–1470
2. Albadi MH, El-Saadany E (2007) Demand response in electricity markets: an overview. In: Proceedings of IEEE power engineering society general meeting, pp 1–5
3. Hartway R, Price S, Woo CK (1999) Smart meter, customer choice and profitable time-of-use rate option. Energy 24:895–903
4. Wu Q, Wang L, Cheng H (2004) Research of TOU power price based on multi-objective optimization of DSM and costs of power consumers. In: Proceedings of 2004 IEEE international conference electric utility deregulation, restructuring, power technologies (DRPT 2004)
5. Kelly L, Rowe A, Wild P (2009) Analyzing the impacts of plug-in electric vehicles on distribution networks in British Columbia. In: Proceedings of IEEE electrical power and energy conference (EPEC), pp 1–9
6. Roe C, Meisel J, Meliopoulos A, Evangelos F, Overbye T (2009) Power system level impacts of PHEVs. In: System Sciences, 2009. HICSS'09. 42nd Hawaii international conference on IEEE, pp 1–10
7. Lopes J, Soares F, Almeida P (2011) Integration of electric vehicles in the electric power system. In: Proceedings of IEEE, vol 99, no 1, pp 168–183
8. Terzidis O, Karnouskos S, Karnouskos P (2007). An advanced metering infrastructure for future energy networks. In: Proceedings of NTMS conference, Paris
9. Chen C, Kishore S, Snyder L (2011) An innovative RTP-based residential power scheduling scheme for smart grids. In: Proceedings of IEEE international conference on acoustics, speech and signal processing (ICASSP), pp 5956–5959
10. Mohsenian-Rad A-H, Wong V, Jatskevich J, Schober R, Leon-Garcia A (2010) Autonomous demand-side management based on game—theoretic energy consumption scheduling for the future smart grid. IEEE Trans Smart Grid 1(3):320–331
11. Clement K, Haesen E, Driesen J (2009) Coordinated charging of multiple plug-in hybrid electric vehicles in residential distribution grids. Power systems conference and exposition, PSCE'09. IEEE/PES, pp 1–7
12. Lopes JP, Soares FJ, Almeida PM, da Silva MM (2009) Smart charging strategies for electric vehicles: enhancing grid performance and maximizing the use of variable renewable energy resources. In proceedings of EVS24 international battery, hybrid and fuel cell electric vehicle symposium, pp 1–11
13. Ma Z, Callaway D, Hiskens I (2010) Decentralized charging control for large populations of plug-in electric vehicles: application of the nash certainty equivalence principle. In: Proceedings of IEEE international conference on control applications (CCA), pp 191–195
14. Gan L, Topcu U, Low S (2013) Optimal decentralized protocol for electric vehicle charging. IEEE Trans Power Syst 28(2):940–951
15. Yang P, Chavali P, Gilboa E, Nehorai A (2013) Parallel load schedule optimization with renewable distributed generators in smart grids. IEEE Trans Smart Grid 4(3):1431–1441
16. Tan Z, Yang P, Nehorai A (2014) Optimal and distributed demand response with electric vehicles in the smart grid. IEEE Trans Smart Grid 5:861–869
17. Tan Z, Yang P, Nehorai A (2013) Distributed demand response for plug-in electrical vehicles in the smart grid. In: IEEE 5th international workshop on computational advances in multi-sensor adaptive processing (CAMSAP), pp 468–471
18. Brey J, Castro A, Moreno E, Garcia C (2002) Integration of renewable energy sources as an optimised solution for distributed generation. In: Proceedings of 28th annual conference on IEEE industrial electronics society (IECON), vol 4, pp 3355–3359
19. Carrasco J, Franquelo L, Bialasiewicz J, Galvan E, Guisado R, Prats M, Leon J, Moreno-Alfonso N (2006) Power-electronic systems for the grid integration of renewable energy sources: a survey. IEEE Trans Ind Electron 53(4):1002–1016

20. Lopes JP, Hatziargyriou N, Mutale J, Djapic P, Jenkins N (2007) Integrating distributed generation into electric power systems: a review of drivers, challenges and opportunities. Elect Power Syst Res 77(9):1189–1203
21. Boyd S, Parikh N, Chu E, Peleato B, Eckstein J (2010) Distributed optimization and statistical learning via alternating direction method of multipliers. Found Trends Mach Learn 3:1–122
22. Yang P, Tang G, Nehorai A (2013) A game-theoretic approach for optimal time-of-use electricity pricing. IEEE Trans Power Syst 28(2):884–892
23. Taylor J, Buizza R (2002) Neural network load forecasting with weather ensemble predictions. IEEE Trans Power Syst 17(3):626–632
24. Xia Y, Yang Y, Ge F, Su J, Yu H (2012) Pattern analysis for load forecasting. In: Proceedings of 2012 8th international conference on computer technology and information management (ICCM), vol 1, pp 339–343
25. Soman S, Zareipour H, Malik O, Mandal, P (2010) A review of wind power and wind speed forecasting methods with different time horizons. In: Proceedings of North American power symposium (NAPS), pp 1–8
26. He M, Murugesan S, Zhang J (2011) Multiple timescale dispatch and scheduling for stochastic reliability in smart grids with wind generation integration. In: Proceedings of IEEE INFOCOM, pp 461–465
27. Golub GH, van Loan CF (1996) Matrix computations, 3rd edn. Johns Hopkins University Press, Baltimore
28. FERC (2011) MISO daily report: electric power markets: Midwest. Retrieved from http://www.ferc.gov/market-oversight/mkt-electric/midwest/miso-archives.asp
29. OPA (2007) Simulated wind generation data. Retrieved from Antario Power Authority, http://www.powerauthority.on.ca/integrated-power-system-plan/simulated-wind-generation-data

Chapter 5
Risk Averse Energy Hub Management Considering Plug-in Electric Vehicles Using Information Gap Decision Theory

Alireza Soroudi and Andrew Keane

Abstract The energy hub is defined as the multi-input multi-output energy converter. It usually consists of various converters like thermal generators, combined heat and power (CHP), renewable energies and energy storage devices. The plug-in electric vehicles as energy storage devices can bring various flexibilities to energy hub management problem. These flexibilities include emission reduction, cost reduction, controlling financial risks, mitigating volatility of power output in renewable energy resources, active demand side management and ancillary service provision. In this chapter a comprehensive risk hedging model for energy hub management is proposed. The focus is placed on minimizing both the energy procurement cost and financial risks in energy hub. For controlling the undesired effects of the uncertainties, the Information gap decision theory (IGDT) technique is used as the risk management tool. The proposed model is formulated as a mixed integer linear programming (MILP) problem and solved using General Algebraic Modeling System (GAMS). An illustrative example is analyzed to demonstrate the applicability of the proposed method.

Keywords Energy hub · Information gap decision theory · Uncertainty modeling · Wind power · Plug-in electric vehicle

---

A. Soroudi is funded through Science Foundation Ireland (SFI) SEES Cluster under grant number SFI/09/SRC/E1780. A. Keane is supported by the SFI Charles Parsons Energy Research Awards SFI/06/CP/E005. This work was conducted in the Electricity Research Centre, University College Dublin, Ireland, which is supported by the Commission for Energy Regulation, Bord Gáis Energy, Bord na Móna Energy, Cylon Controls, EirGrid, Electric Ireland, Energia, EPRI, ESB International, ESB Networks, Gaelectric, Intel, SSE Renewables, and UTRC.

A. Soroudi (✉) · A. Keane
The School of Electrical, Electronic, and Communication Engineering,
University College Dublin, Dublin, Ireland
e-mail: Alireza.soroudi@ucd.ie

A. Keane
e-mail: Andrew.keane@ucd.ie

S. Rajakaruna et al. (eds.), *Plug In Electric Vehicles in Smart Grids*,
Power Systems, DOI 10.1007/978-981-287-302-6_5

Nomenclature

Uc(t,v)	Binary variable indicating the charging state
Ud(t,v)	Binary variable indicating the discharging state
η_{chp}^{ge}	CHP efficiency in converting gas to electricity
η_{chp}^{gh}	CHP efficiency in converting gas to heat
Pc(t,v)	Charged power of vehicle v in time t
η_{v}^{c}	Charging efficiency of vehicle v
$Pd(t,v)$	Discharged power of vehicle v in time t
η_{d}^{v}	Discharging efficiency of vehicle v
Le(t)	Electric load in time t
λ_{t}^{e}	Electricity price
η_{f}^{gh}	Furnace efficiency in converting gas to heat
λ_{g}	Gas price
Lh(t)	Heat load in time t
Pw(t)	Injected wind power
$Pc_{\min/\max}^{v}$	Min/max charging limits of vehicle v
$Pd_{\min/\max}^{v}$	Min/max discharging limits of vehicle v
OF	Objective function
Pg(t)	Purchased gas power
$Pe_{b}(t)$	Purchased electricity power
α_{Le}	Radius of uncertainty for electric load
α_{Lh}	Radius of uncertainty for heat load
α_{w}	Radius of uncertainty for wind power generation
$Pe_{s}(t)$	Soled electricity power
SOC(t,v)	State of charge of vehicle v in time t
$Pg_{chp}(t)$	Share of purchased gas power to feed into CHP
$Pg_{f}(t)$	Share of purchased gas power to feed into furnace
$P_{tr}(t,v)$	Traveling requirement of vehicle v in time t

5.1 Introduction

The concept of energy hub was first introduced in [1]. It is defined as a combination of energy conversion units which satisfy different types of energy demands. Figure 5.1 illustrates an example of energy hub, which provides an interface between the different inputs and outputs energy carriers.

A relevant number of recent researches have proposed some models for energy hub concept. These models describe the energy hubs as a combination of nuclear plants, wind turbines, solar panels, biomass reactors, electrolyzers, fuel cells [2] and energy storage devices. Different optimization techniques are available for solving

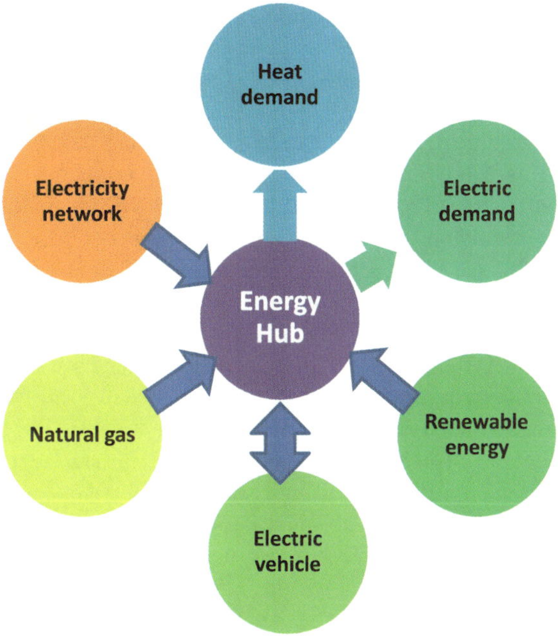

Fig. 5.1 The general concept of energy hub

the optimal management of energy hubs like Simulated Annealing algorithm [3], genetic algorithm [4] and multi-objective goal programming [5]. The optimal operating schedule of an energy hub highly depends on the input parameters of the model. Usually these input parameters are subject to uncertainty due to various reasons. For example, renewable power generations are volatile because of their natural primary resource like wind speed, solar radiation, temperature and etc. Another important uncertain input parameter is the demand whether it is electrical or heat which should be treated properly [6]. It is highly dependent on the consumer behavior which cannot be predicted easily. The last important uncertain parameter is the electricity price which directly affects the payments or benefits of the decision maker. The electricity prices in deregulated electricity markets are uncertain due to various reasons like: competition between the price maker generating companies, contingencies and etc.

There are different types of uncertainty modeling in energy hub management. The most famous technique is stochastic method [7, 8]. The Monte Carlo Simulation (MCS) [9] is used in uncertainty modeling of energy prices [10]. The shortcoming of this technique is that it is computationally expensive and it also requires the probability density functions (PDF) of uncertain parameters. Without them the problem cannot be solved. The second issue is that using the Monte Carlo simulation gives the decision maker the expected value of objective function and also its variance. It's more useful in assessment applications rather than optimization applications. The scenario based modeling which defines some discrete scenarios with specific probabilities and then tries to minimize the expected values. It improves the computational burden significantly compared to MCS.

In [11], this method is used to handle the uncertainty of wind, price and electricity demand. The conditional value at risk (CVaR) [12] is also used for risk controlling. Another uncertainty modeling technique is robust optimization [13]. This technique does not require the PDF of the uncertain parameters. Instead, it uses an interval for uncertain parameters. It tries to find the optimal decision variables while some predefined degree of conservativeness is taken into account. This technique is used for uncertainty modeling of energy prices, energy demand and also the converter efficiencies of energy hubs [14]. To cope with the increasing volatile renewable generation in energy hubs it is possible to use energy storage [15]. Different energy storage technologies have been used in energy hubs such as solid hydrogen storage [3], water electrolyzers for hydrogen production [16], thermal energy storage [17], Hydrogen-Natural Gas Co-Storage [18] and plug-in electric vehicles (PEV) [19]. The PEVs have recently attracted a great deal of attention in energy system management strategies. The advents of these new technologies have changed the original operating philosophy of PEVs from pure transportation into important energy system flexibility providers. They can be used as an energy storage device when not in use for transportation purposes. In this chapter, a risk averse Information Gap Decision Theory (IGDT) [20] framework is proposed for optimal energy management of an energy hub. This technique is exact and does not require the PDF of the uncertain parameters. This hub purchases energy from different resources and converts them to different output forms. It also uses the flexibilities that PEV may provide. The problem is analyzed with the following constraints, decision variables and objective function:

- Decision variables:
  - Electricity purchase from the electricity market
  - Electricity sell to the electricity market
  - Gas purchase from the gas network
  - Operation schedule of energy conversion devices
  - Operation charging and discharging of PEV
- Constraints:
  - Uncertainty of thermal demand
  - Uncertainty of electricity demand
  - Uncertainty of energy production of renewable resources
  - Technical constraints of energy conversion/storage of PEVs
  - Different demand balance
  - Risk of energy management strategy due to different uncertainties

Objective function: it is defined as the total payments regarding the energy management.

5.2 IGDT Based Uncertainty Modeling

The decision makers need some strong tools in order to handle the severe uncertainties. Specially when not enough information is available from the uncertain input parameters (like probability density function or membership function). The information gap decision theory provided such a tool which is computationally efficient and it is robust against the prediction errors. It has been successfully applied on various energy system applications such as:

- Energy procurement in distribution networks [20]
- Risk-constrained self-scheduling of GenCos [21]
- Multi-objective robust transmission expansion planning [22]
- Optimal bidding strategy of generation station in power market [23]

In this chapter, an IGDT based model [8] is proposed to handle the uncertainty of wind power generation, electric load and heat load. The mathematical formulation of risk hedging IGDT framework is as follows:

$$\min_X f(X, \psi) \tag{5.1}$$

$$H_i(X, \psi) \leq 0, \quad i \in \Gamma \tag{5.2}$$

Γ is the set of all constraints. ψ is the vector of input uncertain parameters. In this work, an IGDT based energy management is formulated as:

$$\max_X \hat{\ell} \tag{5.3}$$

$$H_i(X, \psi) \leq 0, \quad i \in \Gamma \tag{5.4}$$

$$\hat{\ell} = \{\max_\ell | f(X, \psi) - \Lambda_c \leq 0\} \tag{5.5}$$

$$\psi \in U(\bar{\psi}, \ell) = \{\psi : |\frac{\psi - \bar{\psi}}{\bar{\psi}}|\} \leq \ell \tag{5.6}$$

Λ_c is the critical value of objective function (for a given value of X) which can be exceeded when the realized values are not the same as forecasted ones. $\bar{\psi}$ is the forecasted value of ψ. ℓ is the unknown radius of uncertainty.

5.3 Problem Formulation

The general operating concept of an energy hub can be described as follows:

$$OF = \sum_t \lambda_\gamma(t) P_\gamma^{in}(t) \tag{5.7}$$

$$P_\gamma^{out}(t) = AP_\gamma^{in}(t) \tag{5.8}$$

where, γ is the set of energy carriers. $P_\gamma^{in}(t)$, $P_\gamma^{out}(t)$ denote the input and output energy carriers of the hub, respectively. $\lambda_\gamma(t)$ is the price of energy carrier γ at time t. The matrix A is the core function of the energy hub which defines the conversion, storage and distribution of different energy carriers. The energy hub under study in this chapter is depicted in Fig. 5.2.

This energy hub has three inputs as the supplying resources namely electric power purchased from electricity market ($Pe_b(t)$), wind power generation ($Pw(t)$) and finally natural gas ($Pg(t)$). The output of energy hub has three different parts namely electric load ($Le(t)$), heat load ($Lh(t)$), power sold to energy market ($Pe_s(t)$) and power charge/discharge for PEV ($Pd(t,v), Pc(t,v)$). The question is how to optimally exploit the energy hub in order to minimize the payments for energy procurement.

The performance of the described energy hub in Fig. 5.2 can be modeled as follows:

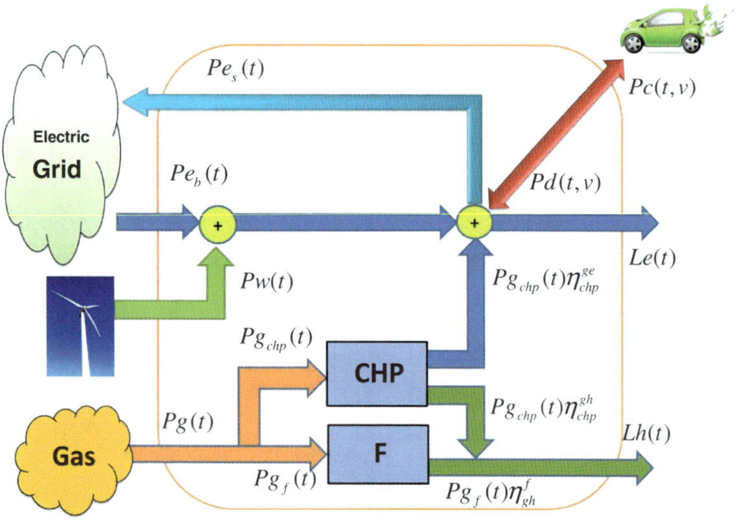

Fig. 5.2 The energy hub under study

$$Pe_b(t) + Pw(t) + Pg_{chp}(t)\eta_{chp}^{ge} + \sum_v Pd(t,v) = Le(t) + Pe_s(t) + \sum_v Pc(t,v) \tag{5.9}$$

$$0 \le Pe_b(t) \le Pe_b^{\max} \tag{5.10}$$

$$0 \le Pe_s(t) \le Pe_s^{\max} \tag{5.11}$$

$$Pw(t) \le (1-\alpha_w)\bar{P}w(t)Cap_w \tag{5.12}$$

$$Le(t) = (1+\alpha_{Le})\bar{L}e(t)Le_{\max} \tag{5.13}$$

$$Lh(t) = (1+\alpha_{Lh})\bar{L}h(t)Lh_{\max} \tag{5.14}$$

$$Pg(t) = Pg_{chp}(t) + Pg_f(t) \tag{5.15}$$

$$Lh(t) = Pg_f(t)\eta_f^{gh} + Pg_{chp}(t)\eta_{chp}^{gh} \tag{5.16}$$

The electric balance is modeled in (5.9). This means that the electric output of the energy hub $(Le(t) + Pe_s(t) + \sum_v Pc(t,v))$ is fed using $Pe_b(t) + Pw(t) + Pg_{chp}(t)\eta_{chp}^{ge} + \sum_v Pd(t,v)$. The third term is the converted gas to electricity in CHP units. The purchased gas $Pg(t)$ is divided into two streams $Pg_{chp}(t), Pg_f(t)$. The $Pg_{chp}(t)$ is fed into the CHP unit and the $Pg_f(t)$ is fed into the furnace unit as described in (5.15). Finally, the heat load $(Lh(t))$ is supplied using furnace and CHP units as given in (5.16).

The operation modeling of PEV is described in (5.17–5.23).

$$\mathrm{SOC(t,v)} = \mathrm{SOC(t-1,\ v)} + \eta_v^c Pc(t,v) - \frac{Pd(t,\mathrm{v})}{\eta_v^d} - \mathrm{P}_{tr}(\mathrm{t,v}) \tag{5.17}$$

$$\mathrm{SOC(t,v)} = \mathrm{E}_v^0 + \eta_v^c Pc(t,v) - \frac{Pd(t,v)}{\eta_v^d} - \mathrm{P}_{tr}(\mathrm{t,v}) \tag{5.18}$$

$$\mathrm{SOC}_{\min}^v \le \mathrm{SOC(t,v)} \le \mathrm{SOC}_{\max}^v \tag{5.19}$$

$$\mathrm{P}_{tr}(\mathrm{t,v}) = \Delta\mathrm{D(t,v)}\,\Omega_v \tag{5.20}$$

$$\mathrm{Pc}_{\min}^v Uc(t,v) \le Pc(t,v) \le \mathrm{Pc}_{\max}^v Uc(t,v) \tag{5.21}$$

$$\mathrm{Pd}_{\min}^v Uc(t,v) \le Pd(t,v) \le \mathrm{Pd}_{\max}^v Uc(t,v) \tag{5.22}$$

$$Uc(t,v) + Ud(t,v) \le 1 \tag{5.23}$$

The state of charge in *vth* PEV at time t $(SOC(t,\ v))$ depends on the state of charge at time t − 1 $(SOC(t-1,\ v))$. as well as the charging/discharging or traveling state of the PEV as modeled in (5.17) and (5.18). The relation between the required energy for traveling of vth PEV (P_{tr}(t,v)) depends on the traveling distance (ΔD(t,v)) and also the efficiency of the vehicle (Ω_v) as described in (5.20). The state of charge should be kept between operating limits as (5.20). The charging and discharging rate of each PEV are limited by technical characteristics as well as the operating state as enforced by (5.22) and (5.23). It is assumed that each PEV is either in charging ($Uc(t,v)=1$)/discharging state $Ud(t,v)=1$ or traveling state ($Uc(t,v)+Ud(t,v)=0$) as described in (5.23).

The objective function is defined as the total payments regarding the energy purchase as follows:

$$OF = \sum_t Pg(t)\lambda^g + \lambda_t^e(Pe_b(t) - Pe_s(t)) \tag{5.24}$$

If the OF is negative in (5.24) it means the energy hub is making profit in the electricity market.

5.4 Simulation Results

The proposed mixed integer linear programming (MILP) model is implemented in GAMS [24] environment solved by CPLEX solver running on an Intel® Xeon® CPU E5-1620 @ 3.6 GHz PC with 8 GB RAM. The predicted hourly electric/heat demand, wind power, electricity price pattern are depicted in Fig. 5.3. The wind

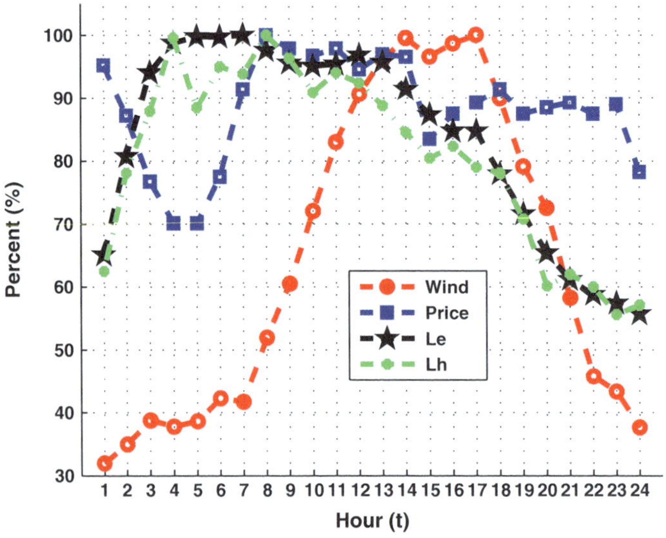

Fig. 5.3 The hourly electric/heat demand, wind power, electricity price pattern

Table 5.1 Energy hub characteristics and data

Parameter	Value	Unit
η_{chp}^{ge}	35	%
η_{chp}^{gh}	45	%
η_{f}^{gh}	75	%
$Pe_{s}^{\max}$	7	kW
$Pe_{b}^{\max}$	7	kW
η_{v}^{d}	93	%
η_{v}^{c}	90	%
E_{v}^{0}	3	kWh
$SOC_{\max}^{v}$	25	kWh
$SOC_{\min}^{v}$	1	kWh
$P(c/d)_{\min}^{v}$	0	kW
$P(c/d)_{\max}^{v}$	12.5	kW
Ω_{v}	$\frac{1}{6}$	$\frac{\text{kW}}{\text{km}}$

capacity is assumed to be $Cap_w = 15\,\text{kW}$. The energy hub characteristics and data are described in Table 5.1. The peak electric and heat load is $Le_{\max} = 5\,\text{kW}$, $Lh_{\max} = 4.5\,\text{kW}$, respectively. The peak value of electric price is $47\,\frac{\$}{\text{kWh}}$. The gas price is assumed to be constant and equal to $\lambda^g = 30\,\frac{\$}{\text{kWh}}$.

The travel pattern of PEV (km) are given in Table 5.2.

In order to demonstrate the applicability and strength of the proposed approach different scenarios are considered as follows:

- Base case (no uncertain parameter exists in the model)
- Wind uncertainty (α_w)
- Electric load uncertainty (α_{Le})
- Heat load uncertainty (α_{Lh})

5.4.1 Base Case

In this case, it is assumed that no uncertain parameter exists in the model. The objective function to be minimized is the energy procurement cost. It is called the base cost (benefit) OF_b. The following optimization is solved:

$$\min_{DV_b} \text{OF}_b = \text{OF} \tag{5.25}$$

Subject to: (5.9–5.24)

Table 5.2 The travel patterns of PEVs (km)

Time (h)	v1	v2	v3	v4	v5	v6	v7	v8	v9	v10
t1	0	0	0	4.6	0	0	0	4	0	0
t2	0	3.6	0	1.8	0	2	0	0	0	0
t3	0	5	0	0	0	2.2	0	0.6	0	2
t4	0	0	0	0	3.6	0	0	1.2	0	0
t5	2.4	0	0	0	1.8	4.2	2.8	3.6	0	4.4
t6	0	4.8	0	0	1.4	1.8	0	0	0	0
t7	0	0	0	0	1.6	2.6	0	0	0	0
t8	4.8	0	1	2	0	3.8	0	0	0	2
t9	0	0	0	0	1.2	3	1.2	0.8	0	1.2
t10	0	2.4	0	4	0	0	3.4	0	0	1
t11	0	0	0	4.6	2.4	0	0	4.4	0	0.4
t12	4	0	0	0	4.2	3	0	1.2	0	0
t13	0	0	0	0	2	0	3.4	0	4.2	0
t14	0	0	0	0	3	0	0	4	0	0
t15	0	0	0	0	0	1.4	0	0	3.8	0
t16	3.6	0	0	4.6	0	0	3.8	0	0	4
t17	0	0	3.6	0	1.6	0	0	3	0	4
t18	0	0	0	4	0	0	0	1.8	0	0
t19	0	0	4	0	2.2	2.6	0	0	2	4
t20	0	0	0	0	3	0	4.2	3.2	2.2	0
t21	0	0	4.8	3.8	0	0	0	2.6	1	1
t22	0	0	0	0	3.8	0	0	0.4	0	0
t23	0	4.8	0	0	0	0	0	0	0	2.2
t24	0	0	0	0	2.2	0	3.2	0	0.4	0

$$\alpha_{Lh} = \alpha_{Le} = \alpha_w = 0 \tag{5.26}$$

$$\mathrm{DV}_b = \left\{ \mathrm{Pe}_b(\mathrm{t}), \mathrm{Pe}_s(\mathrm{t}), \mathrm{Pw}(\mathrm{t}), \mathrm{Pg}_{chp}(\mathrm{t}), \mathrm{Pg}_f(\mathrm{t}), \mathrm{Pd}(\mathrm{t}, \mathrm{v}), \mathrm{Pc}(\mathrm{t}, \mathrm{v}), \mathrm{Ud}(\mathrm{t}, \mathrm{v}), \mathrm{Uc}(\mathrm{t}, \mathrm{v}) \right\} \tag{5.27}$$

The total costs would be $\mathrm{OF}_b = -\$0.480165$. The hourly total charge and discharge pattern of PEVs is shown in Fig. 5.4.

The hourly gas input to CHP and furnace is shown in Fig. 5.5.

The hourly purchased/sold power from/to electric grid is shown in Fig. 5.6.

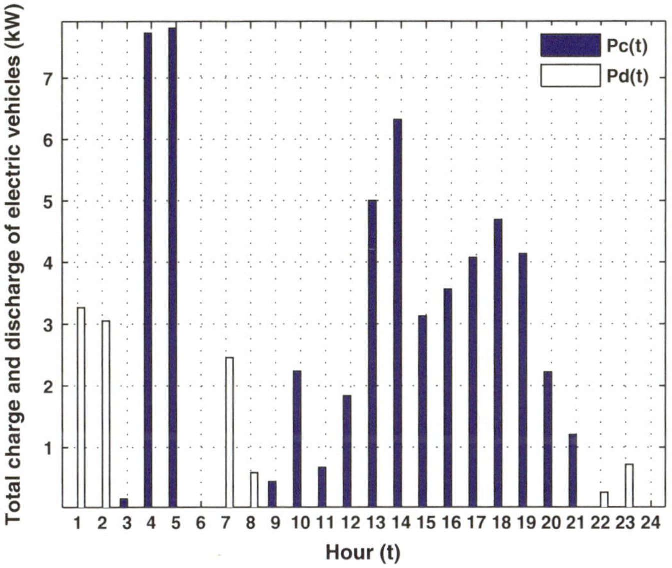

Fig. 5.4 The hourly total charge and discharge pattern of PEVs in base case

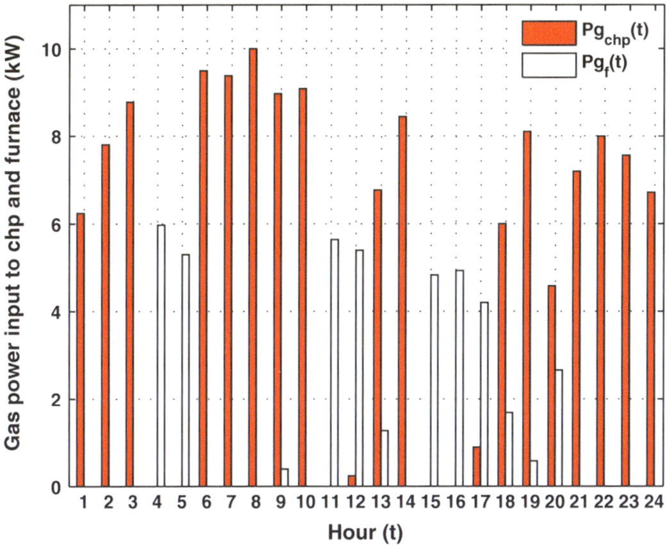

Fig. 5.5 The hourly gas input to CHP and furnace in base case

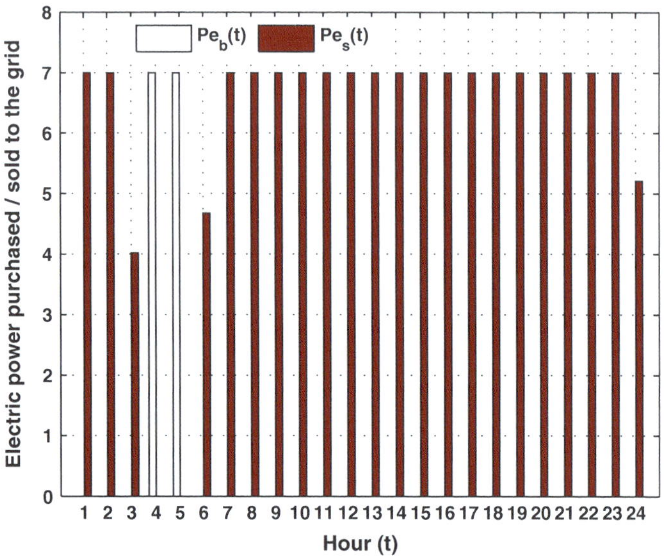

Fig. 5.6 The hourly purchased/sold power from/to electric grid in base case

5.4.2 Uncertain Wind ($\alpha_w \neq 0$)

In this case, it is assumed that the only uncertain parameter existing in the model is wind power generation. The objective function in this case is radius of wind power uncertainty (not the total cost (benefit)). The following optimization is solved:

$$\max_{DV_w} \alpha_w \tag{5.28}$$

$$\text{OF} \leq \text{OF}_b + \beta|\text{OF}_b| \tag{5.29}$$

Subject to: (5.9–5.24)

$$\alpha_{Lh} = \alpha_{Le} = 0 \tag{5.30}$$

$$\text{DV}_w = \{\text{DV}_b, \alpha_w\} \tag{5.31}$$

The interpretation of each β value is simply defined as the relaxation degree of objective function. The objective function is defined as α_w and the decision maker tries to maximize it for a given β value. In this way, the traditional objective function OF would be immune against the wind uncertainty. This means even if the forecasted value of wind doesn't come true, the total payments do not increase more than β percent of the base case costs OF_b. The β is increased from 0 to 1 and the variation of different variables (DV_w) versus β is shown in Fig. 5.7.

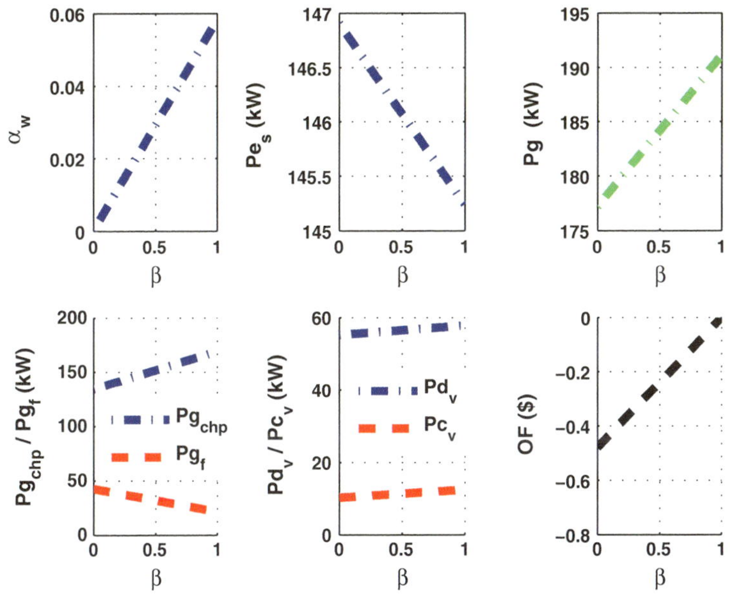

Fig. 5.7 The variation of different variables versus β (uncertain wind)

In this way, the decision maker has a portfolio of the decision variables (DV_w) for each β. The simulation results show that the α_w varies from 0 to 5.829 %. This means that if the total cost is 100 % increased then the decision maker can be immune up to 5.829 % error in wind power prediction. In order to increase the immunity of the objective function against the wind power uncertainty, $Pe_s(t)$, $Pg_f(t)$ are decreased and $Pg_{chp}(t)$ is increased. Both charging and discharging of PEVs ($Pd(t, v)$, $Pc(t, v)$) are increased. For clarification, the decision variables DV_w are given in Table 5.3 for $\beta = 30\,\%$. In this table, the hourly optimal schedule of energy hub $\beta = 30\,\%$ under $Pw(t)$ uncertainty are described. The total payments would be OF = − \$0.3361 and the maximum wind uncertainty that can be tolerated would be $\alpha_w = 1.748692\,\%$.

5.4.3 Uncertainity Electric Load Missing ($\alpha_{Le} \neq 0$)

In this case, it is assumed that the only uncertain parameter existing in the model is electric load. The objective function in this case is radius of electric load uncertainty [not the total cost (benefit)]. The following optimization is solved:

Table 5.3 The hourly optimal schedule of energy hub $\beta = 30\,\%$ under Pw(t) uncertainty

Time (h)	$Pe_b(t)$	$Pe_s(t)$	$Pg(t)$	$Pg_{chp}(t)$	$Pg_f(t)$	$\sum_v Pc(t, v)$	$\sum_v Pd(t, v)$
t1	0	7	6.245	6.245	0	0	3.353
t2	0	7	7.808	7.808	0	0	3.145
t3	0	3.729	8.787	8.787	0	0.36	0
t4	7	0	5.974	0	5.974	7.63	0
t5	7	0	5.306	0	5.306	7.708	0
t6	0	4.569	9.502	9.502	0	0	0
t7	0	7	9.385	9.385	0	0	2.568
t8	0	7	10	10	0	0	0.724
t9	0	7	9.044	8.159	0.885	0	0
t10	0	7	9.091	9.091	0	2.048	0
t11	0	7	8.387	6.866	1.522	2.862	0
t12	0	7	5.544	0	5.544	1.511	0
t13	0	7	7.214	4.718	2.496	4.031	0
t14	0	7	6.536	3.663	2.874	4.378	0
t15	0	7	6.183	3.391	2.792	4.061	0
t16	0	7	8.23	8.23	0	6.186	0
t17	0	7	6.603	4.656	1.947	5.129	0
t18	0	7	8.807	8.807	0	5.439	0
t19	0	7	5.448	0	5.448	1.087	0
t20	0	7	9.011	9.011	0	3.582	0
t21	0	7	6.002	4.207	1.795	0	0
t22	0	7	8	8	0	0	0.389
t23	0	7	7.558	7.558	0	0	0.835
t24	0	5.113	6.713	6.713	0	0	0

$$\max_{DV_w} \alpha_{Le} \tag{5.32}$$

$$\mathrm{OF} \leq \mathrm{OF}_b + \beta|\mathrm{OF}_b| \tag{5.33}$$

Subject to: (5.9–5.24)

$$\alpha_{Lh} = \alpha_w = 0 \tag{5.34}$$

$$\mathrm{DV}_{Le} = \{\mathrm{DV}_b, \alpha_{Le}\} \tag{5.35}$$

The objective function is defined as α_{Le} and the decision maker tries to maximize it for a given β value. In this way, the traditional objective function OF would be immune against the electric load uncertainty. This means that even if the forecasted value of electric load is not equal to the real value, the total payments do not increase more than β percent of the base case costs OF_b. The β is increased from 0 to 1 and the variation of different variables versus β is shown in Fig. 5.8.

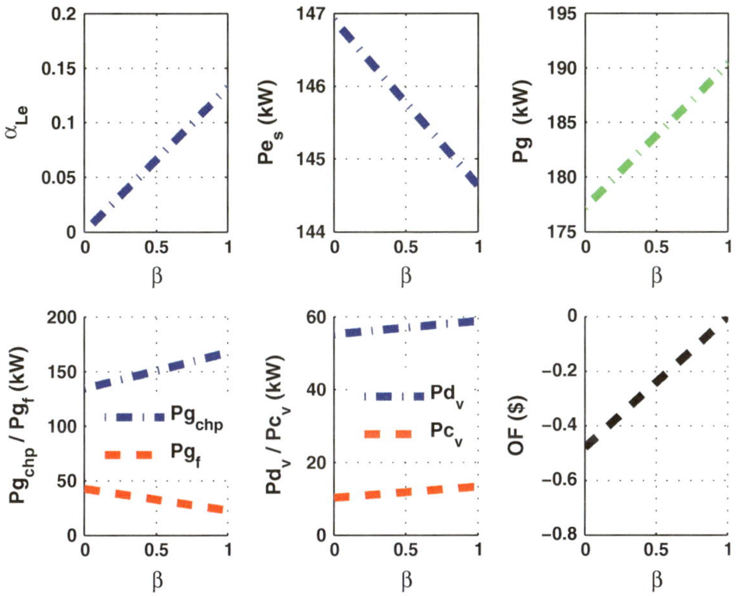

Fig. 5.8 The variation of different variables versus β (uncertain electric load)

In this way, the decision maker has a portfolio of the decision variables (DV_{Le}) for each β. For clarification, the decision variables are given in Table 5.4 for $\beta = 30\,\%$. In this table, the hourly optimal schedule of energy hub $\beta = 30\,\%$ under Le(t) uncertainty are described. The total payments would be OF = − \$0.336115 and the maximum wind uncertainty that can be tolerated would be $\alpha_{Le} = 3.963492\,\%$.

5.4.4 *Uncertain Heat Load* ($\alpha_{Lh} \neq 0$)

In this case, it is assumed that the only uncertain parameter existing in the model is heat load. The objective function in this case is radius of heat demand uncertainty (not the total cost (benefit)). The following optimization is solved:

$$\max_{DV_{Lh}} \alpha_{Lh} \tag{5.36}$$

$$\mathrm{OF} \le \mathrm{OF}_b + \beta|\mathrm{OF}_b| \tag{5.37}$$

Subject to: (5.9–5.24)

Table 5.4 The hourly optimal schedule of energy hub $\beta = 30\,\%$ under Le(t) uncertainty

Time (h)	$Pe_b(t)$	$Pe_s(t)$	$Pg(t)$	$Pg_{chp}(t)$	$Pg_f(t)$	$\sum_v Pc(t,v)$	$\sum_v Pd(t,v)$
t1	0	7	6.245	6.245	0	0	3.398
t2	0	7	7.808	7.808	0	0	3.213
t3	0	3.645	8.787	8.787	0	0.359	0
t4	7	0	5.974	0	5.974	7.533	0
t5	7	0	8.843	8.843	0	10.707	0
t6	0	4.482	9.502	9.502	0	0	0
t7	0	7	9.385	9.385	0	0	2.657
t8	0	7	10	10	0	0	0.781
t9	0	7	9.079	8.246	0.832	0	0
t10	0	7	9.091	9.091	0	2.048	0
t11	0	7	9.402	9.402	0	3.778	0
t12	0	7	5.544	0	5.544	1.557	0
t13	0	7	5.327	0	5.327	2.442	0
t14	0	7	8.452	8.452	0	6.134	0
t15	0	7	4.826	0	4.826	2.955	0
t16	0	7	7.247	5.774	1.473	5.417	0
t17	0	7	4.74	0	4.74	3.594	0
t18	0	7	8.343	7.646	0.697	5.114	0
t19	0	7	7.686	5.595	2.091	3.111	0
t20	0	7	5.407	0	5.407	0.489	0
t21	0	7	7.199	7.199	0	1.079	0
t22	0	7	8	8	0	0	0.385
t23	0	7	7.558	7.558	0	0	0.835
t24	0	5.101	6.713	6.713	0	0	0

$$\alpha_{Le} = \alpha_w = 0 \tag{5.38}$$

$$\text{DV}_{Lh}=\{\text{DV}_b, \alpha_{Lh}\} \tag{5.39}$$

The interpretation of each β value is simply defined as the relaxation degree of objective function. The objective function is defined as α_{Lh} and the decision maker tries to maximize it for a given β value. In this way, the traditional objective function OF would be immune against the heat load uncertainty. This means that even if the forecasted value of heat load is not equal to the real value, the total payments do not increase more than β percent of the base case costs OF_b. The β is increased from 0 to 1 and the variation of different variables (DV_{Lh}) versus β is shown in Fig. 5.9.

In this way, the decision maker has a portfolio of the decision variables (DV_{Lh}) for each β. For clarification, the decision variables are given in Table 5.5 for $\beta = 30\,\%$. In this table, the hourly optimal schedule of energy hub $\beta = 30\,\%$ under

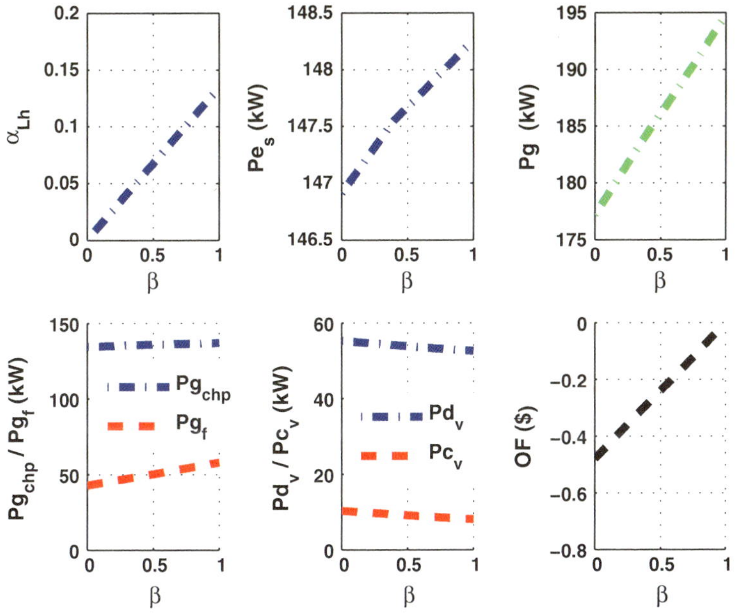

Fig. 5.9 The variation of different variables versus β (uncertain heat load)

Lh(t) uncertainty are described. The total payments would be OF = − \$0.336115 and the maximum wind uncertainty that can be tolerated would be $\alpha_{Lh} = 4.040376\,\%$.

5.5 Comparison and Discussion

In this section, the four assessed cases are compared and discussed. The simulation results show that in order to increase the robustness of the decision variables against the wind uncertainty, some actions should be taken. Selling electricity to the pool should be reduced. This holds also for reducing the undesired impacts of electric demand uncertainty. In contrary to these two cases, the energy selling to the pool market should be increased in order to avoid the financial risks due to uncertainty of heat load. Increasing the amount of gas purchase would have positive impacts on reducing the risks of all uncertain parameters (including wind power generation, electric and heat load). However in order to make the objective function immune to uncertainty of wind and electric load, the share of the natural gas which is fed into the CHP unit is increased and the furnace share is decreased. The decision maker should increase the share of furnace unit to avoid the risks of uncertain heat demand. The amount of PEVs in charging and discharging should be increased in order to handle the uncertainties of wind and electric load in contrary to the

Table 5.5 The hourly optimal schedule of energy hub $\beta = 30\,\%$ under Lh(t) uncertainty

Time (h)	$Pe_b(t)$	$Pe_s(t)$	$Pg(t)$	$Pg_{chp}(t)$	$Pg_f(t)$	$\sum_v Pc(t,v)$	$\sum_v Pd(t,v)$
t1	0	7	6.497	6.497	0	0	3.181
t2	0	7	8.123	8.123	0	0	2.943
t3	0	4.272	9.142	9.142	0	0.042	0
t4	7	0	6.331	0.29	6.041	7.831	0
t5	7	0	5.52	0	5.52	7.809	0
t6	0	4.814	9.886	9.886	0	0	0
t7	0	7	9.764	9.764	0	0	2.326
t8	0	7	10.404	10.404	0	0	0.447
t9	0	7	9.096	7.705	1.391	0	0
t10	0	7	9.458	9.458	0	2.365	0
t11	0	7	5.869	0	5.869	0.677	0
t12	0	7	5.768	0	5.768	1.749	0
t13	0	7	5.542	0	5.542	2.632	0
t14	0	7	8.793	8.793	0	6.435	0
t15	0	7	8.369	8.369	0	6.057	0
t16	0	7	5.137	0	5.137	3.564	0
t17	0	7	8.219	8.219	0	6.639	0
t18	0	7	7.553	5.138	2.415	4.391	0
t19	0	7	8.295	6.565	1.729	3.593	0
t20	0	7	5.625	0	5.625	0.619	0
t21	0	7	6.002	3.771	2.231	0	0
t22	0	7	8.323	8.323	0	0	0.156
t23	0	7	7.864	7.864	0	0	0.615
t24	0	5.307	6.984	6.984	0	0	0

uncertain head load case. The comparison between different cases and the actions to be taken is shown in Fig. 5.10.

Some lines of future research can be concluded from this work, as follows:

- To consider more elements (like energy conversion and storage units) in energy hub
- To consider other uncertain parameters affecting the performance of the energy hub
- To consider the possibility of participating in other markets in addition to energy market
- To assess the Impacts of smart grids on energy hub energy management policies
- To analyze the reliability issues of elements in energy hub
- To develop a model for describing the interaction of multiple energy hubs from technical and economical points of view
- To incorporate the grid (gas and electric) integration constraints of energy hubs

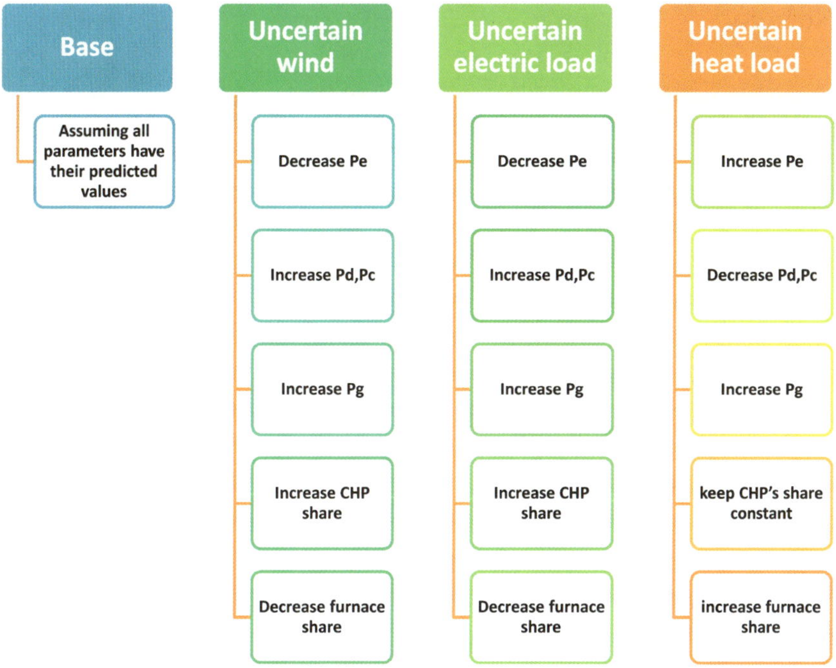

Fig. 5.10 The comparison between different cases and the actions to be taken

5.6 Conclusions

The problem of considering the input uncertainties within the context of the energy hub management has been addressed in this chapter. An IGDT based technique was proposed to obtain the optimal operating strategy of the energy hub. The PEVs have been used as the energy storage device in order to maximize the flexibility of decision making framework. The optimal energy procurement from different resources is determined taking into account the influence of electric/heat demand as well as the wind power generation uncertainties. The obtained results from the proposed risk-averse strategy assures the decision maker that although the predicted values of the uncertain input parameters are not exact, the outcome of the proposed model (payments) would be immune against the prediction error to some controlled extent. The method can be extended to consider the risk seeking behavior of opportunistic decision maker.

Acknowledgments This book chapter is gratefully dedicated to Simin, Shahryar, Mona and Soudeh who taught me how to be a better man.

References

1. Geidl M, Koeppel G, Favre-Perrod P, Klockl B, Andersson G, Frohlich K (2007) Energy hubs for the future. IEEE Power Energy Mag 5:24–30
2. Maniyali Y, Almansoori A, Fowler M, Elkamel A (2013) Energy hub based on nuclear energy and hydrogen energy storage. Ind Eng Chem Res 52:7470–7481
3. Proietto R, Arnone D, Bertoncini M, Rossi A, La Cascia D, Miceli R, Riva Sanseverino E (2014) A novel heuristics-based energy management system for a multi-carrier hub enriched with solid hydrogen storage. In: proceedings of 5th international conference on future energy systems, pp 231–232
4. Barbieri ES, Dai YJ, Morini M, Pinelli M, Spina PR, Sun P, Wang RZ (2014) Optimal sizing of a multi-source energy plant for power heat and cooling generation. Appl Therm Eng 71 (2):736–750
5. La Scala M, Vaccaro A, Zobaa AF (2014) A goal programming methodology for multiobjective optimization of distributed energy hubs operation. Appl Therm Eng 71 (2):658–666
6. Soroudi A (2012) Possibilistic-scenario model for DG impact assessment on distribution networks in an uncertain environment. IEEE Trans Power Syst 27:1283–1293
7. Rabiee A, Soroudi A (2014) Stochastic multiperiod OPF model of power systems with HVDC-connected intermittent wind power generation. Power Deliv IEEE Trans 29:336–344. doi:10.1109/TPWRD.2013.2259600
8. Soroudi A, Amraee T (2013) Decision making under uncertainty in energy systems: state of the art. Renew Sustain Energy Rev 28:376–384
9. Soroudi A (2014) Taxonomy of uncertainty modeling techniques in renewable energy system studies. In: Hossain J, Mahmud A (eds) Large scale renewable power generation. Springer, Singapore, pp 1–17
10. Kienzle F, Ahčin P, Andersson G (2011) Valuing investments in multi-energy conversion, storage, and demand-side management systems under uncertainty. Sustain Energy IEEE Trans 2:194–202. doi:10.1109/TSTE.2011.2106228
11. Soroudi A, Mohammadi-Ivatloo B, Rabiee A (2014) Energy hub management with intermittent wind power. In: Hossain J, Mahmud A (eds) Large scale renewable power generation. Springer, Singapore, pp 413–438
12. Tajeddini MA, Rahimi-Kian A, Soroudi A (2014) Risk averse optimal operation of a virtual power plant using two stage stochastic programming. Energy 73:958–967
13. Soroudi A (2013) Robust optimization based self scheduling of hydro-thermal Genco in smart grids. Energy 61:262–271
14. Parisio A, Del Vecchio C, Vaccaro A (2012) A robust optimization approach to energy hub management. Int J Electr Power Energy Syst 42:98–104
15. Adamek F, Arnold M, Andersson G (2014) On decisive storage parameters for minimizing energy supply costs in multicarrier energy systems. Sustain Energy IEEE Trans 5:102–109. doi:10.1109/TSTE.2013.2267235
16. Sharif A, Almansoori A, Fowler M, Elkamel A, Alrafea K (2014) Design of an energy hub based on natural gas and renewable energy sources. Int J Energy Res 38:363–373. doi:10.1002/er.3050
17. Ferrari ML, Pascenti M, Sorce A, Traverso A, Massardo AF (2014) Real-time tool for management of smart polygeneration grids including thermal energy storage. Appl Energy 130:670–678
18. Peng D (2013) Enabling utility-scale electrical energy storage through underground hydrogen-natural gas co-storage MSc Thesis, Chemical Engineering Department, University of Waterloo, 2013
19. Dargahi A, Ploix S, Soroudi A, Wurtz F (2014) Optimal household energy management using V2H flexibilities. COMPEL Int J Comput Math Electr Electron Eng 33:6–14

20. Soroudi A, Ehsan M (2012) IGDT based robust decision making tool for DNOs in load procurement under severe uncertainty. IEEE Trans Smart Grid 4:886–895
21. Mohammadi-Ivatloo B, Zareipour H, Amjady N, Ehsan M (2013) Application of information-gap decision theory to risk-constrained self-scheduling of GenCos. IEEE Trans Power Syst 28:1093–1102. doi:10.1109/TPWRS.2012.2212727
22. Dehghan S, Kazemi A, Amjady N (2014) Multi-objective robust transmission expansion planning using information-gap decision theory and augmented ε;-constraint method. Gener Transm Distrib IET 8:828–840. doi:10.1049/iet-gtd.2013.0427
23. Nojavan S, Zare K, Feyzi MR (2013) Optimal bidding strategy of generation station in power market using information gap decision theory (IGDT). Electr Power Syst Res 96:56–63
24. Brooke AD, Kendrick AM, Roman R (1998) GAMS: a user's guide. GAMS Development Corporation, Washington, DC

Chapter 6
Integration of Distribution Grid Constraints in an Event-Driven Control Strategy for Plug-in Electric Vehicles in a Multi-Aggregator Setting

Klaas De Craemer, Stijn Vandael, Bert Claessens and Geert Deconinck

Abstract In literature, several mechanisms are proposed to prevent Plug-in Electric Vehicles (PEVs) from overloading the distribution grid [1]. However, it is unclear how such technical mechanisms influence the market level control strategies of a PEV aggregator. Moreover, the presence of multiple aggregators in the same distribution grid further complicates the problem. Often, grid congestion management mechanisms are proposed to solve the potential interference between the technical and market objectives. Such methods come at the expense of additional complexity and costs, which is not beneficial for the large scale application of demand response. In our work, we investigate this problem by combining a simple low level voltage droop controller with an event driven control strategy for the coordination of charging PEVs. The approach is evaluated in different distribution grid settings, using two different market objectives for the aggregator.

6.1 Introduction

In a liberalized electricity market, aggregators are typically seen as the actors who will utilize the flexibility of PEVs. To control their PEVs, an aggregator typically determines a collective charging schedule for the fleet, based on wholesale energy

K. De Craemer · G. Deconinck (✉)
Department of Electrical Engineering (ELECTA - EnergyVille), KU Leuven, PB2445
Kasteelpark Arenberg 10, 3000 Leuven, Belgium
e-mail: geert.deconinck@esat.kuleuven.be

K. De Craemer
e-mail: klaas.decraemer@esat.kuleuven.be

S. Vandael
Department of Computer Science, KU Leuven, Celestijnenlaan 200A, 3000 Leuven, Belgium
e-mail: stijn.vandael@cs.kuleuven.be

B. Claessens
Flemish Research Institute (VITO), Boeretang 200, 2400 Mol, Belgium
e-mail: bert.claessens@vito.be

S. Rajakaruna et al. (eds.), *Plug In Electric Vehicles in Smart Grids*,
Power Systems, DOI 10.1007/978-981-287-302-6_6

prices or its portfolio position. However, charging PEVs are physically connected to a distribution grid, which is inherently constrained by its infrastructure. To assure correct operation of the distribution grid, the Distribution System Operator (DSO) can enforce constraints by using grid congestion management mechanisms.

To integrate both aggregator and DSO objectives in the coordination of PEV charging, we identified two operation levels [2]:

- The **market operation level** entails actions with the objective of following beforehand traded volumes on the wholesale electricity markets, where trading takes place on relatively long-term scale (months, seasons) and amounts are expressed as energy quantities—usually MWh—in time slots of typically 1 h or 15 min.
- The **real-time operation level** entails the actions to comply with instantaneous consumer preferences and respect local grid constraints. Because changes and control are relatively more instantaneous and dynamic at this level, real-time operation (or technical operation) is usually expressed in terms of electrical power, e.g. kW. Granularity is in the range of minutes to seconds. At this level, fast responses are important and the number of exchanged messages will be limited.

The influence between market operation and real-time operation in coordinated charging of PEVs is often overlooked. A large part of research on integration of PEVs is aimed at optimally coordinating charging at the market operation level, facilitating larger shares of renewable energy sources or providing system-wide ancillary services. At the same time, a lot of work in literature has been carried out towards the use of PEVs to avoid distribution grid overloads or reducing losses [3, 4], objectives that are situated in the technical operation level.

However, the market and technical level can come into conflict, which typically occurs when the distribution grid is constrained or overloaded, at which point the technical objectives will intervene in the market objective(s). As market operation is overruled, consumption can deviate from what is intended by the aggregator. Multiple aggregators active in the same distribution grid further complicate this problem.

In this chapter, we analyze the influence of the real-time operation level on the market operation level by simulating both levels in a set of varying distribution grid scenarios with a single aggregator and multiple aggregators. For the market operation level, an existing event-driven market-based control (MBC) for coordinated PEV charging is used. For the real-time operation level, an optional voltage droop controller is used to mitigate local voltage limitations. In our analysis, we quantify the optimality of the aggregator's objective at the market operation level, while using droop controllers.

The contributions of this chapter can be summarized as follows:

1. Analysis of the influence of grid constraints in an event-driven control strategy for PEVs. Attention is paid to the effect of grid constraints on an aggregator's

market-level objectives, optionally with the use of a voltage droop controller to alleviate grid congestion.
2. Analysis of the influence of grid constraints in a multi-aggregator setting.

In Sect. 6.2, existing algorithms and models for both market and real-time operation levels are discussed. In Sect. 6.3, the choice of algorithm for the market operation level is detailed and motivated. Then, in Sect. 6.4, a set of relevant distribution grid scenarios is described, together with an explanation of the models and assumptions for the simulations. In Sect. 6.5, the chosen algorithms are simulated in these predefined scenarios, and the influence of real-time operation on market-level objectives is thoroughly analyzed. Finally, the same scenarios are analyzed for a multi-aggregator setting in Sect. 6.6.

6.2 Background

6.2.1 Market Level Operation

Current research regarding the optimization and coordination of clusters of Demand Response (DR) participants at the market level can roughly be divided according to the way the optimization is performed; distributed, centralized and aggregate and dispatch algorithms. This is illustrated in Fig. 6.1.

Distributed algorithms perform a significant part of the optimization process of allocating energy over the cluster at the participating devices themselves. This way, the computational complexity of finding a suitable solution is spread out over the demand response cluster, typically using an iterative process where information is communicated between the participants. However, the distributed aspect does not exclude the existence of an entity responsible for initiating or coordinating the convergence over the iterations.

One share of distributed algorithms in literature is based around distributed optimization techniques, in which a large optimization problem is divided in smaller parts that can be iteratively and independently solved [5–9]. In particular the use of gradient ascent methods and its derivatives, such as dual decomposition, are common.

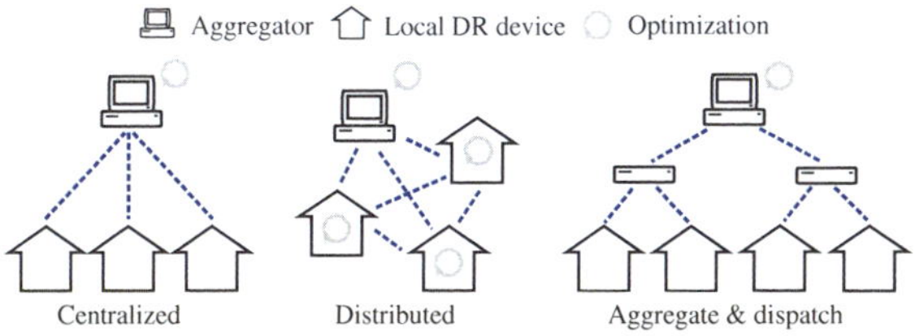

Fig. 6.1 Illustration of the three classes of algorithms and coordination for DR at the market-level

Centralized algorithms are entirely the opposite. A central actor collects information that is sent to it from the DR devices. This information can consist of individual constraints and deadlines or comfort settings. Using the collected knowledge, and possibly including its own additional information such as predictions or stochastic functions, the central coordinator can perform a single optimization that returns an optimal schedule satisfying all the constraints at once. Inherently, this makes centralized algorithms the least scalable, as the optimization process quickly becomes intractable with an increasing number of participating devices. Furthermore, the communication towards and from a single point poses a potential bottleneck. Several solutions are proposed that help to overcome the tractability issue [10, 11]. In [12], focus is on ensuring that vehicle owners truthfully report their value for receiving electricity, willingness to wait and maximum charging rate. Owners could misreport their availability, for example by unplugging early or plugging in the vehicle some time after arrival to try and get a better price.

Inbetween distributed and centralized mechanisms are the **aggregate and dispatch algorithms**. They decouple the optimization of the objective and the dispatch of its outcome, thus alternatively the term 'dispatching mechanism' is equally fitting. An aggregate & dispatch mechanism allows information (such as constraints) from and to the central entity to be aggregated, reducing the complexity of the optimization and improving scalability, but carrying certain compromises or constraints regarding the optimality of the results. The work of [13–15] follows this idea.

While distributed and centralized algorithms can determine an optimal DR schedule given the device's constraints, market data,... they carry some disadvantages regarding computation times, complexity or communication. Aggregate and dispatch mechanisms are a compromise allowing for a scalable and low-cost implementation, at a limited loss in optimality [16]. In our work, we have chosen to work with one aggregate & dispatch algorithm in particular, MBC. We will discuss this method in more detail in Sect. 6.3.

6.2.2 Real-Time Level and Grid Congestion

As the electricity grid cannot get physically congested, the term *grid congestion* refers to a situation where the demand for active power exceeds the nominal power transfer capabilities of the grid [17]. Grid congestion can be mapped to the violation of one or more constraints at its connection points. In the context of this chapter, these will mainly be in the form of power quality problems in distribution grids, and can be attributed to the resistive and unbalanced nature of distribution grids.

6.2.2.1 Grid Congestion Metrics

The European EN 50160 standard, "Voltage characteristics of electricity supplied by public distribution systems" [18], describes, among others, the following important specifications:

- Over- and undervoltage: "The European EN 50160 standard specifies that the 10 min mean RMS voltage deviation should not exceed ± 10 %, measured on a weekly base. For undervoltages, a wider range is allowed in the measurement procedure: −15 to −10 % during maximum 5 % of the week."
- Voltage dip: EN 50160 allows 1,000 voltage dips per year, during which the voltage drops at most to 85 % of its nominal value, for a duration of less than 1 min. Interruptions, defined as lasting less than 180 s, should occur less than 500 times/year.
- Voltage unbalance factor (VUF): When magnitudes of phases or line voltages and the phase angles are different from balanced conditions. "The European EN 50160 standard specifies that the 10 min mean RMS value of the voltage unbalance factor should be below 2 % for 95 % of time, measured on a weekly base." Different ways to compute the VUF exist, and here we will use *True VUF* as shown below. More information on the definitions and calculation of VUF can be found in [19].

$$\begin{aligned} \text{True VUF} &= \frac{\text{negative voltage sequence component } V_n}{\text{positive voltage sequence component } V_p} \\ \text{with } V_p &= \frac{V_{ab} + aV_{bc} + a^2 V_{ca}}{3} \\ \text{and } V_p &= \frac{V_{ab} + aV_{bc} + a^2 V_{ca}}{3} \end{aligned} \tag{6.1}$$

- Harmonics: Caused by the power electronics inside converters such as found inside vehicle chargers or photovoltaic (PV) inverters. Harmonics will not be looked into here, but the use of power electronics such as found inside PEV chargers can create problematic harmonics [20].

6.2.2.2 Congestion Mitigation

A distribution system operator, faced with grid congestion problems, can opt for a number of mitigating strategies.

- **Reactive power and voltage control** to increase the (local) transfer capacity. This is already used in wind generators connected to the medium voltage network. In distribution grids, reactive power and voltage control can be achieved through the use of tap changers and capacitor banks, and their switching is planned using load forecasts. For example, [21] optimizes to limit switching of such devices.
- **Coordinating the power flow** [17] throughput via shifting or curtailment of demand, possible through the implementation of demand response, or through the mandated implementation of voltage droop control.

- **Increasing the transfer capacity** of the local grid by replacing or upgrading equipment (adding or replacing cables, installing a bigger transformer…). While this option is attractive because it limits the involvement of the DSO (retain 'passive' role, no forecasts …), the cost of this option can be substantial and thus only considered when other solutions are exhausted or deemed infeasible.

The first option is already used today. However, in practical operation, low voltage-grid tap changers are usually off-load types and barely used [22]. Tap positions are calibrated and changed only in case of network extension or modification [23]. Automated and remotely controllable on-load tap changers (OLTC) exist, but their use in distribution grids is still reserved to a few test cases [24], due to costs.

The third option is technically attractive for DSOs, since it fits within a predominantly off-line role of installation, maintenance and asset management at the distribution network level.

Adding parallel cables to or upgrading existing lines by using new cables with higher cross sections is considered a straightforward solution [23]. No additional tasks such as day-to-day load forecasting, extensive state estimation and monitoring are required. The high investment costs will likely reserve this to some corner-cases.

In the remainder of this chapter, congestion management will refer to the use of the second option; the coordination of active power demand at congested grid locations. In the light of the real-time and market operation levels, we will now discuss the use of voltage droop control and grid congestion management mechanisms.

6.2.2.3 Voltage Droop Control

As mentioned, lines in distribution grids behave resistively rather than inductively. This causes voltage deviations along the line when large amounts of active power are drawn from or injected into the grid. To avoid such effects, large-scale PV installations in some countries are now required to be able to provide grid services to the DSO. Similarly, small PV installations are required to respond to overfrequency and overvoltage by limiting injected power or temporarily disconnecting [25, 26].

But PV output is determined by the uncontrollable radiation of the sun, whereas charging rates of PEVs can be varied and shifted arbitrarily in time. Thus, in addition to the coordination at the market level, a fast-acting grid-supportive behavior similar as used in PV installations can be implemented inside a charger [27–29]. It is not unthinkable that the use of automatic voltage control for Electric Vehicle Supply Equipment (EVSE) becomes mandatory as well once their impact reaches a significant quantity.

Nonetheless, a droop control scheme is robust and easy to implement because it only requires the measurement of voltages and a way to adjust local active or

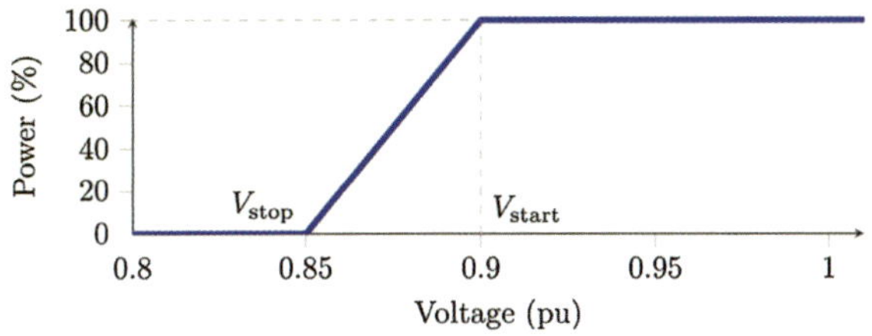

Fig. 6.2 Example voltage droop control characteristic for PEV chargers

reactive power settings. No communication with a central entity is needed. An example of a voltage droop curve for a PEV charger is shown in Fig. 6.2. When the voltage at its connection point drops below 0.9 per unit (pu), power is linearly reduced until 0.85 pu, where charging is completely halted.

On the downside, activation of the droop will almost certainly conflict with market level coordination [30] (Sect. 5.2.4). For example, at some point the fleet manager would send its optimal power set-points or an equilibrium priority to the vehicle agents. But due to local grid problems the EVSE is forced to reduce power. The result is that, even if the real resulting power setting is communicated back to the fleet manager, the deviation holds a disparity from the original optimal market level energy plan. The resulting energy shortfall (negative imbalance) may result in a penalty for the fleet manager.

6.2.2.4 Advanced Congestion Management Mechanisms

The task of a grid congestion management mechanism is to limit the managed loads to the capacity of the distribution grid assets at any time, especially in the presence of multiple competing actors with different objectives. This can be achieved by adding a network cost or penalty for the use of the network during certain times of the day. In [17], algorithms for congestion management are classified according to strategy.

- **Distribution grid capacity market**: In this mechanism, the aggregators involved will start by optimizing the schedule for their PEVs in absence of a network tariff. The schedule is sent to the DSO, which evaluates whether the network constraints are met. If not, the aggregators will receive a price that reflects congestion at each node in the network and are requested to update their schedule.
  The procedure is then repeated until convergence, at which point the network tariff and charging schedules are fixed. As the mechanism is essentially the same as dual decomposition, the use of non-strict convex objective functions can cause problems. In [31, 32], this method is used.
  A capacity market would be complex to implement and the iterations add a lot of computational burden. The DSO could be offloaded by externalizing the

process into a separate capacity market, in which it still has to provide measured and estimated power.

- **Advance capacity allocation system**: The idea behind this mechanism is that the DSO pre-allocates grid capacity at each transformer or line to the aggregators, based on the free capacity remaining at each line or transformer, after inelastic load (mainly household consumption) has been accounted for. The allocation between aggregators would be based on auctioning of this free capacity.
  While relatively straightforward, there are some drawbacks to this method. First of all, the DSO needs to map all its customers' connection points to their respective aggregator. Secondly, there is no way to incorporate the time-dependency of demand; if an aggregator bids for the capacity during certain time, that bid depends on what was allocated before and after that time-period. An iterative approach would solve this, but also increases complexity again.
- **Dynamic grid tariff**: In this case, a time-varying location-dependent grid tariff is determined by the DSO beforehand, based on expected consumption levels at each node in the grid. Predicting loads and estimating price-sensitivity is entirely the responsibility of the DSO. Once the tariffs are published to the aggregators, the latter integrate them into their scheduling. In case of severe deviations from the expected value, the DSO may resort to controlled interruptions in real-time, which in turn also holds a risk for the aggregators. The work of [33] uses this approach. The biggest drawback consists of the high complexity of the problem that needs to be solved by the DSO (predictions, load flow calculations…), let alone when the stochastic properties of inevitable uncertainties are taken into account.

The work of [34] (p. 97) provides an overview and comparison of these 3 types of mechanisms. While all of the mechanisms should lead to the same optimal PEV charging profile, the complexity involved limits their practical implementation. It is also not clear how deviations during the course of the day should be handled, which will inevitably occur as the algorithms are based on the use of allocations in time slots (e.g. 15 min in [32]), besides the last-resort of DSO-controlled interruptions.

In [34] (p. 100) the use of a simpler proxy tariff is proposed, such as a historical *ToU* or real-time tariff, as a compromise. Unfortunately, following simulations, the conclusion suggested that the use of proxy tariffs does not necessarily reduce system peak load, leads to higher costs (approx. +20 %) and distorts the economic signal of the electricity price.

6.3 Market-Level Operation: Market-Based Control for PEVs

The concept of MBC is rooted in the theory of microeconomics, wherein economic activity is modeled as an interaction of individual parties pursuing their private interests [35] (Chap. 4). The market mechanisms that apply provide a way to incentive the parties, referred to as economic agents, to behave in a certain way.

In [36], appliances in a DR cluster are represented by software agents in a multi-agent system (MAS). They have control over one or more local processes (e.g. heating of water or charging of a PEV's battery), but compete for resources (electric power) on an equilibrium market with other agents.

6.3.1 Architecture

The MBC system has been used in a number of field tests and is commercially known as PowerMatcher. The clearing of the market in [36, 37] is operated on a periodic basis, e.g. a time slot length of 15 min, or using events, and is implemented in a hierarchical, tree-like manner [35], as illustrated by Fig. 6.3a.

At the root of the tree is an auctioneer agent, directly connected to a number of *concentrator agents*. The *auctioneer agent* is a special type of concentrator agent and is responsible for the price setting process, just as in the Walrasian auctions. The concentrator agents lower in the tree aggregate the demand functions of their child agents. Because a uniform interface is used between the levels, an unlimited number of such aggregation levels can be used. Eventually, at the bottom of the tree, we find the *device agents* themselves.

The device agents assemble *demand functions* representing their willingness to pay and consume, taking into account the specific constraints of the controlled device. Demand functions are sent upwards and an auctioneer agent performs a matching process with producing agents. An equilibrium price is communicated back to the agents, that start consuming or producing at the equilibrium level.

If equilibrium prices are regarded as a pure control signal, so that there is no direct link to the cost of energy, the MAS MBC mechanism can be viewed as a dispatching method for the aggregator's business case. In such scenario, the demand function data is regarded as input for a scheduling algorithm, and the equilibrium price (or better, *equilibrium priority*) as a level to steer the cluster towards its outcome.

6.3.1.1 Demand Functions for PEV Device Agents

Representative demand functions can be built using various means, but in case of PEVs, a straightforward way is by combining each agent *i*'s requested energy ${}^{i}E_{\mathrm{req}}$, time till departure ${}^{i}\Delta t_{\mathrm{dep}}$ and maximum charging power ${}^{i}P_{\max}$ to create a sloped curve ${}^{i}P_{\mathrm{dem}}$, as shown below each PEV in Fig. 6.3a and also in Fig. 6.3b. In case there is not enough time left to receive the requested energy (t_{critical} occurs before the current time), an inflexible demand function can be used, so that charging happens at maximum power regardless of the control signal.

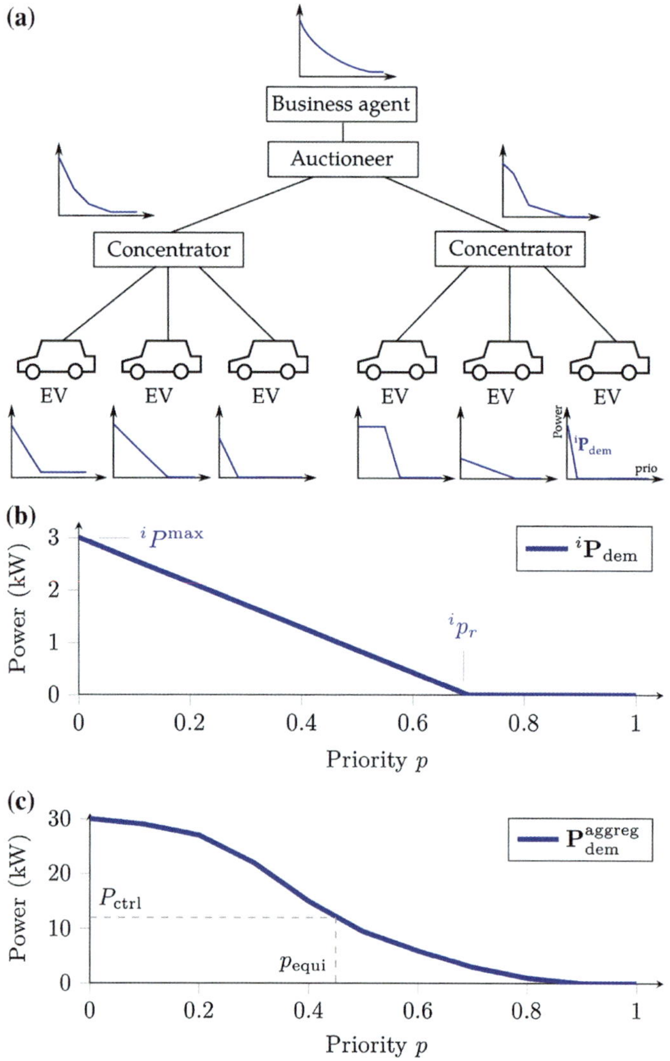

Fig. 6.3 Overview of the control structure in MAS MBC in (**a**). Device agents, pictured as charging PEVs, send demand functions $^iP_{\text{dem}}$, shown in (**b**), upwards. After aggregation of the individual demand functions, equilibrium priority p_{equi} is determined, shown in (**c**), and sent back to the agents

$$^iP_{\text{dem}} = f\left({}^iE_{\text{req}}, {}^i\Delta t_{\text{dep}}, {}^iP_{\max}\right) \tag{6.2}$$

$$^it_{\text{critical}} = t \,|\, {}^iE_{\text{req}} = {}^iP_{\max}\, {}^i\Delta t_{\text{dep}} \tag{6.3}$$

A detailed description of building demand functions for PEVs in this context can be found in [2, 16].

6.3.1.2 Concentrator Agents and Aggregation

At the concentrator agents, the individual demand functions of n agents are aggregated into a single curve $P_{\mathrm{dem}}^{\mathrm{aggreg}}$, shown in Fig. 6.3c. At the auctioneer agent, this aggregated curve is used to find the equilibrium priority p_{equi} that corresponds to a desired power setting P_{ctrl} for the DR cluster.

$$P_{\mathrm{dem}}^{\mathrm{aggreg}} = \sum_{i=1}^{n} {}^{i}P_{\mathrm{dem}} \tag{6.4}$$

$$p_{\mathrm{equi}} = P_{\mathrm{dem}}^{\mathrm{aggreg}} | P_{\mathrm{ctrl}} \tag{6.5}$$

The value for P_{ctrl} has to be determined by the business agent.

6.3.2 MAS MBC Advantages and Drawbacks

Using a multi-agent market based control system (MAS MBC) for demand response, as exemplified by the PowerMatcher, offers several benefits.

- Scalability: In a centralized system, the central entity has to deal with all incoming and outgoing messages, $O(n)$, quickly creating a communication bottleneck. Because of the aggregation on multiple levels in the PowerMatcher, the amount of messages that have to be dealt with per agent can be reduced to $O(\log n)$.
- Low complexity: The construction of demand function data and the matching process itself is straightforward, and is not based on any model. Determining a demand function for a device can be done during its development.
- Openness: Any kind of device can be integrated in the cluster, since operation only depends on the exchange of demand functions and price. Devices without flexibility are represented by an inelastic demand function.
- Privacy: Since demand functions are aggregated there is no central entity that collects all information. Furthermore, the physical processes of devices, bidding strategy and motives of users are all abstracted through their demand functions.

As indicated before, the use of an aggregated model at the auctioneer agent and a heuristic to build the PEV's demand functions will lead to a suboptimal solution. However, a more significant shortcoming compared to other methods presented in this chapter, is the lack of look-ahead functionality.

6.3.3 Addition of Scheduling Functionality and Control Objectives

For loads that can store electric energy, such as PEVs, an *energy constraints graph* can be used to capture the available flexibility over a certain time horizon. This is introduced in the work of [16]. For each PEV i, two vectors ${}^{i}\boldsymbol{E}_{\max}$ and ${}^{i}\boldsymbol{E}_{\min}$ are added to the information ${}^{i}P_{\text{dem}}$ sent from device agents to auctioneer agent.

The vector ${}^{i}\boldsymbol{E}_{\max}$ is the energy path of a PEV agent i, if it were to start charging immediately at maximum power and then (at t_{idle}) stay idle until its departure time t_{dep}. On the other hand, ${}^{i}\boldsymbol{E}_{\min}$ represents the case when charging is postponed as long as possible (up to t_{critical}). This is expressed in the equations below and illustrated in Fig. 6.4a. All area in between ${}^{i}\boldsymbol{E}_{\max}$ and ${}^{i}\boldsymbol{E}_{\min}$ represents the flexibility of the charging process.

$$
\begin{aligned}
{}^{i}\boldsymbol{E}_{\max} &= \left\{ {}^{i}E_{\max}(t) \,|\, {}^{i}E_{\max}(t) = \min\left(t\,{}^{i}P_{\max}, {}^{i}E_{\text{req}}\right) \forall t \in \left\{0, 1, \ldots {}^{i}\Delta t_{\text{dep}}\right\}\right\} \\
{}^{i}\boldsymbol{E}_{\min} &= \left\{ {}^{i}E_{\min}(t) \,|\, {}^{i}E_{\min}(t) = \max\left({}^{i}E_{\text{req}} - \left({}^{i}\Delta t_{\text{dep}} - t\right) {}^{i}P_{\max}, 0\right) \forall t \in \left\{0, 1, \ldots {}^{i}\Delta t_{\text{dep}}\right\}\right\}
\end{aligned}
\tag{6.6}
$$

To represent the battery constraints of an entire PEV fleet of n vehicles, the individual constraints are aggregated into collective battery constraints $\boldsymbol{E}^{\text{aggreg}}_{\max}$ and $\boldsymbol{E}^{\text{aggreg}}_{\min}$, at the intermediate agents and the auctioneer agent. The auctioneer agent can now use the collective energy constraints to determine an optimal path $\boldsymbol{E}_{\text{opt}}$ over the horizon t_{horizon}, according to some objective function C:

$$
\boldsymbol{E}_{\text{opt}} = \underset{\boldsymbol{E}}{\operatorname{argmin}}\, C(\boldsymbol{E}) \tag{6.7}
$$

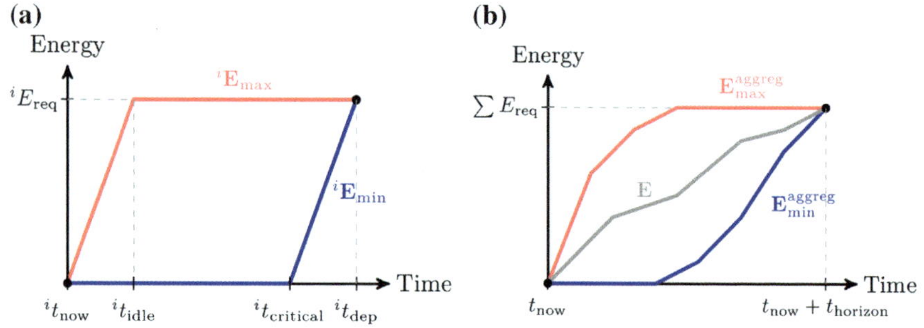

Fig. 6.4 **a** Energy constraints graph for a single vehicle i, and **b** aggregated energy constraints graph and some scheduled path $\boldsymbol{E}$ through it

$$
\begin{aligned}
\text{with:} \quad & E = \{E_t && \forall t \in \{0, 1, \ldots, t_{\text{horizon}}\}\}, \\
\text{subject to:} \quad & P_t \leq P_t^{\text{limit}} && \forall t \in \{0, 1, \ldots, t_{\text{horizon}}\} \\
& E_{\text{min,t}}^{\text{aggreg}} \leq E_t \leq E_{\text{max,t}}^{\text{aggreg}} && \forall t \in \{0, 1, \ldots, t_{\text{horizon}}\} \\
& E_{t+1} = E_t + P_t \Delta t && \forall t \in \{0, 1, \ldots, t_{\text{horizon}} - 1\}
\end{aligned}
$$

Here E_t i s the collective energy of the cluster at time t, and P_t is the power consumed by the cluster during time t to Δt. Any objective $C(\boldsymbol{E})$ can be used to determine a path for the PEV cluster, and in Sect. 6.4.1, two objectives will be discussed.

6.3.4 Event-Driven Approach

Communication takes on an important role in demand side management of PEVs. Charging requirements and constraints need to be communicated to an aggregator, while aggregators need to send control signals back to PEVs in order to steer their charging power towards cluster-wide goals.

In terms of integrating charge coordination algorithms into a realistic "real-world" environment, two challenges are identified: continuous coordination, and messaging limitations.

The first challenge is the need for continuous coordination of the charging process. In energy markets, charging only needs to be optimized in terms of energy volume per hour. However, vehicles arrive and depart continuously, and will want to start charging or depart at asynchronous times.

This means that, ideally, control and coordination actions should also commence immediately, especially for fast-charging applications, and allow for quickly altering the fleet's behavior if the need arises. Consequently, charging needs to be coordinated at two levels: a market level, where time is divided in time slots, and a real-time, event-driven level that is focused on responsiveness. This division applied to the MBC architecture is illustrated in Fig. 6.5.

In this case, event-based interaction allows the PEVs and aggregator to quickly respond to changes in setpoint or in flexibility. If a situation occurs where vehicles have to slow down charging due to distribution grid constraints, the aggregator is informed and will try to use flexibility from other vehicles that do not experience such problems.

The second challenge is related to the exchange of messages between PEVs and an aggregator. In reality, the underlying infrastructure places constraints on the communication, pertaining to packet delays, link reliability or maximum throughput. In the latter case, the exchange of messages should be limited by the coordination mechanism, which is done by caching information from and to the PEVs. More details on the event-driven implementation can be found in [2].

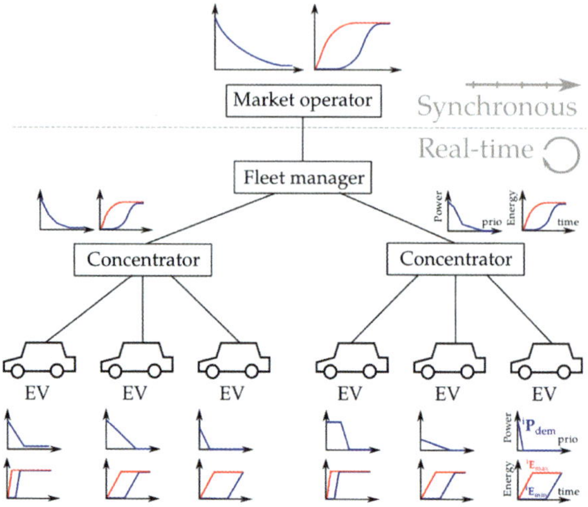

Fig. 6.5 MAS MBC architecture with dual coordination. The real-time part is event-driven, while the market operator works in discrete time intervals (time slots)

6.4 Simulation Objectives and Models

We want to investigate the situation where an aggregator coordinates a cluster of PEVs based on market-level objectives, but a large part or all of the vehicles are situated inside a weak and constrained grid topology. How effective is the use of a voltage droop controller in eliminating or reducing grid congestion problems? To what degree do the technical objectives impact the aggregator's business case?

To answer these questions, a simulation framework was developed; a Java-based part allows to model the interaction between the agents, while the market-level optimization is performed in Matlab using CPLEX. To simulate the effects on the voltages in a distribution grid, a Matlab-based backward-forward sweep load flow solver developed at our research group was also integrated in the framework.

Besides a framework, several models and datasets are required to properly represent the actors and their behavior. In this section, we describe the driving profiles and model for the PEVs, the wind prediction and generation, and the household loads present in the distribution grids.

6.4.1 Aggregator Market-Level Objectives

Two market-level objectives for the auctioneer agent are considered:

- *Time-of-Use (ToU)*, where the aggregator's goal is to minimize the cost of charging a cluster of vehicles, based on a time-varying tariff p_t, and using a Linear Program (LP) optimization:

$$\boldsymbol{E}_{\text{opt}} = \underset{\boldsymbol{E}}{\operatorname{argmin}} \sum_{t=0}^{t_{\text{horizon}}} C(E_t) \quad \text{with} \quad C(E_t) = p_t E_t \tag{6.8}$$

Because this is a linear objective, a sharp on-off control behavior can be expected.

- *Portfolio balancing*, where the goal of the aggregator is to use the flexibility of a fleet of PEVs to limit his portfolio's wind generation exposure to the imbalance markets. This means finding an optimal energy trajectory for the PEVs, E_{PEV}, over a horizon, such that the difference between short-term wind prediction E_{wind} and day-ahead nomination E_{nomin} is minimized:

$$\boldsymbol{E}_{\text{opt}} = \underset{\boldsymbol{E}}{\operatorname{argmin}} \sum_{t=0}^{t_{\text{horiz}}} \left(E_{\text{PEV},t} + E_{\text{wind,t}} + \frac{1}{4} E_{\text{nomin,t/4}} \right)^2 \tag{6.9}$$

In this specific case, the day-ahead nominations are required to be supplied on an hourly-basis, while the short-term wind predictions are on known a quarterly basis, 15 min ahead, and with a horizon of 24 h. The control variables of the PEVs can be on an arbitrary time basis.

Thus, as more accurate wind predictions become available after the nomination, the optimization will try to use the vehicles to limit the difference, and, due to the quadratic term, favor to spread out remaining imbalance in time (Fig. 6.6).

6.4.2 PEV Model

The model of the PEVs (hybrid plug-in or full electric) in the simulations consists of two main parts: a battery model and a usage or driving profile.

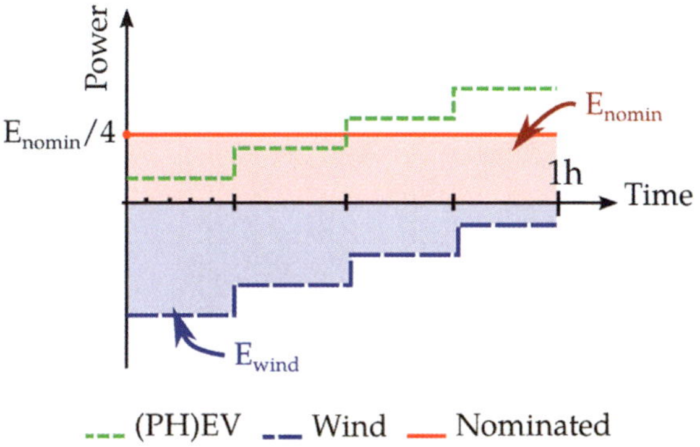

Fig. 6.6 Illustration of the balancing objective

6.4.2.1 Battery and Charging

In literature, a great deal of research has been done on the characterization and use of batteries for electric drive-train applications. The purpose of the envisaged model for the simulations in this context does not include aspects such as aging and depreciation costs, and are subsequently left out in favor of a simple first order approximation of the storage capacity of the battery.

In reality though, the maximum charging current has to decrease before the battery reaches a state of charge (SOC) of 100 %, to avoid damaging the cells. Here, the charging and discharging process in the PEV is simplified to a constant power behavior, and the capacity is chosen such that it corresponds to a depth of discharge (DOD) of 83 %. This is a valid consumption, as the SOC of existing PEVs is also kept within a certain DOD to extend battery life. Summarized, all vehicle instances are equipped with the same usable battery content of 20 kWh. Technically speaking, 20 kWh would then represent a PEV with a total battery capacity of around 24 kWh.

Technical constraints in (European) residential installations limit charging power to around 3.3 kW (corresponding to 16 A at 230 V and 10 % allowed voltage deviation) or 6.6 kW (32 A). In fact, to avoid problems due to inadequate wiring or installations, some car manufacturers only allow the 3.3 kW power level when the vehicle is plugged into a so-called dedicated wall-box. Charging through standard outlets is then typically limited to 2–2.5 kW. In the battery model used here, charging takes place at a variable power level between 0 and 3.3 kW. This may seem to be a slow charging rate, but because of the long standstill times at home, the need for higher charging rates at home is not critical [38].

Also assumed that vehicles want their battery fully charged by departure, as this is the worst case and also more convenient for drivers, not having to enter an expected distance. Vehicle-to-grid scenarios were not considered.

6.4.2.2 Driving Profiles

To complete the PEV model, data about the state of the vehicle during the day (idle at home, driving, unavailable,…) and the energy consumption while driving is required.

In the work of [39], the results of the 3rd Flemish Mobility Study (OVG3) were analyzed. The latter was commissioned by the Flemish government and looks at the transportation behavior of 8,800 drivers during September 2007 and 2008. Recorded data includes the number of trips each day, distances, motives, departure times, … From this, synthetic availability profiles were prepared that can be used in simulations. An example for 2,500 vehicles is shown in Fig. 6.7, where the number of vehicles that is at home, driving or at work over the course of 7 days is plotted. It can be seen that fleet behavior is very periodic and therefore predictable.

Vehicles will only charge at home, so that the amount of energy needed reflects a worst-case scenario for the distribution grid.

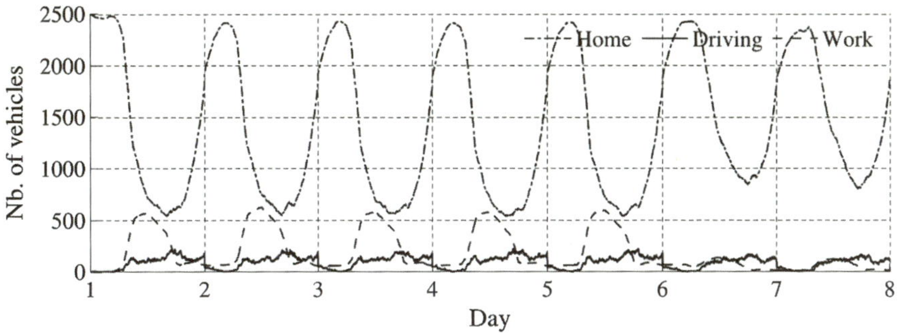

Fig. 6.7 Illustration of the used vehicle availability profiles, cumulative for 2,500 vehicles, over 7 days

For the energy consumption model, required power during acceleration and braking (related to vehicle size, aerodynamics and driver habits) has to be added on top of auxiliaries such as lighting, heating, wipers etc. More information can be found in [30] (Chap. 2), and [39, 40]. From [39], an average driving speed of 42 km/h is combined with an energy consumption of 250 Wh/km.

These numbers result in a theoretical range of 80 km for each simulated vehicle. Figure 6.8 shows the cumulative distribution of the SOC of the battery at arrival time, after a simulation with 1,000 vehicles and over 7 days. From the figure, half of the arrivals happened with a battery of almost 80 % SOC or more. However, for 6.8 % of the simulated trips, 20 kWh was insufficient. Simply increasing the usable battery size to 24 or 26 kWh does not eliminate these occurrences, so these trips are out of range for the average battery electric vehicle (BEV). It will therefore be assumed that these drivers are using a plug-in hybrid electric vehicle (PHEV) to complete their journey.

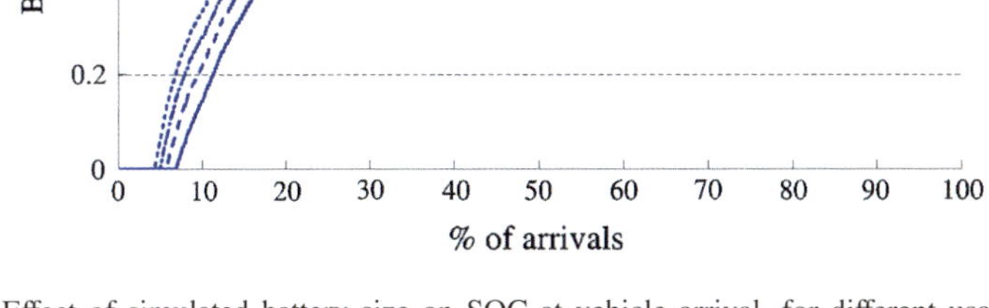

Fig. 6.8 Effect of simulated battery size on SOC at vehicle arrival, for different usable battery contents, obtained for a set of 1,000 vehicles over 7 days

6.4.3 Wind Energy Generation and Predictions

In one objective, renewable energy production from wind turbines is taken into account. For several locations in the Netherlands, both wind speed measurements and predictions are available. The wind speed predictions were calculated by the *Aanbodvoorspeller Duurzame Energie* (AVDE) at ECN [41, 42] and translated to the correct height of the turbine, as winds aloft generally have a higher velocity than winds at ground level.

The resulting wind speed then has to be put alongside the turbine's specified output power. For the turbine specifications, one type from manufacturer Nordex is used, the 2.5 MW peak N80/2500 [43].

Since the available wind speed data consists of predictions that have a horizon of 48 h, and are updated every 6 h, the most accurate predictions that can be submitted for nomination are those generated at 12h00 the day before.

6.4.4 Household Consumption

To be able to simulate the effects on voltage quality in a distribution grid, realistic household consumption profiles are required. Synthetic aggregated profiles, available from the local regulator and used by energy retailers to estimate their customers' consumption, are too generic. In the 'Linear' smart grid project [44], measurements at 100 households were performed over the course of a year, with a resolution of 15 min. When more profiles are needed the available set is rotated. An illustration of 20 of the used profiles is plotted in Fig. 6.9.

It can also be observed that there is a high simultaneity between households consumption in the evening and the arrival of PEVs.

6.4.5 Weak Grid Topology and Agent Architecture

When investigating the effects of coordinated charging on the state of the distribution grid and vice versa, it makes sense to focus on weak grid configurations, where problems are more likely to occur. The question then arises what specific topology should be used as grid model. We are focusing on the grid situation in Belgium, but from discussions with experts, information on the current state of distribution grids seems to be lacking. During planning and deployment of new grid segments, DSOs selected appropriate values for the cable sizes and lengths, and individual connection points were spaced out evenly over the phases. Decades later, sections have been added, reconfigured, new connections points have been attached to "random" phases, etc. This makes the occurrence of virtually any situation possible in practice and with the increasing share of PV installations on

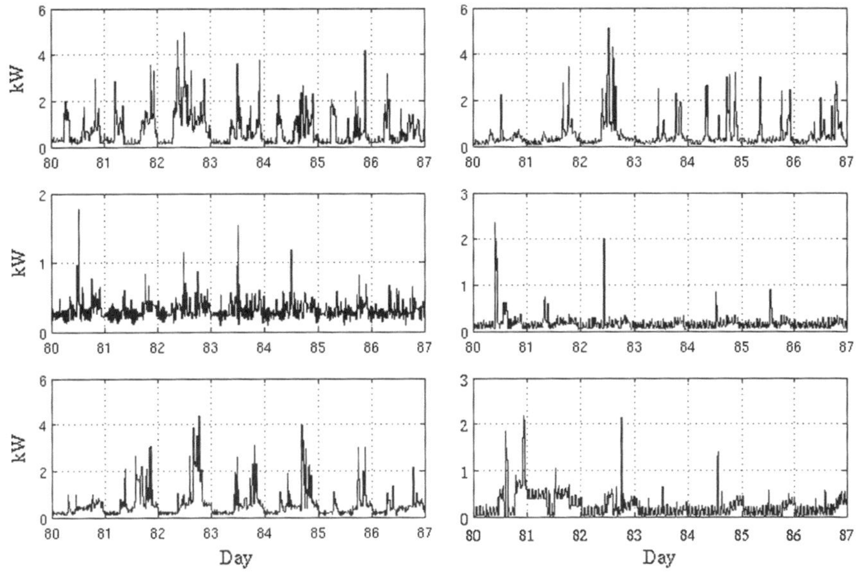

Fig. 6.9 Examples of the household profile data used, for 6 individual households in Belgium, one week starting at day 80 of the year

the roof of households, power quality problems have already started to appear. More information regarding PV and power quality problems in Belgian grids can be found in [22].

Nonetheless, indicative simulations and other work [45, 46] (Chap. 3) suggests that power quality problems in distribution grids due to charging PEVs only comes into view at larger penetration levels of over 30–50 %, and then mostly in weak grids, with unfavorable cable types and lengths. Since we want to study the interference between technical and market level objectives, the focus in the next sections will be on specific cases that represent constrained grids, and not on some average grid situation (if that even exists).

6.4.5.1 Base Physical Grid Structure

Figure 6.10 shows the base topology used in the simulations. A 400 kVA transformer supplies several parallel feeders. Each feeder then supplies a number of household loads, bringing the equivalent transformer load up to 191 households. This is within the limits of the DSO; in a document published by the VREG [47], a maximum occurrence of 220 connections per transformer cabin can be derived. Unfortunately, there is no mention of the rating of the corresponding transformer. The resulting topology is similar to the urban setting used by [45], also with a PEV penetration level of 100 %, but here no PV installations are added, since it was found that they do not cause major changes in the occurrence of undervoltages due

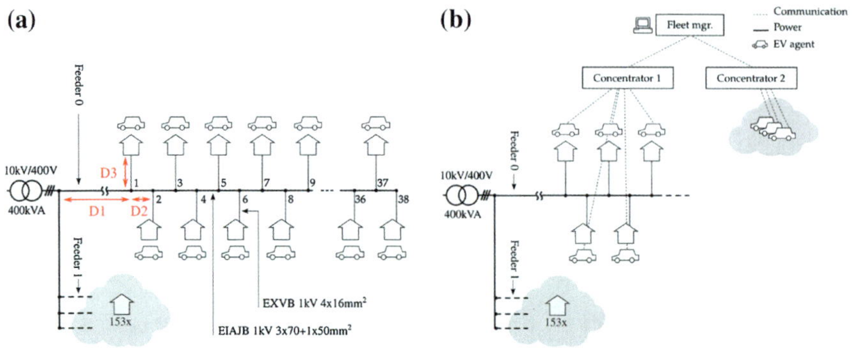

Fig. 6.10 *Left* single instance of the physical grid topology. *Right* agent topology in relation to the physical grid in the single aggregator scenario

to charging. This can be attributed to the non-coincidence of PV production and PEV availability.

One of the feeders, Feeder0, is linked to a line-segment supplying 38 single-phase household connections. These are alternatingly attached to phases 1 to 3 and spaced apart by distance D2. The distance from the transformer to the first household connection is D1. From each connection point, a cable with length D3 runs from the line to the household's supply terminals. In the simulated model, the other feeders and loads (153 households) connected to the transformer are lumped together into one single entity (Feeder 1), as their impact is not studied in detail.

Cable parameters are taken from the design specifications of the standard for underground distribution cables, NBN C33-322 [48]. Cable type EIAJB 1 kV $3 \times 70 + 1 \times 50\ \text{mm}^2$ is used for the main feeder and line (D1, D2), while cable type EXVB 1 kV $4 \times 16\ \text{mm}^2$ is used to connect the household's supply terminals to the main cable (D3).

Table 6.1 shows the variations on this topology that are evaluated in the next sections. Case NS and NL have a relatively short cable between the transformer and the first household terminal (100 m). Case NL and FL represent scenarios with rather long total cable lengths (914 and 805 m), due to longer distances between the household connection points.

Table 6.1 Variations of the physical base topology, representing various weak grids

Case name	Abbreviation	D1 (m)	D2 (m)	D3 (m)	Total length (m)
NearTransf ShortCable	NS	100	15	20	655
NearTransf LongCable	NL	100	22	20	914
FarTransf ShortCable	FS	250	7	20	509
FarTransf LongCable	FL	250	15	20	805

Table 6.2 Variations of the agent topology, representing different amounts of vehicles situated in weak grids

Case name	EVs inside weak grid
x-38	38
x-114	3 × 38
x-380	10 × 38
x-760	20 × 38

6.4.5.2 Agent Structure

The organization of the software agents that represent the charging vehicles is independent of the grid topology from the previous section. But, in our simulations, it is assumed that all agents for vehicles that are physically connected to the same transformer are grouped under a single concentrator agent.

At the same time, for the market operation at the fleet manager to function properly, more flexibility than what is provided by the 38 vehicles in the base topology should be available in the cluster. To that end, the cluster is extended so that, depending on the scenario, a total of 200 or 1,000 vehicle agents takes part in the coordinated charging. These additional agents are not part of the load flow calculations. The right side of Fig. 6.8 shows the resulting agent topology.

To test additional shares of PEVs inside weak distribution grids, additional variations of the agent structure are created by having multiples of the base topology. These are shown in Table 6.2. The suffix after the case number determines the share of agents used in the topology.

6.5 Single Aggregator Simulations and Results

In this section, the effect of coordinated charging using market-level objectives on local grid congestion, in the distribution grid scenarios from Sect. 6.4.5, will be examined.

Besides the MAS MBC event-based implementation that was outlined before, we will also include an uncoordinated or dumb charging scenario, during which vehicles plug in and start charging upon arrival at their maximum rated power P_{max}.

6.5.1 Aggregator with ToU Cost Objective

The objective of the fleet manager during a ToU scenario is to respond on a 24-hour horizon ToU tariff in such a way as to minimize the charging cost of the vehicle fleet. The 24-hour tariff is based on the wholesale energy price of the hourly BELPEX day-ahead market. It should be noted that using the price profile of a day-ahead market is not fully representative of a future ToU tariff as it could be implemented by utilities. Still, prices on the day-ahead market do reflect real-world peak and off-peak periods on an hourly basis, which is what is needed in these simulations.

Because of the seasonal effects of household consumption and tariffs, distribution grid problems are correlated to the time of the year. To limit the influence of the choice of day on the results and get a global picture, randomized sets of scenario parameters are generated and tested. The randomized parameters consist of the day of the year for the tariff, vehicle driving profiles and household load profiles.

The result of 100 randomized parameter sets for each case and coordination option regarding voltage problems according to the EN 50160 standard are shown in Fig. 6.11.

6.5.1.1 Real-Time Level Results

Looking at the household-only (HHOnly) results of Fig. 6.11 indicates that the chosen topologies are sufficient as long as no PEVs are introduced, although voltage regularly fluctuated within the EN50160 specifications. With charging PEVs, the voltage problems are outside the EN 50160 specifications by a wide margin, confirming that the grid topologies qualify as 'weak grid'. Voltages regularly drop below 0.9 pu for more than 5 % of the time, and events where the voltage drops below 0.85 pu are quite common. The problems will no doubt turn for the worse in situations with unbalanced phase connections, higher charge currents (such as future 6.6 kW chargers) and increasing household loads.

Still, the severity of distribution grid problems strongly depends on the grid topology, shown as cases NS, NL, FS and FL. Having the longest cable sections to the loads, case FL leads to the highest amount of voltage magnitude and VUF problems, while case NS and FS experience the least problems.

However, the observed trend is the same: uncoordinated charging is responsible for a peak in the evening that overlaps with the peak of household loads. Charging coordination based on ToU cost minimization objectives leads to only a little less voltage problems. The reason is that, while the coincidence of household loads and charging has disappeared, all available vehicles are now asked to commence charging at one or two points during the day. This creates a new peak that is in itself sufficient to create voltage problems.

To illustrate, Fig. 6.12 shows the power through the feeder and the voltage profile at the worst node for one specific simulated week inside case FL-38, for the event-driven MAS MBC implementation. The situation has the potential to be a lot worse, were the low wholesale prices to correspond to the household evening peak.

It is also immediately visible that the severity of voltage deviations for the implementation with voltage droop controllers is reduced. However, because the voltage droop control only activates below 0.9 pu, the measured values for 0.9 pu deviations are still often outside the 5 % specifications of the EN 50160 standard. Looking at the 0.85 pu results reveals that such occurrences are entirely solved by the use of the voltage droop controller. By tuning the setpoints of the controller so

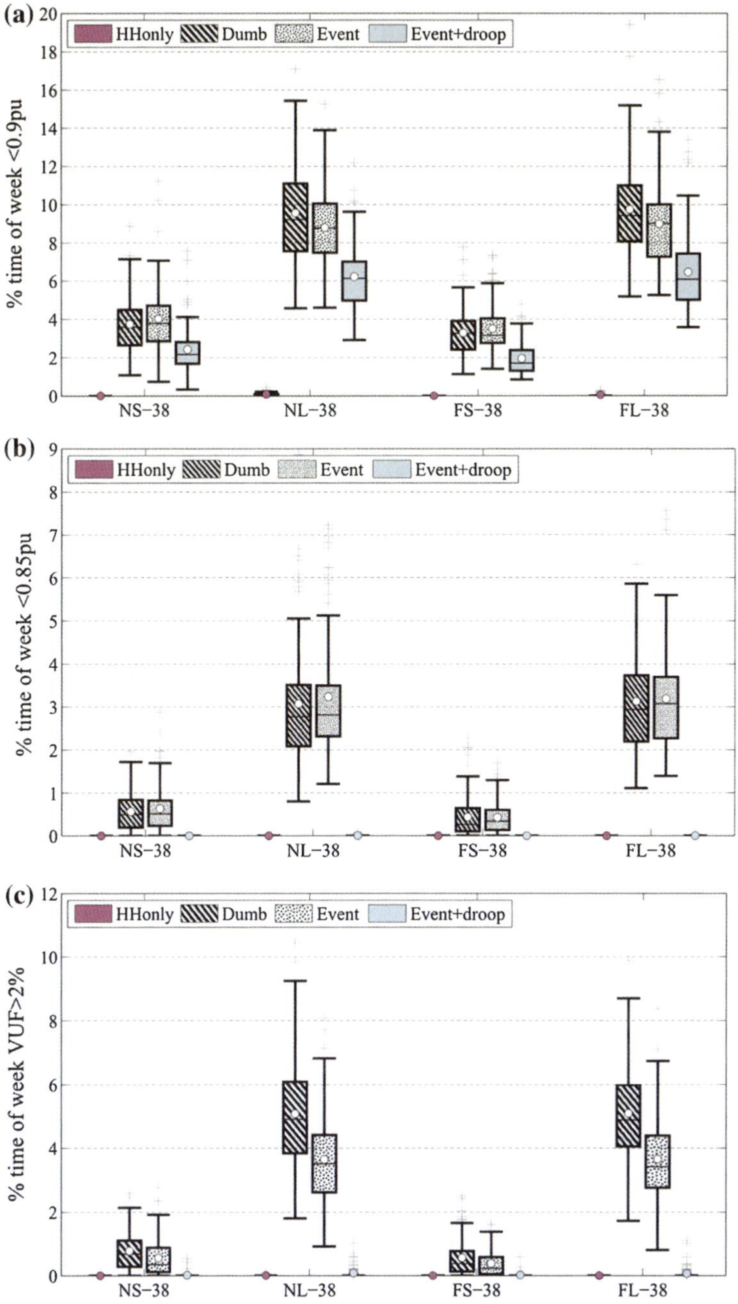

Fig. 6.11 EN 50160 voltage magnitude and unbalance problems, over the course of 7 days, for 100 randomized days

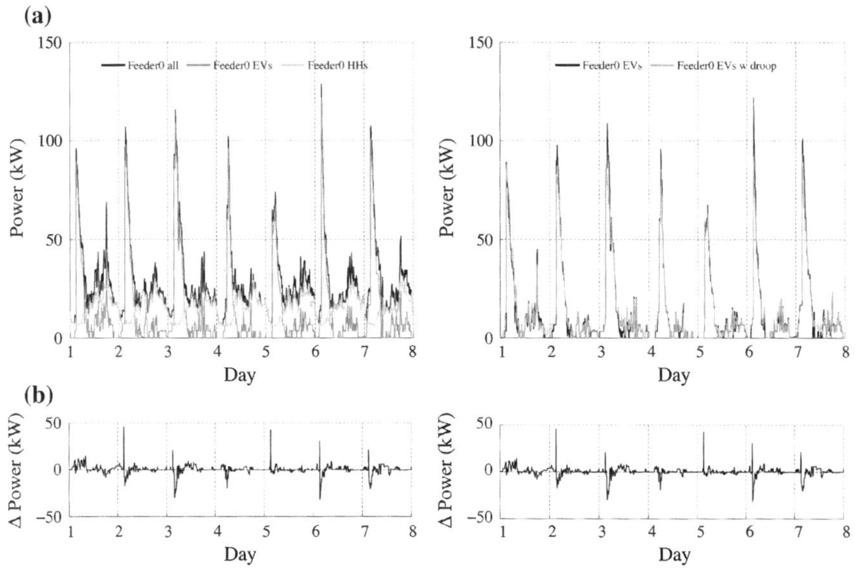

Fig. 6.12 Single simulation instance of case FL-38, for the event-based MAS MBC algorithm, week starting at day 16; **a** power profiles, **b** difference between (non)-droop enabled chargers, **c** voltages in the 3 phases of the Feeder0-line and **d** tariff used for ToU objective

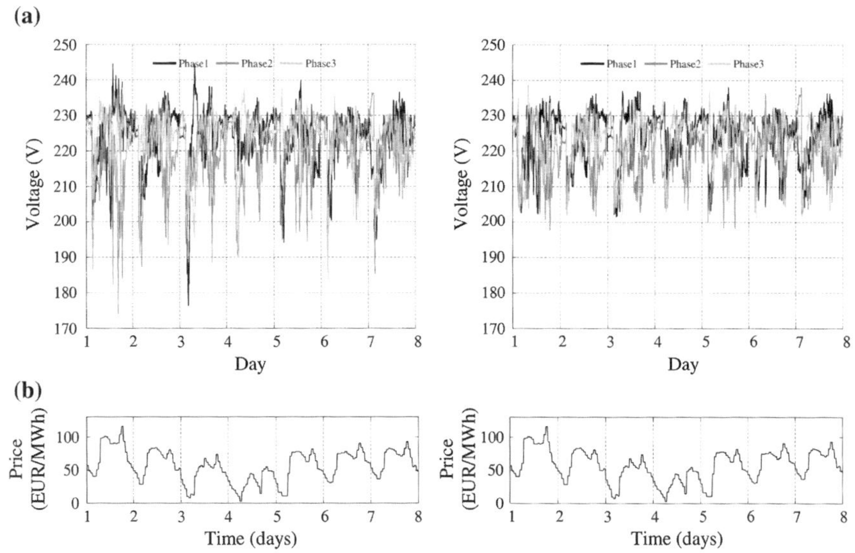

Fig. 6.12 (continued)

that it intervenes sooner, the weak grids can be brought into full EN 50160 compliance.

The difference between the power profiles for the case with and without voltage droop controller is also shown in Fig. 6.12. It is visible that, initially, power during the peak is lower, but immediately afterwards part of the 'lost' energy is recovered.

6.5.1.2 Market-Level Results

Table 6.3 shows the cost of charging for a cluster of 200 PEVs. Due to technical constraints, the 8 cases were simulated in separate batches. Because different random parameter sets were generated for each batch, the total cost values between the cases cannot simply be compared.

During droop control intervention, some vehicles can end up with an incompletely charged battery at departure time t_{dep}. Since this influences the cost numbers, a cost has to be attached to the resulting *energy deficit*. E_{deficit} equals the difference between the requested battery level and the level by which the vehicle departed:

$$E_{\text{deficit}} = \sum_i \left(\sum_i {}^iE_{\text{req},t} -{}^i E_{\text{batt},t}|t = {}^i\, t_{\text{dep}} \right) \tag{6.10}$$

A cost of €50/MWh is assigned to this energy deficit. Of course, the amount of deficit is directly related to the amount of vehicles that can suffer from distribution grid problems. Vehicles outside of weak distribution grids will obviously never end up with lost energy.

Table 6.3 Cost results for the ToU scenarios, and the difference due to the use of voltage droop control in the PEV chargers

Case name	Dumb	Event MBC	Event MBC + droop		Cost diff. due to V droop (%)
			w/o E_{deficit}	w. E_{deficit}	
NS-38	€805.46	€595.00	€596.71	€598.03	+0.28
NL-38	€795.83	€589.72	€594.25	€600.02	+0.77
FS-38	€814.26	€604.59	€606.71	€608.20	+0.35
FL-38	€806.75	€603.77	€609.84	€616.50	+1.00
NS-114	€792.28	€580.16	€585.32	€589.99	+0.89
NL-114	€823.00	€611.03	€626.54	€645.26	+2.50
FS-114	€816.26	€614.34	€620.61	€625.55	+1.02
FL-114	€819.28	€610.95	€630.91	€653.93	+3.27

The cost difference due to the undelivered energy is also shown

While the droop controller has a positive effect on the occurrence of voltage problems, it also increases the cost of charging the fleet, as more energy is consumed during unfavorable periods. Without taking into account the energy deficit at departure time, there is already a small cost increase of 0.6 % for the a-cases, and almost 2 % for the b-cases, where close to 60 % of the PEVs are situated in weak distribution grids. Taking into account E_{deficit}, this cost increase is doubled, and the cumulative battery deficit volume takes up to 1.15 % of the total delivered energy.

6.5.1.3 Conclusions on the ToU Scenario

From the results, it is apparent that *ToU* based controlled charging of PEVs has the potential to create significant power quality problems, because of the tendency to synchronously switch a large amount of the controlled loads when market prices are low, thereby creating large power peaks.

The effect on the state of the distribution grid can be even worse than when no coordinated charging is used (dumb charging). In fact, there were two mitigating factors in the simulations; the household connection points' phases were alternatingly distributed along the line and the price profiles used by the aggregator kept the power peak of the vehicles out of the household's evening peak. If the latter two were not the case, the EN 50160 results would be even worse.

One could argue that, once the penetration level of PEVs reaches a significant share, peak periods will be reflected in the ToU prices, which in turn will favor the spreading of charging load. However, problems in distribution grids can arise much earlier, due to clustering effects, meaning we have large penetration levels in a relatively small geographic area due to demographics. Additionally, when the share of variable renewable energy sources increases, the wholesale price will become more decorrelated from the instantaneous load. E.g. when wind or solar generation is peaking, electricity prices could be low even though the distribution grids are experiencing high load. Influencing distribution grid congestion through ToU tariffs will need carefully designed tariffs [49, 50].

On the positive side, the use of a simple voltage droop controller can practically solve the encountered power quality issues and is able to bring relatively weak distribution grids back into EN 50160 compliance, with some tuning. However, the use of a droop controller has a negative impact on the business case of the aggregator, as the cost of charging goes up and a small number of vehicles do not get their required charge at departure time. But quantitatively speaking, the differences only start to become significant (>2 %) when a large share (>50 %) of an aggregator's fleet is situated inside weak grids (Fig. 6.13).

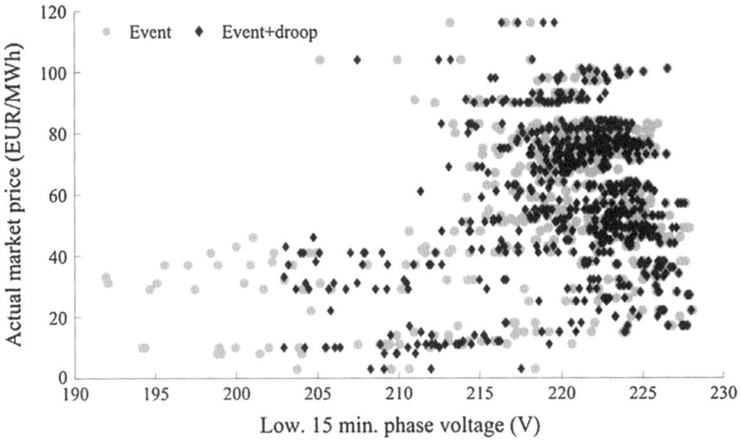

Fig. 6.13 Worst phase voltages observed in Feeder0 versus actual market prices over 7 days, for the MAS MBC algorithm with the ToU-objective, during case FL-38, both active and passive distribution grid. A correlation can be seen between low market prices and the occurrence of low voltages

6.5.2 Aggregator with Balancing Objective

In the previous sections, the objective for the coordinated charging at the market-level has been the cost of charging for the whole fleet. The outcome of an optimization over a *ToU* tariff of the next 24 h and the constraints of the vehicles results in a charging schedule. While a well-established generic objective, it does not entirely represent the potential of coordinated charging for fleet aggregators.

Alternatively, an aggregator could use the flexibility of a fleet to reduce the uncertainty on his portfolio after day-ahead commitments are made, to limit his exposure to the balancing market. In Europe, balancing services are traded on separate markets than wholesale energy [51]. While the prices for these services are correlated to those of the energy markets, they tend to be more expensive. The responsibility and the costs of balancing are usually attributed to an Access Responsible Party (ARP), which will prefer to reschedule their own generation portfolio rather than being exposed to the balancing market.

For wind farms, for example, wind predictions are used to build estimated production profiles and the required day-ahead nominations. Since the predictions are not perfect, real output will deviate from the day-ahead prediction during the day itself, and without intervention this difference leads to a positive or negative imbalance. An example is shown in Fig. 6.14a. By using the energy flexibility of the charging vehicles, an aggregator could try to reduce this wind imbalance.

The main difficulty in compensating for wind prediction errors with PEVs, however, is that large imbalances require the shifting of a considerable share of the fleet's available flexibility. Because the driving behavior of a fleet has a 24 h

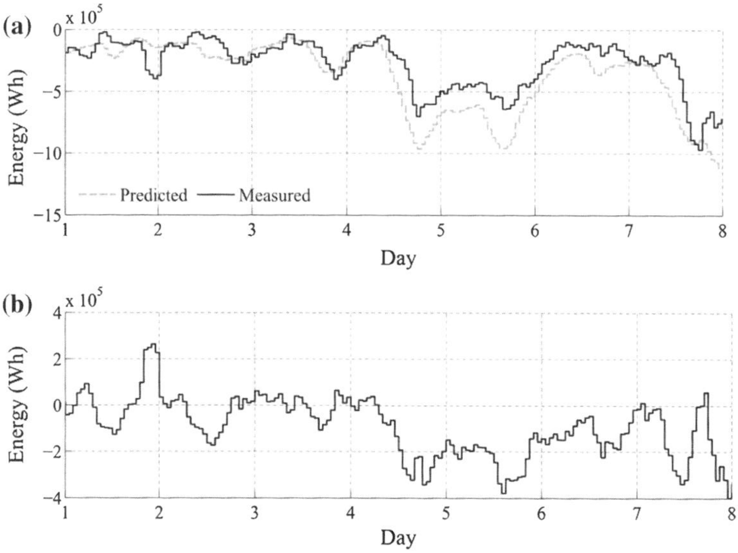

Fig. 6.14 **a** shows predicted and nominated versus measured hourly wind energy for 1.25 MW peak wind power ($W = 0.5$), for one week during March of 2008, and **b** resulting hourly prediction error over the simulated days

periodicity (as seen on Fig. 6.7) and remains relatively constant over time, so is the amount of charging energy per day. At the same time, wind prediction errors do not equal each other out over the course of a day and persist for longer times. Therefore, using all the vehicles' flexibility early in the day means any unexpected imbalance later that day cannot be compensated anymore. A possible solution could consist of incorporating stochastic optimization [52] and intra-day prediction updates to refine the scheduling process.

Another possible source of imbalance lies in the time resolution of the nominations; nominations for the day-ahead market in Belgium require energy values on an hourly basis [53]. However, imbalance volumes are settled on a 15 min basis. Even if an ARP has predictions on his portfolio with high resolution and accuracy, imbalance will still occur as nominated values are averaged per hour.

The description of the optimization problem was already provided in Sect. 6.4.1.

$$\boldsymbol{E}_{\text{opt}} = \underset{\boldsymbol{E}}{\operatorname{argmin}} \sum_{t=0}^{t_{\text{horiz}}} \left(E_{\text{PEV},t} + E_{\text{wind,t}} + \frac{1}{4} E_{\text{nomin,t/4}} \right)^2 \tag{6.11}$$

$$E_{\text{nomin,t}} = \text{E}_{\text{PEV,nomin,t}} + \text{E}_{\text{wind,nomin,t}} \tag{6.12}$$

The nominated energy E_{nomin} consists of a nomination for the PEV fleet and the day-ahead wind power prediction with a resolution of 1 h, for 24 h. Such nominations have to be determined by the ARP or aggregator, for example from

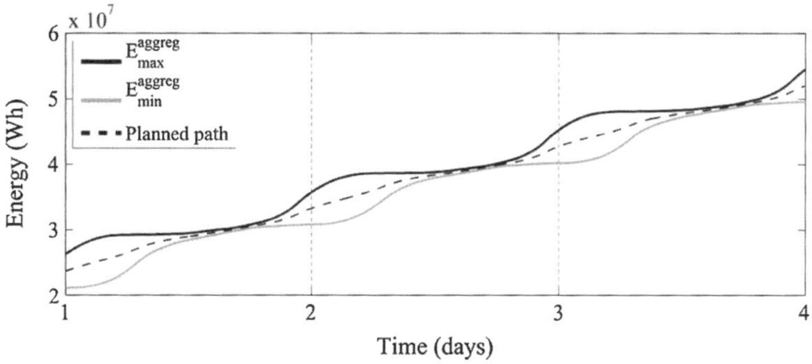

Fig. 6.15 Example PEV nominated energy based on historic aggregated energy constraints data

historical records or estimates. Because the driving behavior of an entire fleet behaves stable and predictable over time, it can be justified to use the power profile of a previous day or week as nomination for the fleet.

When historic energy constraints graphs are used, the amount of flexibility at any given time can be maximized by following an energy path through it according to a fixed ratio of e.g. 1/2 or 1/3 in between $\boldsymbol{E}_{\max}^{\text{aggreg}}$ and $\boldsymbol{E}_{\min}^{\text{aggreg}}$. Figure 6.15 illustrates such a planned path. The power values that correspond to the path can then be translated to hourly energy values to compose $E_{PEV,nomin,t}$.

An extra decay-term, γ can be added to reduce the influence of long-term information in the objective function.

$$\boldsymbol{E}_{\text{opt}} = \underset{\boldsymbol{E}}{\text{argmin}} \sum_{t=0}^{t_{\text{horiz}}} \gamma^{t/t_{\text{horiz}}} \left(E_{\text{PEV},t} + E_{\text{wind,t}} + \frac{1}{4} E_{\text{nomin,t/4}} \right)^2 \tag{6.13}$$

A $\gamma < 1$ will assign a higher optimization cost to the quarter hour imbalance values that are closest in time. In the limit, a $\gamma \rightarrow 0$ will mean that the system will act *myopic*, as no information on the future is taken into account. It behaves as the MAS MBC algorithm without planning and minimize instantaneous imbalance.

In order to evaluate the benefit of using this objective, a new ‘dumb’ scenario is added during which the fleet manager only tries to keep the energy consumption as close as possible to the nomination (referred to as *tracking* the nomination with the fleet). All scenarios use the event-based MAS MBC system to coordinate the fleet, but in the ‘tracking’ scenario, no optimization to minimize the difference with the nomination using short-term wind data takes place.

6.5.2.1 Simulation Scenarios and Performance Metrics

Due to the relatively long simulation times, the need to prepare nomination data for the wind and PEVs and an exponentially increasing set of parameters, a fixed simulation case is chosen for the simulations, in which the wind and vehicle profiles start at day 112 of the year. This was chosen because the first 3 days of the consequent week had relatively little wind imbalance and the last 3 relatively large. To end up with a significant amount of energy flexibility, the PEV cluster consists of 1,000 vehicles, instead of 200 for the previous case. Similar to the ToU scenarios, different shares of vehicles can be inside weak distribution grids, according to Table 6.2.

The main performance indicator consists of the total energy volume of remaining quarter hourly imbalance and the resulting cost. For the latter, real market data on the positive and negative imbalance price from the Transmission System Operator (TSO) Elia is used. It should be noted that the price data used dates from 2012, because the operating principle of the imbalance settlement was changed from then onwards, while the wind data available is from 2008.

While the total remaining imbalance volume accumulated during a simulation gives a good idea about the performance, it does not tell anything about its distribution during the day. From Fig. 6.14a, it can be seen that during the first 3 days, nominated and measured wind energy values are reasonably balanced over the course of a day. However, during the last 4 days, the difference between prediction and measured energy exists for the whole period. This is apparent from Fig. 6.14b, where the resulting prediction error during each hour of the simulation is shown. Unless the ratio of energy flexibility to wind power is very high, it is difficult to end up without imbalance under such conditions.

But the quadratic nature of the objective will favor to spread out imbalance as much as possible, so that a relatively flat imbalance profile should be obtained in the case of $\gamma = 1$. Therefore, looking solely at the remaining imbalance volume as a measure of performance would not capture the intent of the algorithm's objective. In fact, a myopic algorithm, instantly matching imbalance figures with the flexibility of PEVs, will perform better regarding the remaining imbalance volume.

Because the ability to smoothen or influence the occurrence of imbalance can be very beneficial for an ARP, it makes sense to look at the "variability" of the imbalance profiles. The spectral content of the imbalance profile is obtained by taking the sum of FFTs over a sliding window of 32 profile samples. Then the mean value is subtracted to get rid of the DC component, and the surface under the spectral plot is kept, expressed in kW Hz. The higher this value, the more variability there is on the remaining imbalance's power profile.

To evaluate the effects at the real-time level, the EN 50160 specifications and performance indicators from the ToU case, are used here as well.

6.5.2.2 Market-Level Results

In a first simulation, only the behavior at the market level is investigated, disregarding the distribution grid completely. In Fig. 6.16a, the 15 min imbalance volumes are plotted for different values of γ, for a simulation covering the 7 days from Fig. 6.14. It is visible that the event-based balancing successfully reduces the amount of imbalance with the nomination. Smaller γ values lead to aforementioned

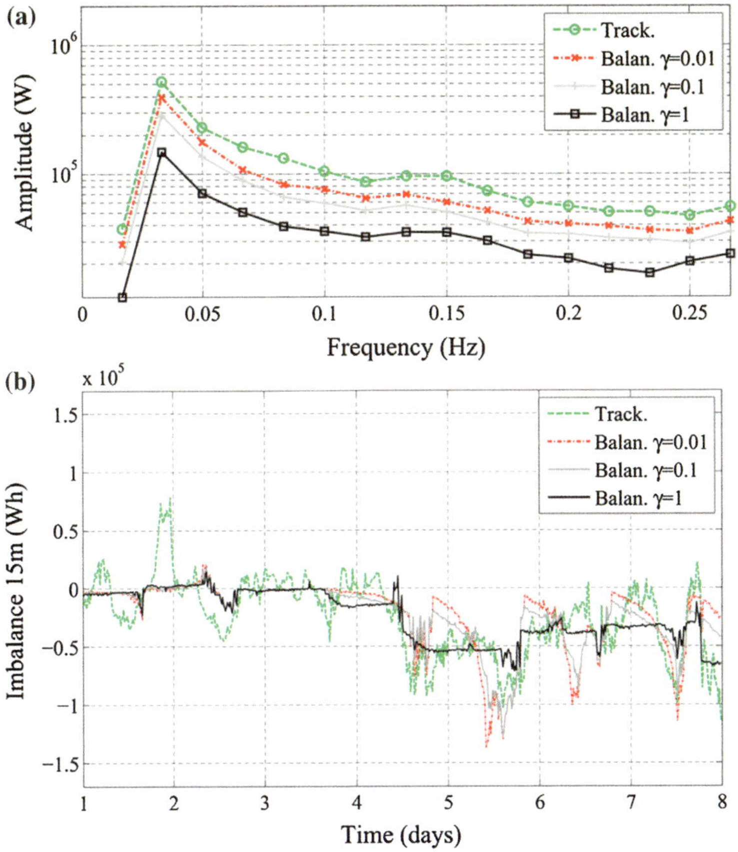

Fig. 6.16 Imbalance scenario, **a** remaining imbalance profile for different values of γ over the course of 7 days and for a peak wind output of 1.25 MW (W = 0.5), together with the tracking scenario and **b** spectral plot of the power profiles expresses variability of the remaining imbalance

'myopic' behavior and force the imbalance profile close to zero, until of course the aggregator runs out of short-term flexibility.

In Fig. 6.16b, the Fourier transformed imbalance volume is plotted. This figure thus shows its frequency components. In case of the balancing optimization scenarios, it is visible that their imbalance profiles contain less high-frequency components then when no balancing optimization is done. This confirms what can be seen in Fig. 6.16a, namely that the case with the balancing optimization for $\gamma = 1$. is able to better spread out the remaining imbalance.

In the above scenario, the wind nominations and measurements were scaled with a factor $W = 0.5$, to obtain a peak wind output of 1.25 MW. Varying ratios of wind and vehicles have also been examined, of which the results are shown in Table 6.4.

The improvement in remaining imbalance volume over the tracking case is between 20 and 30 %. Smaller γ values lead to slightly less remaining imbalance

Table 6.4 Balancing case simulation results for 7 consecutive days and a cluster 1,000 PEVs, for different values of the wind scaling parameter W and discount factor γ

W = 0.05 (0.125 MWp)	Imbal Volume (MWh)	Imbal Cost	Volume diff. (%)	Spectr. (kW Hz)	Spectr. diff (%)
Tracking nomin.	2.543	€171.2	0	2.7	0
Balancing $\gamma = 1$	2.087	€128.7	−17.9	2.5	−7.4
Balancing $\gamma = 0.1$	1.988	€120.9	−21.8	3.1	+14.8
Balancing $\gamma = 0.01$	1.967	€117.1	22.7	3.5	+29.6
W = 0.2 (0.5 MWp)					
Tracking nomin.	8.633	€580.3	0	9.3	0
Balancing $\gamma = 1$	6.832	€434.3	−20.7	3.3	−64.5
Balancing $\gamma = 0.1$	6.322	€397.8	−26.8	6.2	−33.3
Balancing $\gamma = 0.01$	6.131	€379.7	−28.9	8.0	−14.0
W = 0.5 (1.25 MWp)					
Tracking nomin.	21.056	€1413	0	23.2	0
Balancing $\gamma = 1$	16.680	€1091	−20.8	7.6	−67.2
Balancing $\gamma = 0.1$	15.775	€1014	−25.1	13.2	−43.1
Balancing $\gamma = 0.01$	15.313	€989.4	−27.3	16.9	−27.2
W = 0.7 (1.75 MWp)					
Tracking nomin.	29.364	€1970	0	32.5	0
Balancing $\gamma = 1$	23.888	€1570	−18.6	12.2	−62.5
Balancing $\gamma = 0.1$	22.860	€1471	−22.1	18.9	−41.2
Balancing $\gamma = 0.01$	22.216	€1443	−24.3	23.8	−23.8
W = 1.0 (2.5 MWp)					
Tracking nomin.	41.830	€2806	0	46.4	0
Balancing $\gamma = 1$	35.112	€2324	−16.1	20.8	−55.2
Balancing $\gamma = 0.1$	33.762	€2186	−19.3	29.2	−37.1
Balancing $\gamma = 0.01$	33.141	€2150	−20.8	34.9	−24.8

over 7 days compared to $\gamma = 1$. However, since the objective of the optimization is related to the quadratic imbalance over the optimization horizon, the conclusion that a myopic algorithm performs better based on the total remaining imbalance would be misleading. It has to be looked at together with the 'spreading' of the remaining imbalance, expressed by the spectral content on the 'Volume difference' column of Table 6.4.

For larger wind scaling factors and thus larger wind prediction error volumes, the improvement regarding remaining imbalance decreases to 16–21 %. A similar effect is observed for the spectral content values. It can be deduced that, based on this balancing method, around 1–1.25 MW of wind power can be properly compensated per 1,000 PEVs. Higher or lower shares of wind power decrease the efficiency of this system.

6.5.2.3 Real-Time Level Results

For the effects at the distribution level, the different cases and its variations again come into play. From the tested parameters in the previous section, we keep the wind scaling of $W = 0.5$, since this parameter led to the best performance at the market level, and a γ of 1, as this is the most generic application.

The EN 50160 results of the passive distribution grid scenarios are grouped together with the active distribution grid scenarios in Fig. 6.17, to improve clarity and avoid duplication. These plots show the results for the FL case, but the household-only results have been omitted, since their results are the same as in the previous section.

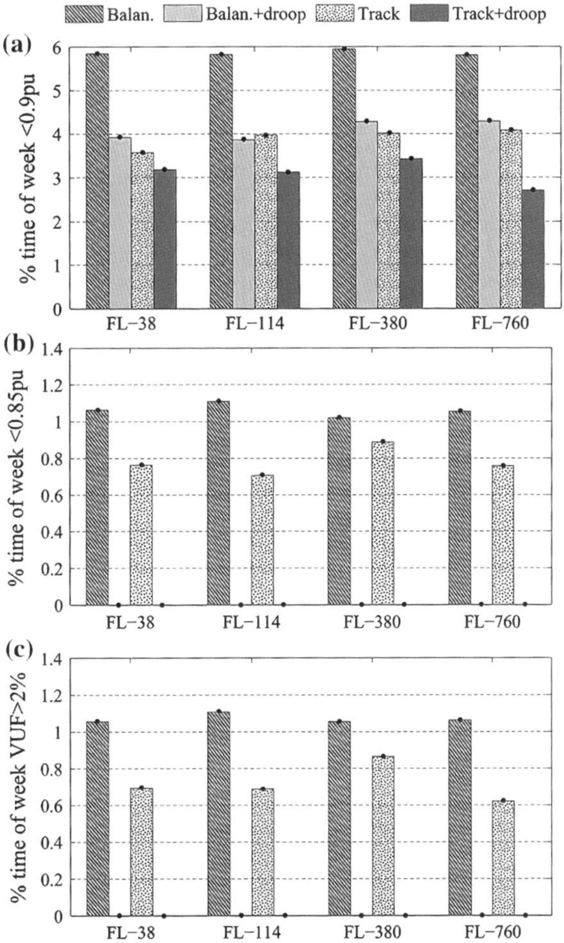

Fig. 6.17 EN 50160 voltage magnitude stats for the single-aggregator balancing scenarios; **a** V < 0.9 pu, **b** V < 0.85 pu and (**c**) VUF > 2 %

Compared to the ToU results, problems are a lot less worse, but voltages still drop below 0.85 pu. The use of voltage droop control reduces the limited remaining voltage problems to below the EN 50160 specifications.

Since the tracking scenario already tries to follow the nomination, which is a smooth path through the aggregated energy constraints graph for the PEVs, the reduction in voltage deviations are relatively small when voltage droop controllers are introduced, in comparison to the balancing case.

6.5.2.4 Impact of Droop Control on Market-Level Objectives

The amount of vehicles that is affected by voltage droop activation is expected to influence the business case at the market level. It would follow that moving from case FL-38 to FL-760 will increase the remaining imbalance, as less and less peak flexibility is available to the fleet manager. Figure 6.18 shows that the imbalance volume is constant for case FL-38 and FL-114, having respectively 4 and 21 % of all the PEVs inside of a weak grid.

For case FL-380, with 38 % of the fleet inside the weak distribution grids, a small increase of 2.4 % in the imbalance volume is noticeable, and finally, for the case FL-760 with 76 % of the PEVs located inside the constrained grids, the observed increase in imbalance volume is 10.3 %. During the latter, the 'dumb' tracking scenario also suffered slightly with a minor 0.95 % increase.

6.5.2.5 Conclusions on the Balancing Case

The balancing concept was successfully tested on a portfolio consisting of wind generation and charging PEV's. The optimization reduces both the imbalance that

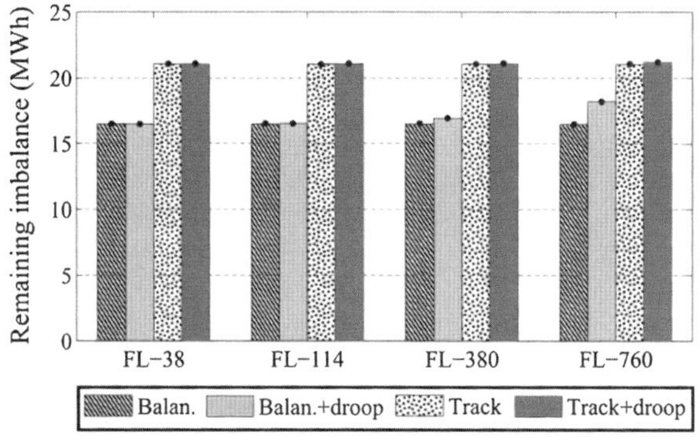

Fig. 6.18 Total remaining imbalance after 7 days, for different shares of PEVs in weak distribution grids. At larger shares, an effect on the remaining imbalance is noticeable, as the aggregator fails to compensate for the activation of the droop controllers

originates from the hourly discretization of the day-ahead nomination, and the imbalance that exists because of imperfect wind speed predictions. Using short-term information on the wind production, the imbalance can also be intentionally spread in time. This can be beneficial for the aggregator, as the remaining imbalance could then be countered by other generation units in its portfolio.

Additionally, the effect of varying the discount factor γ was shown. By including γ as variable into the optimization, one could move the remaining imbalance towards points in time where this has economical benefits, such as by using stochastic information on the imbalance market prices.

Regarding grid constraints, the use of the (quadratic) balancing objective puts less load on the grid compared to the (linear) ToU objective, because flexibility of the PEVs is intentionally spread out when creating the nomination of the charging energy and therefore not enabled all at once.

As in the ToU case, voltage droop controllers inside PEV chargers are successful in mitigating weak grid constraints. Some tuning of its parameters may be needed to find a setting where the grid state at all nodes is within the EN 50160 specifications during all the time.

Unless a very large share of PEVs of a coordinated charging fleet manager is located inside weak grids, the business case is practically unaffected by the addition of local voltage droop control, using the coordination system that was implemented in this work. That means being event-based for fast response and having a compensation loop at the fleet manager. The combination of both ensures that, when droop controllers activate, the equilibrium priority is changed quickly enough so that the flexibility of other vehicles is dispatched to compensate for the 'loss' in expected energy over time.

6.6 Multi-aggregator Simulations and Results

In many cases, when studying coordinated charging of PEVs, there is only a single fleet manager or aggregator. However, if the business case of using the energy flexibility of vehicles takes off, it can be anticipated that multiple competing services will become available. This leads to the question what problems can arise if multiple aggregators are active within the same distribution grid, as illustrated by Fig. 6.19.

In case of problems, is there a need for additional congestion management mechanisms, to ensure that capacity inside individual grids is allocated to the aggregator's objective that has the highest value, or is the use of a voltage droop controller that intervenes when problems arise sufficient?

The advantage of a voltage droop controller lies in its simplicity of operation and the fact that it does not rely on communication with external actors. More complex grid congestion management systems, briefly touched upon in Sect. 6.2.2.2, assign

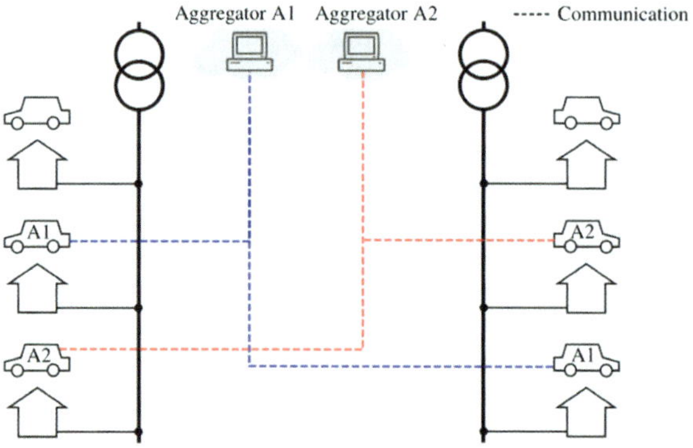

Fig. 6.19 Multi-aggregator grid situation. Two aggregators, A1 and A2, control a number of charging EVs that are connected to the same distribution grid transformer

an active role to the DSO, that must perform ahead-of-time capacity allocation and/or check iteratively whether all aggregators' schedules are feasible (advance capacity allocation). This would be required for every grid segment wherein aggregators are active. Or, a DSO could set up dynamic ToU network tariffs based on location and projected network load.

6.6.1 Aggregators with ToU Cost Objective

Since the use of a ToU objective implies that aggregators use the same actual market prices, a multi-aggregator version of the ToU scenario of Sect. 6.5.1 does not perform any different than its single-aggregator counterpart. Therefore, the simulations and results have been omitted.

However, in case one aggregator is serving mostly customers that are located at the beginning of a line and the other aggregator mainly ones at the end, the latter will be at a disadvantage. A similar situation will occur if the phase connections are heavily correlated with the aggregator assignment.

6.6.2 Aggregators with Balancing Objective

During the balancing case, different aggregators can base their optimizations on different predictions or portfolios, and the expected results are not as straightforward to derive as in the ToU cases.

In the simulations, both aggregators will be using an identical portfolio, again consisting of a fleet of 1,000 PEVs combined with 1.25 MW of peak wind generation. To have a realistic case that represents wind generation in a geographically shared region, the wind predictions should at least be correlated, which is taken care of by adding one day of difference for the second aggregator.

To ensure that aggregators each have the same fleet size, the total amount of vehicles in the simulations has to be doubled. Again, different cases represent varying shares of vehicles that are inside the weak distribution grids. Case FL-38 has been left out, since at less than 4 % of PEVs inside a weak grid, the effects during the balancing scenario are practically zero, as previously shown.

6.6.2.1 Real-Time Level Results

On Fig. 6.20, the EN 50160 results are plotted for cases x-114, x-380 and x-760 (respectively with 114,380 and 760 of 1,000 PEVs inside weak distribution grids). Compared to the single-aggregator scenario, the severity of the voltage deviations is a lot less. This can be entirely attributed to the reduced coincidence of the objectives of both aggregators.

6.6.2.2 Market Level Results

The total remaining imbalance for both aggregators after one simulated week is shown in Fig. 6.21. Just as with the single aggregator case in Fig. 6.18, the imbalance volume is only affected by the droop controllers at high shares of PEVs in weak grids. The absolute volume of aggregator A2 is lower because its wind profile starts one day earlier than that of A1, thereby avoiding a day with large prediction error.

6.6.2.3 Conclusions on the Multi-aggregator Case

From the results, it can be concluded that settings where two aggregators are active within the same part of the distribution grid do not show problematic behavior. This is due to the fact that the wind profiles are not the same for both aggregators, so that access to PEVs' flexibility in one distribution grid is spread out and less voltage deviations appear. Therefore, in the worst case, voltage deviations would be similar to those of the single-aggregator case.

With these results in mind, and based on the implemented DR algorithm with voltage droop controllers inside the presented grid configurations, the necessity of additional grid congestion management mechanisms can be questioned. The

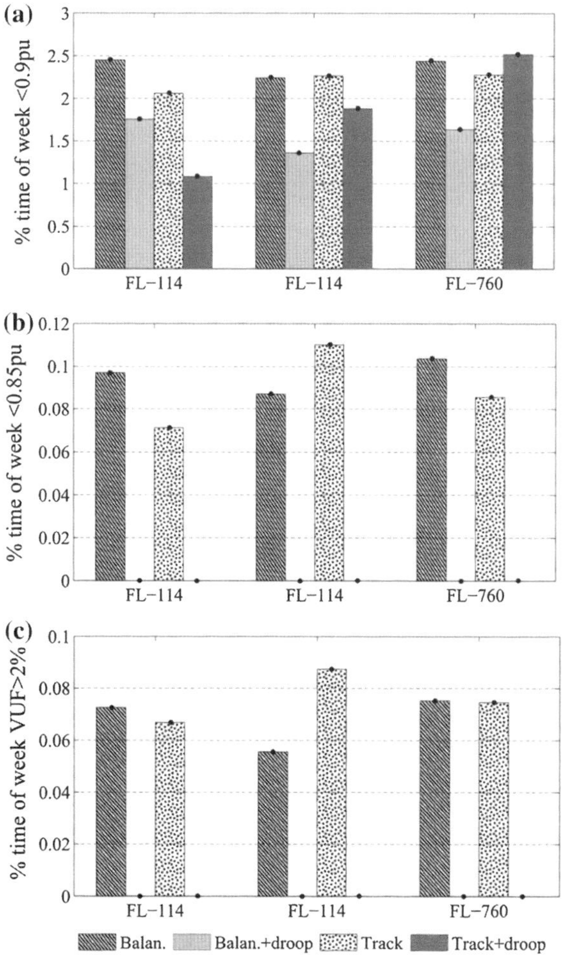

Fig. 6.20 EN 50160 voltage magnitude stats for the multi-aggregator balancing scenarios; **a** V < 0.9 pu, **b** V < 0.85 pu and **c** VUF > 2 %

complexity introduced by such solutions, computationally and from a responsibility perspective, are hard to justify with the amount of gains that can be achieved.

It was however assumed that the PEVs were evenly assigned to both aggregators. In the situation where one aggregator controls all the PEVs at the beginning of a grid and the other all the PEVs towards the end of the line, the latter will be subjected to more droop activations and be at a disadvantage compared to the other aggregator. But again, the limited energy deficits this causes might not warrant the deployment of grid congestion management mechanisms (e.g. capacity markets in cooperation with the DSO, Sect. 6.2.2.2).

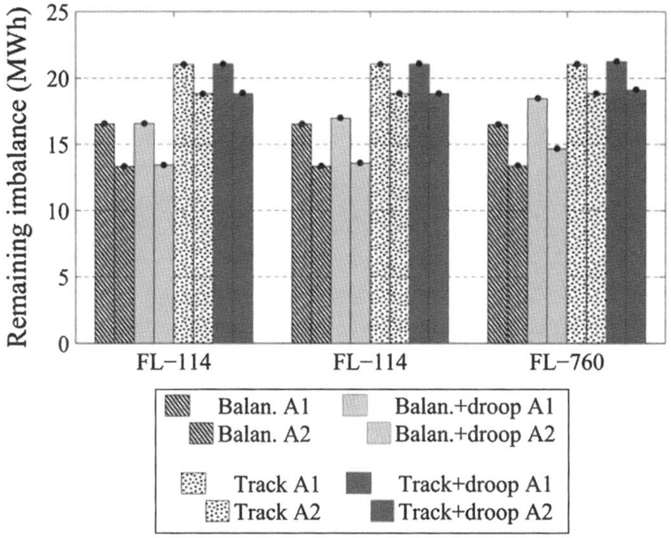

Fig. 6.21 Total remaining imbalance after 7 days. *Left side* of the bar represents aggregator A1, *right side* aggregator A2

6.7 Conclusions

In the light of the challenges that were discussed in the introduction, we can summarize the results and contributions as follows:

- The separation between two demand response operation levels was identified; the market operation level is responsible for the business case of a fleet of PEVs and operates synchronous with the energy markets. The technical or real-time operation level uses the setpoints determined by the business case and uses an event-driven architecture to efficiently dispatch constraints from and control signals to the charging PEVs. At the market level, an algorithm based on MBC was adapted for the coordination of PEVs, and at the technical level, a voltage droop controller is integrated to be able to respect the local grid constraints.
- The effect of using market-level objectives on congestion in weak distribution grids has been examined. Especially the use of ToU cost minimization objectives has a negative effect on the occurrence of undervoltages, with respect to the EN 50160 standard. Synchronization of large amounts of controllable loads is to be avoided in DR.
- Besides a ToU cost minimization objective, it has been shown that a cluster of fast-responding PEVs can be used to limit an aggregator's exposure to the balancing market. An optimization at the market level determines setpoints for the fleet such that the remaining imbalance between predicted and nominated wind output and more recent short-term predictions is spread out in time. This can be beneficial for the aggregator, as the remaining imbalance could be then

be countered by other generation units in its portfolio. Additionally, one could include γ as variable into the optimization to express a preference of having the remaining imbalance occur when this has economical benefits.

- A straightforward and common way of mitigating grid congestion is the use of a voltage droop controller. While fast, inexpensive and able to act independently from any central coordinator, its activation will intervene in the the business case. In literature, the overruling of the market operation level by technical objectives is often presented as a major challenge to be addressed. The results in Sects. 6.5 and 6.6 show that, unless very large shares of the PEV fleet are located inside weak grids, the effects of the activation of voltage droop controllers on the business case remains relatively modest. This is due to the limited amount of scheduled energy that is 'lost' and the possibility to compensate for by other parts in the DR cluster, with the event-driven approach.
- Additionally, situations where multiple aggregators are active within the same distribution grid were also looked at. Based on the assumptions made and using the presented DR algorithm, it can be stated that the use of voltage droop controllers only is already effective in mitigating grid congestion problems without significantly disturbing the aggregators' business case. The need for additional grid congestion management algorithms, e.g. a capacity market in cooperation with the DSO, might better be reserved to a few corner cases where increasing the transfer capacity of the grid is (economically or otherwise) infeasible.

References

1. Geth F, Leemput N, Van Roy J, Buscher J, Ponnette R, Driesen J (2012) Voltage droop charging of electric vehicles in a residential distribution feeder. In: 3rd IEEE PES innovation smart grid technology Europe, (ISGT Europe) IEEE, pp 1–8
2. De Craemer K, Vandael S, Claessens B, Deconinck G (2013) An event-driven dual coordination mechanism for demand side management of PHEVs. IEEE Trans Smart Grid pp 1–10. doi:10.1109/TSG.2013.2272197
3. Clement-Nyns K, Haesen E, Driesen J (2010) The impact of charging plug-in hybrid electric vehicles on a residential distribution grid. IEEE Trans Power Syst 25:371–380. doi:10.1109/TPWRS.2009.2036481
4. Shao S, Pipattanasomporn M, Rahman S (2012) Grid integration of electric vehicles and demand response with customer choice. IEEE Trans Smart Grid 3:543–550. doi:10.1109/TSG.2011.2164949
5. Gatsis N, Giannakis GB (2012) Residential load control: distributed scheduling and convergence with lost AMI messages. IEEE Trans Smart Grid 3:770–786. doi:10.1109/TSG.2011.2176518
6. Gatsis N, Giannakis GB (2011) Cooperative multi-residence demand response scheduling. In: 45th Annual conference on information sciences and systems, IEEE, pp 1–6
7. Fan Z (2012) A distributed demand response algorithm and its application to PHEV charging in smart grids. IEEE Trans Smart Grid 3:1280–1290. doi:10.1109/TSG.2012.2185075
8. Weckx S, Driesen J, D'hulst R (2013) Optimal real-time pricing for unbalanced distribution grids with network constraints. In: IEEE Power and energy society general meeting, IEEE, pp 1–5

9. Ma Z, Callaway D, Hiskens I (2010) Decentralized charging control for large populations of plug-in electric vehicles. 49th IEEE conference on decision and control, IEEE, pp 206–212
10. Anderson RN, Boulanger A, Powell WB, Scott W (2011) Adaptive stochastic control for the smart grid. Proc IEEE 99:1098–1115. doi:10.1109/JPROC.2011.2109671
11. Wong VWS (2011) An approximate dynamic programming approach for coordinated charging control at vehicle-to-grid aggregator. In: International conference on smart grid communications, IEEE, pp 279–284
12. Robu V, Stein S, Gerding E, Parkes D, Rogers A, Jennings N (2012) An online mechanism for multi-speed electric vehicle charging. In: 2nd International conference on auctions, market mechanisms, and their applications. Springer, Heidelberg, pp 100–112
13. Galus MD, La Fauci R, Andersson G (2010) Investigating PHEV wind balancing capabilities using heuristics and model predictive control. In: IEEE power and energy society general meeting, IEEE, pp 1–8
14. Biegel B, Andersen P, Pedersen TS, Nielsen KM, Stoustrup J, Hansen LH (2013) Smart grid dispatch strategy for on/off demand-side devices. Control Conference (ECC), Europe, pp 2541–2548
15. Koch S, Mathieu JL, Callaway DS (2011) Modeling and control of aggregated heterogeneous thermostatically controlled loads for ancillary services. In: Proceedings of 17th Power Systems Computation Conference
16. Vandael S, Claessens B, Hommelberg M, Holvoet T, Deconinck G (2013) A scalable three-step approach for demand side management of plug-in hybrid vehicles. IEEE Trans Smart Grid 4:720–728. doi:10.1109/TSG.2012.2213847
17. Bach Andersen P, Hu J, Heussen K (2012) Coordination strategies for distribution grid congestion management in a multi-actor, multi-objective setting. In: 3rd IEEE PES innovation smart grid technology Europe, (ISGT Europe) IEEE, pp 1–8
18. CENELEC (2010) EN 50160—Voltage characteristics of electricity supplied by public electricity networks, July 2010
19. Pillay P, Manyage M (2001) Definitions of voltage unbalance. IEEE Power Eng Rev 21:50–51. doi:10.1109/39.920965
20. Seljeseth H, Henning T, Solvang T (2013) Measurements of network impact from electric vehicles during slow and fast charging. In: 22nd International Conference on Electricity Distribution
21. Liu MB, Canizares CA, Huang W (2009) Reactive power and voltage control in distribution systems with limited switching operations. IEEE Trans Power Syst 24:889–899. doi:10.1109/TPWRS.2009.2016362
22. Gonzalez C, Geuns J, Weckx S, Wijnhoven T, Vingerhoets P, De Rybel T, Driesen J (2012) LV distribution network feeders in Belgium and power quality issues due to increasing PV penetration levels. In: 3rd IEEE PES innovation smart grid technology Europe, (ISGT Europe) IEEE, pp 1–8
23. Efkarpidis N, Gonzalez C, Wijnhoven T, Van Dommelen D, De Rybel T, Driesen J (2013) Technical assessment of on-load tap-changers in flemish LV distribution grids. International work integrative solar power into power systems
24. Kester CPJ, Heskes JMP, (Sjaak) Kaandorp JJ, (Sjef) Cobben JFG, Schoonenberg G, Malyna D, De Jong ECW, Wargers BJ, (Ton) Dalmeijer AJF (2009) A smart MV/LV-station that improves power quality, reliability and substation load profile. In: 20th International conference on electricity distribution, pp 8–11
25. DIN VDE Std. VDE-AR-N 4105 (2011) Erzeugungsanlagen am Niederspannungsnetz, Technische Mindestanfor-derungen für Anschluss und Parallelbetrieb von Erzeugungsanlagen am Niederspannungsnetz
26. Synergrid C10/11 (2012) Specifieke technische voorschriften voor decentrale productie-installaties die in parallel werken met het distributienet
27. Loix T (2011) Participation of inverter-connected distributed energy resources in grid voltage control. KU Leuven

28. Clement-Nyns K, Haesen E, Driesen J (2011) The impact of vehicle-to-grid on the distribution grid. Electr Power Syst Res 81:185–192. doi:10.1016/j.epsr.2010.08.007
29. Peças Lopes JA, Polenz SA, Moreira CL, Cherkaoui R (2010) Identification of control and management strategies for LV unbalanced microgrids with plugged-in electric vehicles. Electr Power Syst Res 80:898–906. doi:10.1016/j.epsr.2009.12.013
30. Garcia-Valle R, Peças Lopes JA (2013) Electric vehicle integration into modern. Power Netw. doi:10.1007/978-1-4614-0134-6
31. Biegel B, Andersen P, Stoustrup J, Bendtsen JD (2012) Congestion management in a smart grid via shadow prices. In: 8th IFAC symposium on power plant and power system control, pp 518–523
32. Sundstrom O, Binding C (2012) Flexible charging optimization for electric vehicles considering distribution grid constraints. IEEE Trans Smart Grid 3:26–37. doi:10.1109/TSG.2011.2168431
33. O'Connell N, Wu Q, Østergaard J, Nielsen AH, Cha ST, Ding Y (2012) Day-ahead tariffs for the alleviation of distribution grid congestion from electric vehicles. Electr Power Syst Res 92:106–114
34. Verzijlbergh R (2013) The power of electric vehicles—exploring the value of flexible electricity demand in a multi-actor context. Dissertation, TU Delft
35. Kok K (2013) The powermatcher: smart coordination for the smart electricity grid, p 314. Vrije Universiteit, Amsterdam
36. Kok K, Warmer K, Kamphuis R (2005) PowerMatcher: multiagent control in the electricity infrastructure. In: Proceedings of the 4th international joint conference on autonomous agents and multiagent systems (AAMAS 05'). New York, pp 75–82
37. Kok JK, Scheepers MJJ, Kamphuis IG (2010) Intelligence in electricity networks for embedding renewables and distributed generation. Intelligent infrastructures, Springer, pp 179–209
38. Van Roy J, Leemput N, Geth F, Salenbien R, Buscher J, Driesen J (2014) Apartment building electricity system impact of operational electric vehicle charging strategies. IEEE Trans Sustain Energy 5:264–272. doi:10.1109/TSTE.2013.2281463
39. Van Roy J, Leemput N, De Breucker S, Geth F, Peter T, Driesen J (2011) An availability analysis and energy consumption model for a flemish fleet of electric vehicles. European Electrical Vehicle Congr Brussels, p 12
40. Raab AF, Ellingsen M, Walsh A (2011) Mobile energy resources in grids of electricity—WP 1 Task 1.6 Deliverable D1.4—learning from EV FIeld Tests
41. Brand AJ, Kok K (2003) Aanbodvoorspeller duurzame energie. https://www.ecn.nl/avde/
42. Brand AJ (2008) Wind power forecasting method AVDE. China Glob Wind Power
43. Nordex (2009) Datenblatt N80/2500 (2.5 MW). http://www.nordex-online.com/en/produkte-service/wind-turbines/n80-25-mw/product-data-sheet-n80-25mw.html
44. Dupont B, Vingerhoets P, Tant P, Vanthournout K, Cardinaels W, De Rybel T, Peeters E, Belmans R (2012) Linear breakthrough project: large-scale implementation of smart grid technologies in distribution grids. In: 3rd IEEE PES innovation smart grid technology Europe, (ISGT Europe) IEEE, pp 1–8
45. Leemput N, Geth F, Van Roy J, Delnooz A, Büscher J, Driesen J (2014) Impact of electric vehicle on-board single-phase charging strategies on a flemish residential grid. IEEE Trans Smart Grid, 5(4)1815–1822. doi: 10.1109/TSG.2014.2307897
46. Clement-Nyns K (2010) Impact of plug-in hybrid electric vehicles on the electricity system. KU Leuven
47. VREG (2013) RAPP-2013-06—De kwaliteit van de dienstverlening van de elektriciteitsdistributienetbeheerders in het Vlaamse Gewest in 2012
48. NBN (1975) C33-322—Kabels Voor Ondergrondse Aanleg, met Synthetische Isolatie en Versterkte Mantel (Type 1 kV)
49. Shao S, Zhang T, Pipattanasomporn M, Rahman S (2010) Impact of TOU rates on distribution load shapes in a smart grid with PHEV penetration. IEEE PES transmission and distribution, IEEE, pp 1–6

50. Dupont B, De Jonghe C, Olmos L, Belmans R (2014) Demand response with locational dynamic pricing to support the integration of renewables. Energy Policy 67:344–354. doi:10.1016/j.enpol.2013.12.058
51. Baritaud M (2012) Securing power during the transition. Generation investment and operation, IEA (International Energy Agency), pp 47
52. Vaya MG, Andersson G (2013) Integrating renewable energy forecast uncertainty in smart-charging approaches for plug-in electric vehicles. IPowerTech (POWERTECH), 2013 IEEE Grenoble, pp 1–6, IEEE
53. Elia (2014) Nominated capacity: Belgium—Netherlands. http://www.elia.be/en/grid-data/interconnections/nominated-capacity-bel-neth. Accessed 20 Jan 2014

Chapter 7
Distributed Load Management Using Additive Increase Multiplicative Decrease Based Techniques

Sonja Stüdli, Emanuele Crisostomi, Richard Middleton, Julio Braslavsky and Robert Shorten

Abstract Due to the expected increase in penetration levels of Plug-in Electric Vehicles (PEVs), the demand on the distribution power grid is expected to rise significantly during PEV charging. However, as PEV charging in many cases may not be time critical, they are suitable for load management tasks where the power consumption of PEVs is controlled to support the grid. Additionally, PEVs may also be enabled to inject power into the grid to lower peak demand or counteract the influence of intermittent renewable energy generation, such as that produced by solar photovoltaic panels. Further, PEV active rectifiers can be used to balance reactive power in a local area if required, to reduce the necessity for long distance transport of reactive power. To achieve these objectives, we adapt a known distributed algorithm, Additive Increase Multiplicative Decrease, to control both the active and reactive power consumption and injection. Here, we present this algorithm in a unified framework and illustrate the flexibility of the algorithm to accommodate different user objectives. We illustrate this with three scenarios, including a domestic scenario and a workplace scenario. In these scenarios the various objectives allow us to define a type of "fairness" for how the PEVs should

S. Stüdli (✉) · R. Middleton
School of Electrical Engineering and Computer Science, The University of Newcastle, Callaghan, NSW 2308, Australia
e-mail: sonja.stuedli@uon.edu.au

R. Middleton
e-mail: richard.middleton@uon.edu.au

E. Crisostomi
Department of Energy, Systems, Territory and Constructions Engineering, University of Pisa, Pisa, Italy
e-mail: emanuele.crisostomi@unipi.it

J. Braslavsky
Energy Flagship, CSIRO, 10 Murray Dwyer Circuit, Mayfield West, NSW 2304, Australia
e-mail: julio.braslavsky@csiro.au

R. Shorten
Optimization, Control and Decision Sciences, IBM Research Ireland, Dublin, Ireland
e-mail: robshort@ie.ibm.com

S. Rajakaruna et al. (eds.), *Plug In Electric Vehicles in Smart Grids*,
Power Systems, DOI 10.1007/978-981-287-302-6_7

adapt their power consumption, i.e. equal charging rates, or charging rates based on energy requirements. We then validate the algorithms by simulations of a simple radial test network. The simulations presented use the power simulation tool OpenDSS interlinked with MATLAB.

Keywords Power sharing · Load management · AIMD · Reactive power compensation · PEV charging · G2V · V2G

7.1 Introduction

The increased integration of renewable energy into the electricity grid has stimulated significant interest in designing "demand side management" and "load management" strategies to support the distribution grid [4]. In this context, "peak shaving" and "load tracking" are two important support services required for proper functioning of the grid with highly variable renewable power generation; see for example [18] and more recently [22].

Peak shaving is an instance of time-shifting power demand. Peaks in aggregate demand experienced in the power grid may be reduced by shifting the power usage of controllable loads to other times of the day. Load tracking is a network service where controllable loads are driven to follow a given varying power signal. Load tracking is particularly useful to follow the fluctuating power generated from renewable sources, and to limit the use of electricity from more polluting power plants at times when demand exceeds the power available from renewable sources.

There is a general consensus that should widespread adoption occur, Plug-in Electric Vehicles (PEVs), will play an important role in demand side management. In fact, PEVs can be often treated as controllable loads and their charging can be postponed to some later time of the day (unless the owner has some urgency in using the vehicle) see [23, 31]. With the capability to act as energy storage, PEVs may also be utilized to inject power into the grid, often referred to as vehicle to grid (V2G) operation. Especially during day-times, such functionality can be used to flatten the peak demand and to help regulate the grid frequency, see for example [1, 20]. When injecting power into the grid, it is important to take into account the needs of the PEV owner such that their energy requirements are met, see for example [30]. This can be implemented in several ways, for example by limiting the energy that is allowed to be used for V2G operation [1]. In that way it is guaranteed that the battery of the PEV is not completely depleted when the vehicle is needed, due to providing V2G services.

In addition, PEV charging infrastructure can also support the power grid, by exchanging reactive power with the grid, as mentioned in for example [5]. In this case, the PEVs can either consume or inject reactive power into the grid to compensate for the reactive power required locally where they are connected. The advantage of using

PEVs for reactive power balance, is that in some cases this can be done without affecting the charging process, see for instance [5].

The aim of this chapter is to explore how PEVs may be integrated into the distribution grid for peak shaving, load tracking and reactive power support purposes without imposing a significant impact on the existing infrastructure. In doing so, we use algorithms well-known to the communication community, namely, additive increase multiplicative decrease (AIMD) algorithms, and adapt them for the PEV charging problem. Preliminary work in this direction is reported in [28, 29]. In the present chapter we combine the previous work in a unified framework, and evaluate the performance of the proposed approach in a more realistic simulation setting.

7.2 PEV Charging Problem Description

We formulate the active load management task as the task of sharing a limited resource (here, the power) among several loads. Some of the loads may be controllable (e.g., PEVs, where we use the term 'agent' to describe local management of charge rate) and some uncontrollable (i.e., lights, televisions, and other appliances whose power consumption cannot be shifted to a later time without inconvenience to the user). We denote the aggregate power limit at time step k by $P(k)$. Note that this power is generally time-varying, as it depends itself on the power generated from renewable sources, and is subject to some physical constraints, such as limitations at power lines or distribution transformers. Further, note that we operate in discrete time with the time index denoted by k.

We denote by $p_i(k)$ the active power drawn by the ith controllable load. This power can be either positive or negative, as we assume that PEVs can both draw power from the grid and inject power into the grid. We say that when the PEV absorbs real power, it operates in grid to vehicle (G2V) mode, and when it injects power into the grid, it operates in the V2G mode. The admissible power drawn or injected into the grid is also subject to constraints such as a limit on the apparent power, i.e., $-\bar{s}_i \leq p_i(k) \leq \bar{s}_i$ for all k. This limit may arise, for example, due to inverter current limitations where we assume there is negligible variation in supply voltage. In addition, we denote by $\tilde{p}(k)$ the aggregated demand of the uncontrollable loads. Then, neglecting losses in the local distribution grid, the sum of the (controllable and uncontrollable) loads should be smaller than the available power, i.e.,

$$\sum_{i=1}^{N} p_i(k) + \tilde{p}(k) \leq P(k) \qquad \forall k. \tag{7.1}$$

In most cases, we treat Eq. (7.1) as a hard constraint, though in some cases, we allow minor transient excursions beyond this limit. This is allowed, for instance, if the power limit is due to thermal constraints at a power distribution transformer.

By inspecting Eq. (7.1) it is clear that in some cases the problem can be infeasible depending on the values of $P(k)$ and of the demand by uncontrollable loads $\tilde{p}(k)$. This happens, for instance, if the demand of the uncontrollable loads is greater than the available power plus the maximum power that may be injected by the PEVs into the power grid. In such cases, we will be interested in a "best-effort" solution, where the PEVs will be required to provide as much power as possible to mitigate the effects of the power mismatch. The exchange of active power between the grid and the PEVs can be used to implement the peak-shaving and load-tracking functionalities described previously.

As the most important service remains the charging of the vehicles, the first part of the PEV charging problem that we consider here is to govern the active power consumption of the vehicles such that the constraint in Eq. (7.1) is not violated, while maximizing the energy transferred to the vehicles. This can be expressed by

$$\begin{array}{lll} \max_{p_1(k),\dots,p_N(k)} & \sum_{i=1}^{N} p_i(k) & \\ \text{s.t.} & -\bar{s}_i \le p_i(k) \le \bar{s}_i & \text{for all } i, k \\ & \sum_{i=1}^{N} p_i(k) + \tilde{p}(k) \le P(k) & \text{for all } k, \end{array}$$

which represents the first part of the PEV charging problem.

This objective assumes that PEV owners permit the reduction of their charge rates in order to lessen the stress on the distribution grid. To encourage the owners of PEVs to participate in such a program energy distributors may give incentives to the owners in form of electricity price reductions.

As the primary objective of the PEVs is to use them as a mode of transportation, it is important that the batteries of the vehicles have enough stored energy to accommodate the needs of the owners. In the case that the PEV is a hybrid model which combines a combustion engine with an electric motor, the combustion engine can be used to compensate for the required energy, possibly given to the grid. In this case, it is important that the customers are compensated for the inconvenience or for the costs incurred after using the combustion engine (e.g., fuel costs). If the PEVs do not have a combustion engine, the missing energy could even prevent the owners from doing a trip, and this situation is clearly very undesirable. In some studies, it was shown that the average traveled distance per day usually lies around 40 km [9, 24]. The charging levels define a power rate of 2.4 kW as Level 1 [2] which corresponds to an approximate charging time of 200 min for 40 km. This implies that if we assume as in [9] that most vehicles charge at least while staying at home where they are connected for 7–8 h [9, 24] less than half of the time is required to charge the 40 km that are required for the next day. The range of fully electric vehicles lies between 90 and 395 km with battery sizes ranging from 9 to 53 kWh [2]. Hence, for most drivers it is not necessary to fully charge their PEV batteries every day.

Obviously, there are some exceptions, as when the PEVs are required to travel long distances. In such cases, we allow the owners of the PEVs to temporarily stop their participation to the energy exchange program and treat the PEVs as uncontrollable loads. They would then pay a higher tariff if charging occurs at peak times.

On the other hand, when the owners decide to participate in an energy exchange program with the grid, they automatically allow signals from the grid to influence their local power consumption. Also, note that it is reasonable to believe that the problems regarding the limited range of PEVs will likely lessen in the future as the technology progresses and charging points become available at additional places than just residential houses, such as at the parking lots at shopping centers, restaurants, working places, and many others. In [24], it is assumed that vehicles stay parked for an average duration of 3 h at work places, while parking at shopping facilities lasts 2 h on average. If charging facilities are available at such places, then the necessity to fully charge the battery at each of these locations is clearly reduced. This further means that we can safely assume that even with the limitations imposed by our algorithm, the PEV will have enough power to complete the next trip.

It remains to define how the power should be shared among the vehicles. As in principle there are many ways in which the power can be shared among the connected PEVs, in Sect. 7.3 we define different scenarios that give rise to different ways of sharing the available active power.

In addition, the PEVs can also exchange reactive power with the grid to balance the reactive power required locally. Reactive power management is particularly attractive when a large fleet of PEVs is connected to the grid in close proximity to industrial areas where a large amount of reactive power is required. We denote by $q_i(k)$ the reactive power drawn by the ith PEV at time step k. Accordingly, the upper bound on the active and reactive power injected or drawn from the grid becomes in practice

$$\sqrt{p_i(k)^2+q_i(k)^2} \quad \leq \bar{s}_i. \tag{7.2}$$

Ideally, the PEVs can be used to balance all the reactive power consumed in the area of interest, for instance by the uncontrollable loads, i.e.,

$$\sum_{i=1}^{N} q_i(k) + \tilde{q}(k) = 0 \qquad \forall k, \tag{7.3}$$

where $\tilde{q}(k)$ denotes the total reactive power consumed by the uncontrollable loads in the area of interest at time k. However, we will not consider Eq. (7.3) as a hard constraint. In fact, if it is not possible to achieve a full balance, generators and other devices nowadays used for reactive power balancing in the grid can be used to balance the residual reactive power, though at the price of transporting reactive power over a longer distance. Not being a critical task, we give a lower priority to reactive power management compared to active power management. However, if needed by the grid, it is possible to exchange the priorities of active and reactive power management.

The reactive power management is the second part of the PEV charging problem.

7.3 Charging Scenarios

We assume that PEVs can modulate the active and reactive power exchanged with the grid to accommodate their own charging needs, the energy needs of the other PEVs connected, and also the needs of the distribution grid itself. Hence, at each time step the PEVs are supposed to adjust their active and reactive power consumption such that ideally the constraints in Eqs. (7.1)–(7.3) are fulfilled.

In the future with a higher penetration of PEVs, it will be possible to recharge the vehicles at a variety of different locations, such as at homes, work places, fast charging stations, shopping centers, hospitals, and airports. Accordingly, the needs and desires of PEV owners and the providers of the charging service, for example at a shopping center, should be considered and the PEV charging problem should reflect the different scenarios of interest. In this section we will illustrate three specific scenarios, which give rise to three different concepts of "fairness" among the participating vehicles. These scenarios are:

- Power Fairness (PF): In this scenario each connected PEV will receive (or provide) exactly the same share of the available power, hence the power consumption (or injection) is equalized. This is fair in the sense that each PEV receives exactly the same quantity of power;
- Energy Fairness (EF): In this scenario the power consumption is proportional to the expected needs of the users, while the power injection is inversely proportional (e.g., smaller power is given to the PEVs that will not need to travel anytime soon). Hence, the scenario is fair in regard to the energy needed;
- Time Fairness (TF): In this scenario the PEVs are categorized depending on the time they have already been connected. Thus PEVs that have been connected for a long time consume (or inject) a smaller amount of power compared to recently connected PEVs. Hence, this scenario is fair in a way that short connection times are not penalized.

In the remainder of this section, we illustrate each scenario in greater detail.

7.3.1 *Power Fairness (PF) Scenario*

The most obvious and simple way to share the available active power among a fleet of PEVs is to give exactly the same amount of power to each PEV. Such a solution can, for instance, be adopted in a domestic charging scenario, where it might be too complicated, and also unfair, to give higher charging rates to some particular vehicles. Note that the power shared among the connected vehicles will change during the day; either due to the presence of a high power demand from uncontrollable loads in the same area, or as a consequence of a reduction of energy produced from renewable energy generation (e.g., using the energy produced by solar panels on top of buildings).

In this scenario, all PEVs should on average consume, or inject, the same amount of power. We mathematically model the complete PEV charging problem as a prioritized optimization problem. This means that we have an objective with very high priority, denoted $O_1(k)$, which is solved first. This objective represents rapid charging while maintaining the constraint on the active power demand by the grid. If the solution of this objective allows additional degrees of freedom, an objective with lower priority, denoted $O_2(k)$, is solved. This objective represents the power fairness condition imposed in this scenario. If further flexibility is available (for example if the total power available means that many chargers are not operating at their individual complex power limits), the third objective, which has even lower priority, denoted $O_3(k)$, is then solved to balance the reactive power. These prioritized optimizations can be represented by

$$
\begin{aligned}
O_1(k) = &\max_{p_1(k),\dots,p_N(k)} \sum_{i=1}^{N} p_i(k) \\
\text{s.t.} \quad & -\bar{s}_i \le p_i(k) \le \bar{s}_i \quad \text{for all } i,k \\
& \sum_{i=1}^{N} p_i(k) + \tilde{p}(k) \le \mathrm{P} \quad \text{for all } k \\
O_2(k) = &\min_{p_1(k),\dots,p_N(k)} \|B\boldsymbol{p}(\boldsymbol{k})\|_1 \\
\text{s.t.} \quad & -\bar{s}_i \le p_i(k) \le \bar{s}_i \quad \text{for all } i,k \\
& \sum_{i=1}^{N} p_i(k) = O_1(k) \quad \text{for all } k \\
O_3(k) = &\min_{q_1(k),\dots,q_N(k)} \left| \sum_{i=1}^{N} q_i(k) + \tilde{q}(k) \right| \\
\text{s.t.} \quad & \sqrt{p_i(k)^2 + q_i(k)^2} \le \bar{s}_i \quad \text{for all } i,k \\
& \sum_{i=1}^{N} p_i(k) = O_1(k) \quad \text{for all } k, \\
& \|B\,\boldsymbol{p}(\boldsymbol{k})\|_1 = O_2(k) \quad \text{for all } k,
\end{aligned}
\tag{7.4}
$$

where $\boldsymbol{p}(\boldsymbol{k})$ is the vector containing the power consumption of each PEV and B is a matrix containing 1, −1, and 0 such that $B\boldsymbol{p}(\boldsymbol{k})$ contains the difference between the power consumptions. For example

$$B = \begin{bmatrix} 1 & -1 & 0 & \cdots & 0 \\ 0 & \ddots & \ddots & \ddots & \vdots \\ \vdots & \ddots & \ddots & \ddots & 0 \\ 0 & \cdots & 0 & 1 & -1 \\ 1 & 0 & -1 & \cdots & 0 \\ \ddots & \ddots & \ddots & \ddots & \vdots \\ 1 & 0 & \cdots & 0 & -1 \end{bmatrix}$$

is such a matrix.

Note that in the above mathematical formulation, we implicitly assumed that all PEVs participating are also able to participate in the V2G program. In some cases this might not be true and only a subset of the participating PEVs would allow reverse power flows. However, Eq. (7.4) can easily be modified to include such situations by making the lower bound on the real power absorbed zero.

The previous optimization problem explicitly requires that the charge rates of two vehicles must be exactly the same at every time step. However, our interpretation of "fairness" in this scenario can also be relaxed if we require only the running average of the power consumption by each PEV to be equal, which in the following will be denoted by ρ_i. The average can either be computed from the beginning of the charging procedure by

$$\rho_i(k) = \frac{1}{k}\sum_{l=0}^{k} p_i(l)$$

or over the past τ time steps (e.g., the last few minutes of charging) by

$$\rho_i(k) = \frac{1}{\tau}\sum_{l=k-\tau}^{k} p_i(l).$$

Accordingly, the second optimization objective in Eq. (7.4) can be reformulated to

$$\begin{aligned} O_2(k) = & \min_{p_1(k),\ldots,p_N(k)} \|B\,\boldsymbol{\rho}(\boldsymbol{k})\|_1 \\ \text{s.t.} \quad & -\bar{s}_i \le p_i(k) \le \bar{s}_i \qquad \text{for all } i,k \\ & \sum_{i=1}^{N} p_i(k) = O_1(k) \qquad \text{for all } k, \end{aligned} \tag{7.5}$$

where $\boldsymbol{\rho}(\boldsymbol{k})$ is the vector containing the average power consumption of each vehicle and B is the same matrix as above.

Note that there is no objective in Eq. (7.5) that states how the reactive power should be provided by the vehicles if there are multiple possibilities. However, extensions of this work could be pursued in the future to introduce a type of "fairness" for the reactive power compensation as well as for the real power. For example if the PEV owners get paid according to the amount of reactive power they compensate, it is important to impose a "fairness" on the reactive power compensation. This is simply possible by adding a forth objective $O_4(k)$ with even lower priority that represents a "fairness" notion for the reactive power compensation.

7.3.2 Energy Fairness (EF) Scenario

The PF scenario of Sect. 7.3.1 shares the instantaneous power among the participating PEV in the same way, independently from the actual power requirements of the PEVs connected. As a consequence, some PEVs might be fully charged long before they are actually needed, while other PEVs might not be fully recharged by the time their owners require the vehicle for transportation. To include different energy needs by the owners in this scenario we design charging strategies that prioritize the PEVs according to the time they are connected to the grid for charging their batteries, and their energy requirements. Note that such a solution cannot be implemented in a competitive scenario where all the PEV owners are only interested in their own needs. Hence, it is more realistic to implement it in a scenario like a work place, where employees are not in competition with one another.

In this scenario, let us assume that at time step k a PEV requires a certain amount of energy $E_i(k)$ to fully charge its battery to a desired level. Note that the required energy is non-negative. In addition, let $T_i(k)$ denote the remaining time the PEV is expected to remain connected to the grid at time step k, before it is used again. Then, the objective in the EF scenario is to give an amount of energy $E_i(k)$ to the ith PEV within time $T_i(k)$. This corresponds to a desired average charging rate $\hat{p}_i(k)$ and is computed as

$$\hat{p}_i(k) = \min\left(\frac{E_i(k)}{T_i(k)}, \bar{s}_i\right), \tag{7.6}$$

namely, that rate to allow the PEV to finish the charging procedure in the desired time. Note that we explicitly bound the desired charging rate by the maximum power consumption that is physically allowed by the electrical power outlet and the charger. This upper bound makes it impossible to obtain an unrealistically large amount of energy in a small time. Note that the desired charge rate $\hat{p}_i(k)$ has 0 as natural lower bound, since the required energy is larger than or equal to 0. In that case, it might still make sense to connect the vehicle to the grid to perform V2G services, or to exchange reactive power.

In this scenario, we prioritize the vehicles according to their desired charge rates, i.e., vehicles with a high desired charge rate $\hat{p}_i(k)$ actually receive more power than the ones with a lower desired charge rate. The PEV charging problem related to this scenario can be formulated as in the previous scenario where we order in total three objectives depending on their priorities. This leads to

$$
\begin{aligned}
O_1(k) = & \max_{p_1(k),\ldots,p_N(k)} \sum_{i=1}^{N} p_i(k) \\
\text{s.t.} \quad & -\bar{s}_i \leq p_i(k) \leq \bar{s}_i \quad \text{for all } i,k \\
& \sum_{i=1}^{N} p_i(k) + \tilde{p}(k) \leq \mathrm{P} \quad \text{for all } k \\
O_2(k) = & \min_{p_1(k),\ldots,p_N(k)} \|B\zeta(\boldsymbol{k})\|_1 \\
\text{s.t.} \quad & -\bar{s}_i \leq p_i(k) \leq \bar{s}_i \quad \text{for all } i,k \\
& \sum_{i=1}^{N} p_i(k) = O_1(k) \quad \text{for all } k \\
O_3(k) = & \min_{q_1(k),\ldots,q_N(k)} \left| \sum_{i=1}^{N} q_i(k) + \tilde{q}(k) \right| \\
\text{s.t.} \quad & \sqrt{p_i(k)^2 + q_i(k)^2} \leq \bar{s}_i \quad \text{for all } i,k \\
& \sum_{i=1}^{N} p_i(k) = O_1(k) \quad \text{for all } k, \\
& \|B\zeta(\boldsymbol{k})\|_1 = O_2(k) \quad \text{for all } k,
\end{aligned}
\tag{7.7}
$$

where B is a matrix as in Sect. 7.3.1 and $\zeta(\boldsymbol{k})$ is a vector with the jth element

$$
\zeta_j(k) = \frac{p_j(k)}{\tilde{p}_j(k_0)}
$$

whenever the vehicles consume power from the grid and

$$
\zeta_j(k) = p_j(k)\tilde{p}_j(k_0),
$$

whenever the PEVs inject power into the grid.

In the above notation, k_0 corresponds to the initial (or intermediate) time step at which the desired charge rate is computed, according to Eq. (7.6).

As for the PF scenario, the second objective $O_2(k)$ can be relaxed by using the running average $\rho_i(k)$ instead, as in Eq. (7.5). The running average can again be taken over the whole connection period, or over a smaller time window τ.

Another similarity with the PF scenario is that we are not interested in how much reactive power each PEV consumes (or injects) as long as the aggregated reactive power consumption (injection) compensates the reactive power by uncontrollable loads in the region. As mentioned before, additional, lower priority reactive power objectives can easily be incorporated by adding an additional objective $O_4(k)$.

7.3.3 Time Fairness (TF) Scenario

The two scenarios presented in Sects. 7.3.1 and 7.3.2 accommodate situations where the PEVs are connected for long periods of time. However, during the day a lot of situations arise where the vehicle is parked for short periods of time, for example in shopping centers, restaurants, cafes, parking lots in a city center, etc. While it is currently unlikely to find charging facilities at such locations, the increasing amounts of PEVs on the roads can increase the desire for them. In addition, local authorities or shopping mall owners may wish to provide incentives for PEV owners to visit there and therefore provide charging infrastructure.

In such a framework, one possible way to charge the PEVs of the customers to encourage short connections and avoid excessively long stays (e.g., to encourage people to leave as soon as they have finished shopping, so they make their parking spot available for new customers). In this case, we suggest allowing a higher power consumption (or injection) to PEVs that are connected more recently than the PEVs that have been connected for a longer period.

First, we categorize the vehicles into groups with different priorities depending on the length of their connection. Accordingly, upon connection a PEV automatically joins the group with highest priority. Then after a predefined period has elapsed, the PEV is moved into a group with lower priority. The PEV is then repeatedly shifted to a group with lower priority, until, after another predefined period of time, it ends up in the group with lowest priority. We assume that there are L groups, where group 1 has priority 1 (the lowest) and group L has priority L (the highest). Also, we assume that the time period before a PEV is moved into another group is constant and the same for all PEVs. In the following, we denote such a time period by κ and it is measured in time steps. Further, let T_i be the time step at which the ith PEV connects to the power grid. Then, the priority of the ith PEV at time step k, denoted $\Upsilon_i(k)$, can be computed as

$$\Upsilon_i(k) = \min\left(L - 1, \left\lfloor \frac{k - \mathrm{T_i}}{\kappa} \right\rfloor\right), \tag{7.8}$$

where $\lfloor x \rfloor$ denotes the integer part of x.

Now, the power should be shared according to user priority. This can again be formulated as a prioritized optimization problem similar to the ones before,

$$
\begin{aligned}
O_1(k) = &\max_{p_1(k),\dots,p_N(k)} \sum_{i=1}^{N} p_i(k) \\
\text{s.t.} \quad & -\bar{s}_i \leq p_i(k) \leq \bar{s}_i \quad \text{for all } i, k \\
& \sum_{i=1}^{N} p_i(k) + \tilde{p}(k) \leq P \quad \text{for all } k \\
O_2(k) = &\min_{p_1(k),\dots,p_N(k)} \|B\,\Xi\,(\boldsymbol{k})\|_1 \\
\text{s.t.} \quad & -\bar{s}_i \leq p_i(k) \leq \bar{s}_i \quad \text{for all } i, k \\
& \sum_{i=1}^{N} p_i(k) = O_1(k) \quad \text{for all } k \\
O_3(k) = &\min_{q_1(k),\dots,q_N(k)} \left| \sum_{i=1}^{N} q_i(k) + \tilde{q}(k) \right| \\
\text{s.t.} \quad & \sqrt{p_i(k)^2 + q_i(k)^2} \leq \bar{s}_i \quad \text{for all } i, k \\
& \sum_{i=1}^{N} p_i(k) = O_1(k) \quad \text{for all } k \\
& \|B\,\Xi\,(\boldsymbol{k})\|_1 = O_2(k) \quad \text{for all } k,
\end{aligned}
\tag{7.9}
$$

where B is the same as in the previous sections and $\Xi(\boldsymbol{k})$ is the vector with ith element

$$\Xi_i(k) = \frac{p_i(k)}{\Upsilon_i(k)}.$$

Analogously to the previous two scenarios it is possible to relax objective $O_2(k)$ by using the running average of the power consumption $\rho_i(k)$.

Finally, in the current formulation we are not interested in the reactive power consumed or injected by each vehicle, but only in the aggregated reactive power.

7.4 The Additive Increase Multiplicative Decrease Algorithm (AIMD)

To solve the PEV charging problem for the different scenarios defined in Sect. 7.3, we propose a distributed algorithm. Distributed algorithms are attractive for a number of reasons. Firstly, such solutions are known to be robust against possible failures. Secondly, the requirements for distributed algorithms usually place a smaller burden than centralized algorithms on the communication infrastructure. Finally, distributed solutions sometimes lead to "plug-and-play" type functionalities which could be convenient in the future where a large but unknown number of PEVs connect to the same power distribution grid and compete for power. Note that after the preliminary work [29], other papers have been published to solve the charging problem in a distributed fashion, see for instance [3, 11, 32].

The PEV charging problem formulated in Sect. 7.2 is a typical resource-sharing problem, where several agents compete to acquire their share of the resource (in this case, power). This is similar to what occurs in the Internet, where the connected devices compete with each other to obtain as much bandwidth as possible. The similarities between the power distribution network and the communication network have already been observed by a number of authors [17, 19]. Algorithms developed for the transmission control protocol (TCP), namely additive increase multiplicative decrease (AIMD) type algorithms, have recently been used in power networks [8], and in the PEV charging problem [3, 19, 28, 29]. AIMD based algorithms are known to be flexible and reliable, require a small amount of communication between a central management unit and agents, such as PEVs, and have been extensively investigated and tested in the past 20 years [6, 15, 16, 25–27].

The AIMD algorithm is a distributed algorithm that relies on a central management unit to broadcast a binary control signal. The PEVs autonomously react to this control signal by changing their power consumption and injection in a stochastic manner. As the PEVs, or the charger outlet they are connected to, are themselves in command of their reaction, it is possible to accommodate for the individual needs of the PEVs. Thus, AIMD-like algorithms are perfectly suitable for distributed resource allocation problems found in smart grid applications.

As mentioned in Sect. 7.2, the PEV charging problem involves management of active and a reactive power. Normally, the PEVs will draw power (active and/or reactive) from the grid. However, in some cases it might be desired to reverse one or both of the power flows, and make the vehicles inject power into the distribution grid. This situation occurs if, for instance, the grid at a given moment does not have enough power to supply the uncontrollable loads.

The AIMD algorithm can be extended to allow management of active and reactive power exchange and both G2V and V2G power flows. We call such an algorithm double (prioritized) AIMD (in the following, DAIMD). The DAIMD algorithm comprises an active power AIMD algorithm, which manages the active power consumption, and a reactive power AIMD algorithm, which governs the reactive power consumption. Each of these AIMD sub algorithms is able to operate in two modes: the G2V mode, in which the PEV draws power from the grid, and the V2G mode, in which the PEV injects power into the grid.

In the following, we first present the active power AIMD algorithm as it operates in G2V mode, which is the most basic form of the algorithm. Afterwards, we show how we can extend this algorithm, so that the PEVs can also operate in V2G mode, and how the PEVs can autonomously determine in which mode they should operate. We then illustrate how the reactive power AIMD algorithm works and how it can be implemented to obtain the DAIMD algorithm. In Sects. 7.4.1, 7.4.2 and 7.4.3 we illustrate three ways to tune the DAIMD to accommodate for the three charging scenarios presented in Sect. 7.3.

While the PEVs operate in G2V mode, the active power AIMD algorithm controls the active power consumption of each PEV by switching between two distinct phases. The first phase is the additive increase (AI) phase, where the PEVs gently increase their charging rate, according to equation

$$p_i(k+1) = p_i(k) + \alpha_i(k)\bar{\alpha}. \tag{7.10}$$

The additive increase is scaled by a fixed scalar $\bar{\alpha}$, which is identical for all PEVs. This allows for some control over the increase from a central management unit, if necessary, where occasional broadcast of $\bar{\alpha}$ may be desirable. Further, the charging rate cannot exceed some value $\bar{s}_i$ (given by the physical constraints of the individual charging infrastructure).

The second phase is called the multiplicative decrease (MD) phase which occurs when Eq. (7.1) is violated, i.e., when the sum of the consumed power by all connected PEVs and the demand by uncontrollable loads exceeds the maximum amount of power allowed by the power grid. We will refer to such an event as a capacity event (CE). When a CE occurs, the central management unit notifies all the connected PEVs of such an event by broadcasting a binary feedback signal. In response, the PEVs decrease their charge rate by a multiplicative factor $\beta_i^{(1)}(k)p_i(k)$ with probability $\lambda_i(k)$, or by another multiplicative factor $\beta_i^{(2)}(k)p_i(k)$ with residual probability $1-\lambda_i(k)$. In addition, when a PEV decreases its power consumption from an already small value, we force the decrease to be greater than a fixed threshold ψ. In this way, it is possible to handle the situations where the power is near zero e.g., when transitioning between V2G and G2V modes.

If the PEVs operate in the V2G mode, the AIMD algorithm described above is inverted. This means that upon receiving a CE the PEVs increase their power injection additively, which corresponds to an actual decrease of the power consumption. Similarly, when no CE is received the PEVs decrease their power injection multiplicatively, which corresponds to an increase in power consumption.

The PEVs can automatically recognize at which point they need to change the operating mode (i.e., from V2G to G2V or vice versa) in the following way. The switch from G2V to V2G mode occurs after a CE if the actual power consumption is very small. Let $m_i(k)$ indicate whether the ith PEV operates in G2V mode at time step k, i.e. $m_i(k) = 1$ if at time step k vehicle i is in G2V mode and $m_i(k) = 0$ if at time step k vehicle i is in V2G mode. Then, the indicator is updated after a CE by

$$m_i(k+1) = \begin{cases} 1, & \text{if } p_i(k+1) > \varepsilon \text{ and } m_i(k) = 1, \\ 0, & \text{if } p_i(k+1) \leq \varepsilon \text{ and } m_i(k) = 1, \end{cases}$$

where ε is a positive scalar parameter. The return from V2G mode to G2V mode occurs when no CE is received, and the indicator changes as

$$m_i(k+1) = \begin{cases} 1 & \text{if } \ p_i(k+1) > -\varepsilon \text{ and } m_i(k) = 0 \\ 0 & \text{if } \ p_i(k+1) \leq -\varepsilon \text{ and } m_i(k) = 0 \end{cases}.$$

Figure 7.1 illustrates the active power AIMD algorithm executed by the vehicles for both G2V and V2G mode operation.

Different values of the parameters $\alpha_i(k)$ in the AI phase, and $\lambda_i(k)$, $\beta_i^{(1)}(k)$ and $\beta_i^{(2)}(k)$ in the MD phase give rise to different solutions, and this flexibility will be used in the Sects. 7.4.1, 7.4.2 and 7.4.3 to handle the different scenarios presented in Sect. 7.3.

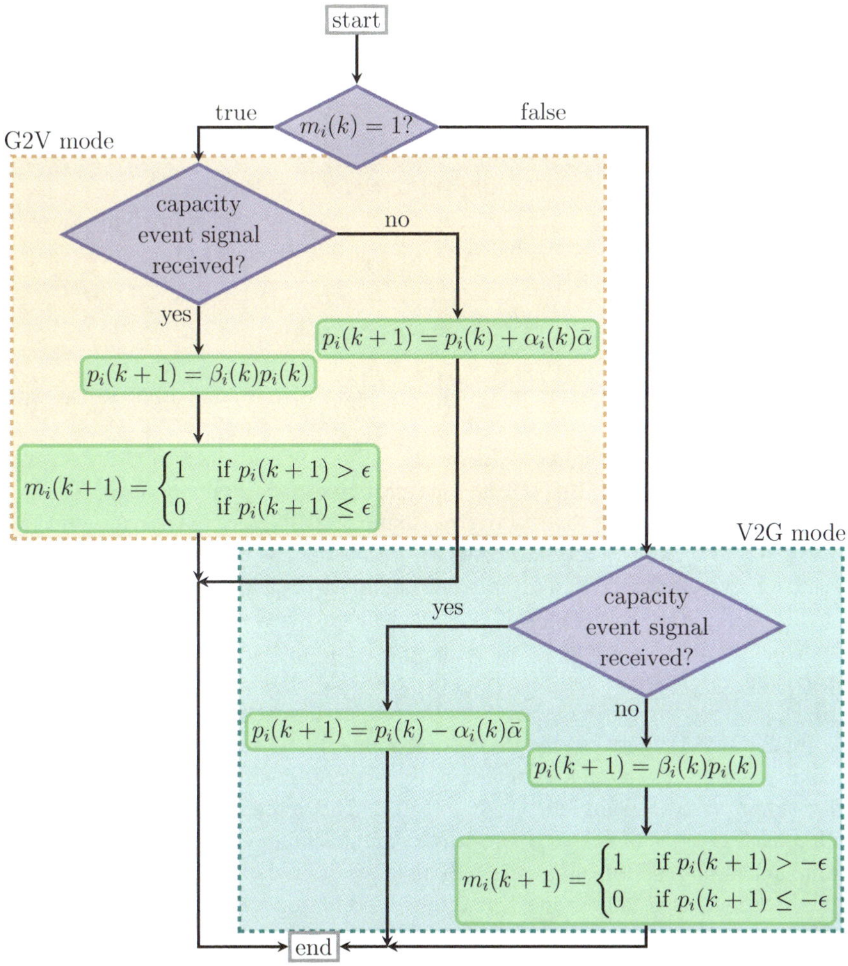

Fig. 7.1 Illustrative diagram of the active power AIMD for PEVs

To handle both the active and the reactive power exchange with the grid, the above active power AIMD algorithm is embedded in a DAIMD, which also includes a reactive power AIMD algorithm. Figure 7.2 shows a flow chart implementing such a DAIMD. In a first step the active power AIMD algorithm is executed as previously explained, and as illustrated in Fig. 7.1. Afterwards, a second AIMD algorithm, the reactive power AIMD algorithm, is executed, where the PEVs aim at computing the value of the reactive power $q_i(k)$ to be exchanged with the grid.

The reactive power AIMD algorithm depends on reactive CEs. Such events occur whenever the reactive power at a defined measuring point, for example placed at a transformer, is larger than 0. This indicates that all reactive power in the area has been compensated and additional consumption of reactive power would lead to over-compensation. Similarly to the active power AIMD, the PEVs are able to draw or inject reactive power depending on the requirements of the power grid. The reactive power consumption (or injection) additively increases if no reactive CE occurs, and multiplicatively decreases otherwise. Figure 7.3 illustrates this algorithm in detail. To distinguish between the parameters used in the reactive power AIMD from the ones used in the active power AIMD, we denote the additive parameter $a_i(k)$, its additive scaling factor $\bar{a}$, the two multiplicative factors $b_i^{(1)}(k)$ and $b_i^{(2)}(k)$, the associated probability $\gamma_i(k)$, and the indicator $z_i(k)$.

Naturally, at all times, the charger outlet gives a maximum bound on the apparent power that can be exchanged between the vehicles and the grid

$$\sqrt{p_i(k)^2 + q_i(k)^2} \leq \bar{s}_i.$$

Regarding this bound it is important to note that we first bound the active charging rate $p_i(k+1)$, and then we bound the reactive power consumption $q_i(k+1)$, see Fig. 7.2. Thus, we give a higher priority to the active power exchange rather than to the reactive power exchange. This is deliberate and based on the assumption that charging PEVs is more important than satisfying some ancillary services for the grid (i.e., exchanging reactive power). If necessary the priorities can easily be reversed, giving reactive power exchange first priority and active power a lower priority.

One of the main advantages of AIMD-like algorithms is that they can be easily implemented in a distributed way with small communication constraints. In particular, the basic active and reactive power AIMD algorithms as described above only require a central management unit to broadcast binary CE signals. In this case, no communication between the PEVs, or from the PEVs to the central management unit is required. Other studies that deploy a central controller require more communication, see for examples [10, 21, 23, 33]. This requires higher investments to equip each charger outlet with two-way communication capabilities and stringent communication requirements, especially in larger scale deployments, to mitigate the effects of increased delays and signal loss. In addition, two-way communications may face user resistance from PEV owners who might not be willing to share all the required data with the central controller.

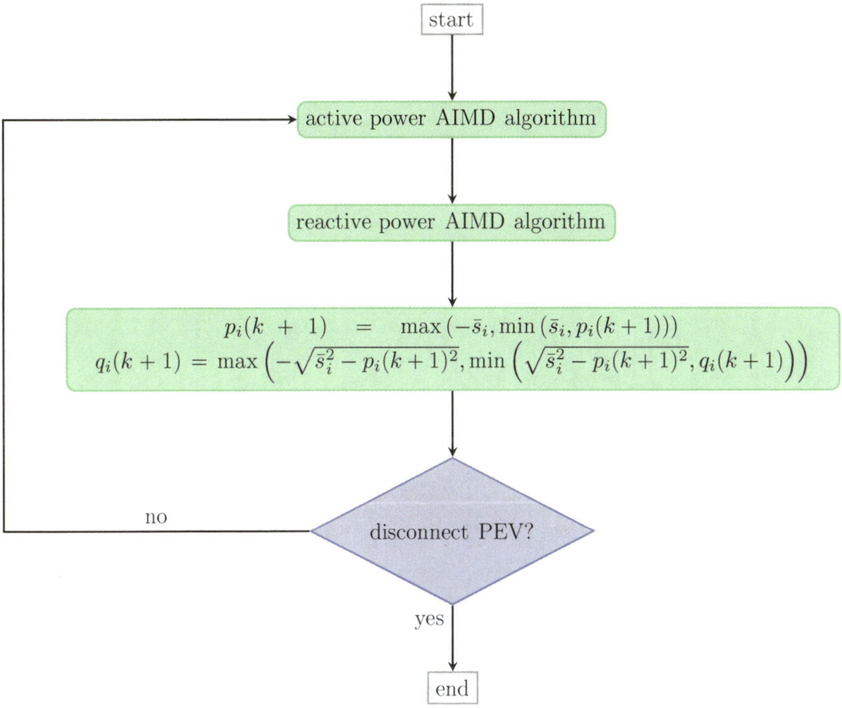

Fig. 7.2 Illustrative diagram of the DAIMD, where the active power AIMD is illustrated in detail in Fig. 7.1 and the reactive power AIMD in Fig. 7.3

7.4.1 Algorithm for the Power Fairness Scenario

In [26] it is shown that an equal share of the available active power can be achieved by setting the AIMD parameters identical for all participating vehicles. This means that $\alpha_i(k) = \alpha$, $\beta_i^{(1)}(k) = \beta^{(1)}$, $\beta_i^{(2)}(k) = \beta^{(2)}$, and $\lambda_i(k) = \lambda$ for all i and k. This can be done if the infrastructure informs the PEVs of the values of the parameters before beginning the charging procedure, which would require additional communication. Another possibility is to have static parameters that are coded in the charger, or dynamic ones that are broadcasted to the vehicles during CEs (in this way, different parameters can be used in different situations).

Since we are not primarily interested in reactive power exchange, the parameters of the reactive power AIMD can be selected with more freedom. Hence, we choose the reactive power AIMD parameters to be equal for all connected PEVs, i.e. $a_i(k) = a$, $b_\mathrm{i}^{(1)}(k) = b^{(1)}(k)$, $b_\mathrm{i}^{(2)}(k) = b^{(2)}(k)$, and $\gamma_i(k) = \gamma(k)$. In this way, the PEVs draw (or inject) equal amounts of reactive power.

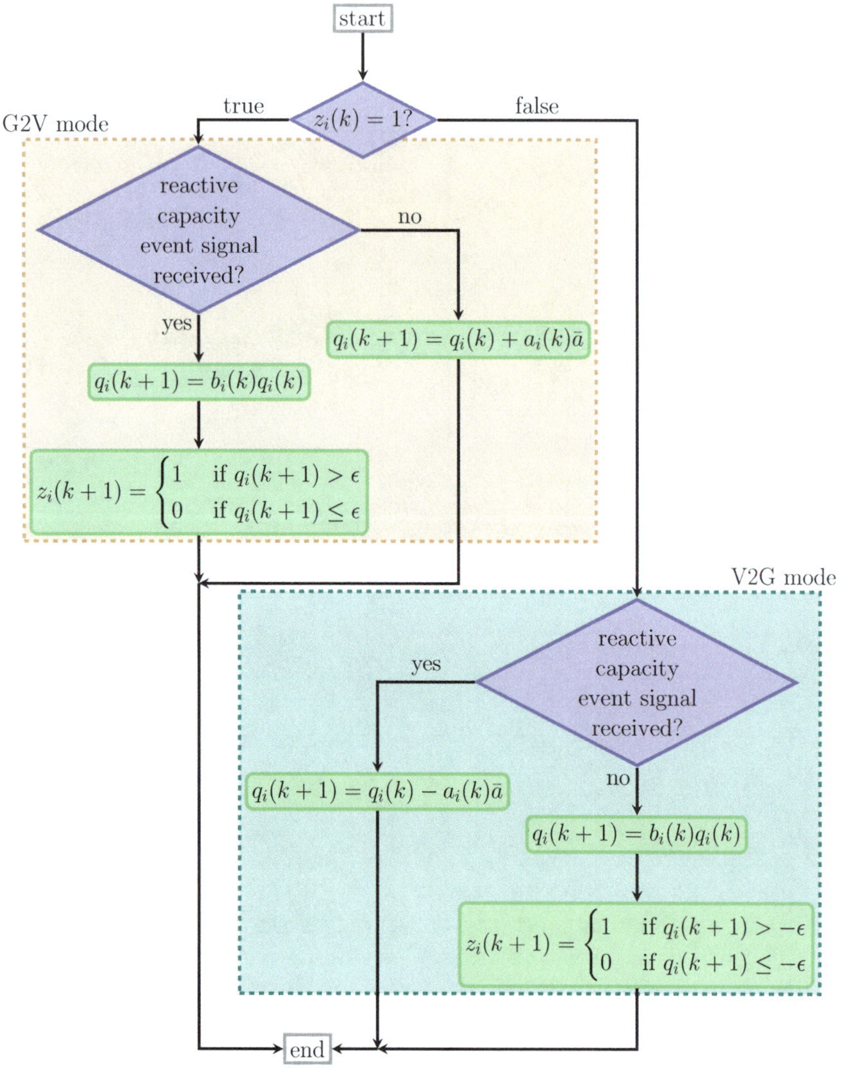

Fig. 7.3 Illustrative diagram of the reactive power AIMD for PEVs

7.4.2 Algorithm for the Energy Fairness Scenario

In some cases, one is interested in sharing the power directly (inversely) proportionally to the desired charge rate in the G2V (V2G) mode. This objective can be achieved by appropriately changing the parameters of the AIMD algorithm. In the G2V mode, the parameters have to be changed such that

$$\frac{\alpha_i(\mathrm{k})}{\beta_i^{(1)}(k)\lambda_i(k) + \beta_i^{(2)}(k)(1 - \lambda_i(k))}$$

is proportional to the desired charge rates $\tilde{p}_i(k)$.

In this regard, it does not matter which of the AIMD parameters are adapted to obtain such a result. However, such a choice affects the behavior of the algorithm. In fact, adapting the additive parameter $\alpha_i(k)$ influences the ability of the demand to increase, while adapting the multiplicative factors $\beta_i^{(1)}(k)$ and $\beta_i^{(2)}(k)$, or the probability $\lambda_i(k)$, influences the ability to decrease the demand. In this section, we only adapt the additive parameter $\alpha_i(k)$ to achieve objective $O_2(k)$ in this scenario, while all the other parameters are chosen identical for all the connected PEVs. Therefore, the additive parameter is adjusted as

$$\alpha_i(k) = \frac{\hat{p}_i(k)}{\bar{s}_i} \tag{7.11}$$

in the G2V mode, and as

$$\alpha_i(k) = \frac{\bar{s}_i}{\hat{p}_i(k)} \tag{7.12}$$

in the V2G mode.

This scenario requires that the PEVs are informed of the value of the other parameters $\bar{\alpha}$, $\beta^{(1)}$, $\beta^{(2)}$, and λ. As for the PF scenario, this information can be transmitted along with the CEs or be coded in the charger to avoid additional communication requirements.

As in the PF scenario, we use identical parameters for all PEVs for the reactive power AIMD. Hence, the reactive power drawn or injected by the PEVs should be equal.

7.4.3 Algorithm for the Time Fairness Scenario

For the TF scenario the power has to be proportional to the priority $\Upsilon_i(k)$ assigned to the PEVs. Hence, as in the EF scenario, we adapt the parameters such that $\Upsilon_i(k)$ is proportional to

$$\frac{\alpha_i(k)}{\beta_{\mathrm{i}}^{(1)}(k)\lambda_i(k) + \beta_{\mathrm{i}}^{(2)}(k)(1 - \lambda_i(k))}.$$

Here, we adapt only the additive parameter $\alpha_i(k)$, while the remaining parameters of the active power AIMD are kept identical for all connected PEVs. The parameter is updated at each time step according to

$$\alpha_i(k) = \frac{\Upsilon_i(k)}{L}. \tag{7.13}$$

Note that we scale the priority $\Upsilon_i(\mathrm{k})$ with L, i.e., the number of available groups, such that the additive parameter remains in the interval $[0, 1]$. While it is not essential that the additive parameter lies in that interval, this is a useful property, since then $\bar{\alpha}$ is the upper limit for the increase per time step and PEV.

7.5 Simulations

In this section we illustrate the behavior of our algorithms in a simulated scenario, using a customized OpenDSS-Matlab simulation platform. In particular, Matlab was used to compute the power consumption according to the different algorithms, while OpenDSS, a power simulation tool developed by [13], was used to simulate the power grid.

We tested our algorithms on a revised version of the power distribution system based on the IEEE37 bus test feeder found among the OpenDSS examples [12]. This is depicted in Fig. 7.4. Note that Fig. 7.4 only shows the interconnections between the loads, buses, and the transformer, and is not meant to depict the real dimensions of the distribution grid. We also assumed that the actual power was measured at the transformer that connects the loads with the external grid, and we assumed a power limit at this transformer of 180 kW for the active power. The transformer is depicted in Fig. 7.4 as the square block with label "SubXF". Additionally, the algorithm controls the reactive power flow to compensate it completely at the transformer, i.e., the reactive power at the transformer should be equal to 0 VAR.

Overall there are 25 uncontrollable three-phase loads, indicated by inverted triangles in Fig. 7.4, connected to different buses. Such loads follow a pre-specified load pattern over a day. For illustrative purposes, we made the assumption that the peak load of the uncontrollable loads would overlap the connection time of the PEVs. While this assumption is clearly not always true (PEVs could be recharged at night time when the load curve is lower), it still allows us to investigate a worst-case scenario. Also, note that there are some studies that predict that domestic charging is likely to partly occur during the evening load peak, directly after work, and this last scenario is consistent with this assumption. For example in [7] three charging periods are identified: during daytime, during the night, and during the evening. Similarly, [14] assumes that without control the charging starts around 6 p.m. and identifies such a scenario as a worst-case, which is consistent with our simulation.

In our simulation, up to 20 PEVs can connect at the locations specified in Fig. 7.4 with ellipses. We assume that the PEVs are connected with uniform probability between hours 7 and 10 of the daily simulation. The connection is single phase, where each charger has a maximal apparent power capacity of 3.7 kVA.

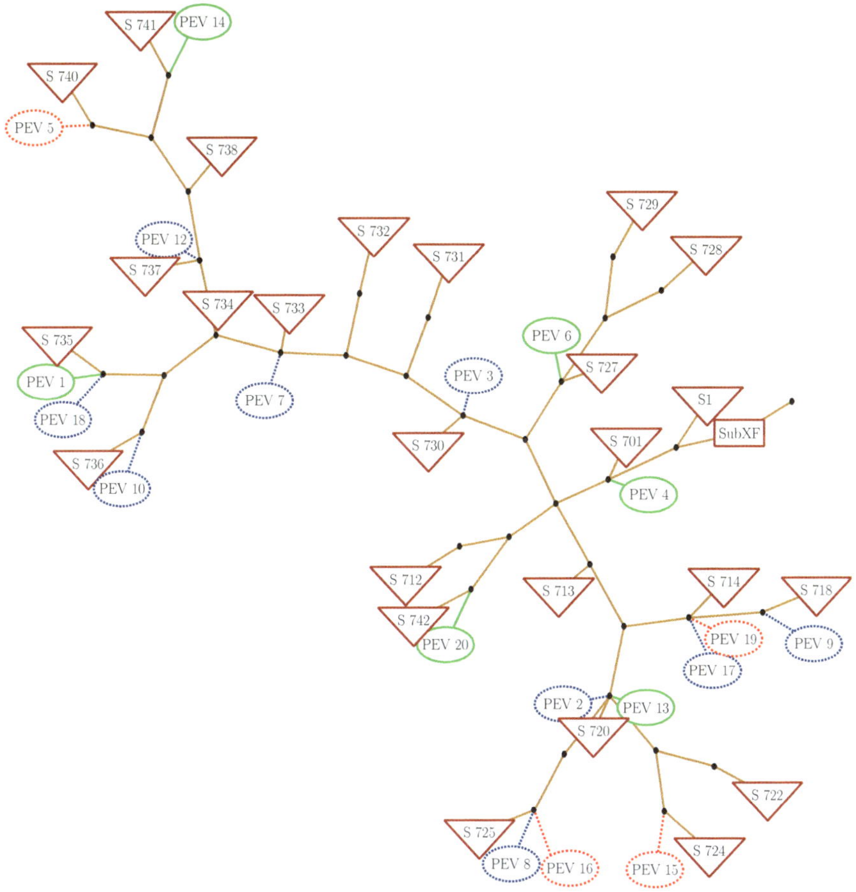

Fig. 7.4 Connection graph of the IEEE37 test feeder including PEVs. *Inverted triangles* symbolize uncontrollable three-phase loads. The *square block* represents the transformer that connects this part of the distribution grid to the external grid. The *ellipses* represent connected PEVs and the line style (*solid*, *dashed*, *dotted*) indicates to which phase the PEV is connected to

In Fig. 7.4 the different line styles (solid, dashed, dotted) of the ellipses indicate the different phases the PEVs are connected to. The required energy is assigned using a uniform distribution between 15 and 20 kWh. We also assume that the PEV is automatically disconnected when it is charged to the desired level. Note that this means that fully charged vehicles do not participate in reactive power balancing. Also, we assume that PEVs can be disconnected after a predefined time, independently from their charging state. Further, we only simulate one scenario per time. This means that all PEVs are in the same situation and therefore deploy the same algorithm, corresponding to the simulated scenario. In the simulation, the PEVs react synchronously to the CE, and possible communication delays have not been taken into account.

The results obtained using the proposed DAIMD algorithm to control the active and reactive power consumption are compared with:

(i) the case where there are no PEVs connected at all, to evaluate the possibly different utilizations of the uncontrollable loads; and with
(ii) the case where PEVs are charged with the maximum charge rate until they are fully charged, i.e., $p_i(t) = \bar{s}_i$ for all i.

7.5.1 Simulation of the Power Fairness Scenario

In this section the PEVs connected to the distribution grid, shown in Fig. 7.4, use the PF scenario algorithm described in Sect. 7.4.1. This means that they should be recharged with the same average charge rate, without exceeding the maximum active power allowed by the transformer. The active power AIMD parameters are identical for all the PEVs: $\alpha = 1$, $\bar{\alpha} = 0.1\,\frac{\text{kW}}{\text{s}}$, $\beta^{(1)} = 0.75$, $\beta^{(2)} = 0.99$, $\lambda = 0.7$, and $\psi = 0.15$. The values for the reactive power AIMD are also identical for all PEVs and identical to the ones used for the active power AIMD, i.e. $a = 1$, $\bar{a} = 0.1\,\frac{\text{kVA}}{\text{s}}$, $b^{(1)} = 0.75$, $b^{(2)} = 0.99$, $\gamma = 0.7$, and $\psi = 0.15$.

Figure 7.5 shows the active and reactive power consumption at the transformer and Fig. 7.6 shows the active power consumption of four randomly selected PEVs. The results are filtered using a moving average filter with a window length of 600 time steps which corresponds to 10 min.

Note that the load demand exceeds the allowed limit by a small margin for a brief period of time near 9 h. However, when the PEVs are connected to the distribution grid, they are able to inject power into the grid and reduce the total demand to below the limit. On the other hand, if the PEVs are not controlled, then there is a peak demand which exceeds the power limit by a large margin. By appropriately controlling the charge rates, it is possible to mitigate the peak, though the overall charging time obviously increases. Furthermore, the PEVs can also support the grid with reactive power management, and successfully push the reactive power at the transformer towards zero. This is helpful both in terms of reduced grid transmission losses and local voltage support. Finally, note that in our set-up the PEVs disconnect as soon as they are charged to the desired level (e.g., fully charged).

7.5.2 Simulation of the Energy Fairness Scenario

In this section, the charge rates of the PEVs are determined using the modified algorithm illustrated in Sect. 7.4.2. To make a comparison with the previous charging strategy, we use a similar setting as previously described. While the

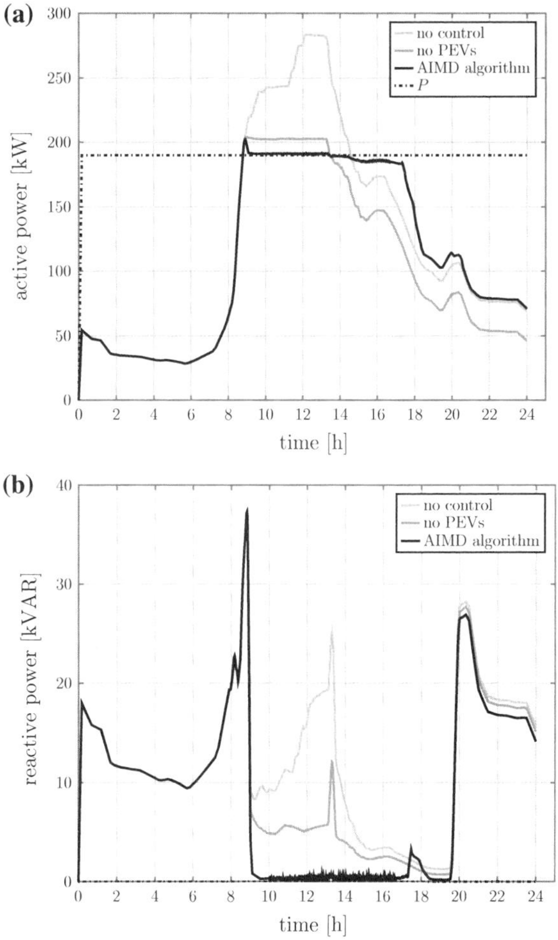

Fig. 7.5 The power consumption at the transformer "SubXF" filtered using a moving average filter with a window length of 600 time steps, while the connected PEV apply the PF scenario. The active power consumption is depicted in (**a**), and the reactive in (**b**)

additive parameter α_i is determined by the Eqs. (7.11) and (7.12), respectively, the remaining active power AIMD parameters are chosen identically to those in Sect. 7.5.1 for the previous simulation, i.e. : $\alpha = 1$, $\bar{\alpha} = 0.1\,\frac{\text{kW}}{\text{s}}$, $\beta^{(1)} = 0.75$, $\beta^{(2)} = 0.99$, $\lambda = 0.7$, and $\psi = 0.15$. The values for the reactive power compensation are also chosen identically to the previous scenario, i.e. $a = 1$, $\bar{a} = 0.1\,\frac{\text{kVA}}{\text{s}}$, $b^{(1)} = 0.75$, $b^{(2)} = 0.99$, $\gamma = 0.7$, and $\psi = 0.15$.

Each PEV has to know the expected time it will be connected to the power grid in advance, in order to compute the additive parameter as in Eqs. (7.11) and (7.12). In this simulation, we assumed that every PEV is expected to stay connected for 9 h. Then the desired charge rate is computed once at connection of the vehicle to the grid according to Eq. (7.6). This desired charge rate is then used to continuously

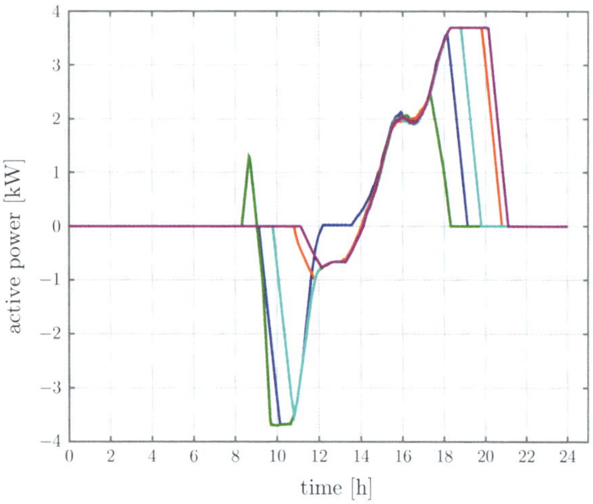

Fig. 7.6 The active charge rates of four randomly selected PEVs applying the PF scenario filtered using a moving average filter with a window length of 600 time steps

update the additive parameter $\alpha_i(k)$ using Eqs. (7.11) and (7.12) while the PEVs operate in G2V and V2G mode, respectively.

Figure 7.7 depicts the active and reactive power at the transformer. Again, we show a comparison of the results relative to the case of no connected vehicles, and to the case of uncontrolled charge rates.

The second objective $O_2(k)$ in this scenario is to share the power proportionally to the desired charge rate (i.e., more power to those who need more energy in a shorter time). To investigate whether this objective is fulfilled, the ratio between the desired charge rate and the actual average power consumption is plotted in Fig. 7.8. As before, the power consumption is filtered using a moving average filter with a window size of 10 min.

7.5.3 Simulation of the Time Fairness Scenario

We repeat the simulation to simulate the TF scenario. As mentioned in Sect. 7.3.3 the power consumption or injection should be proportional to an assigned priority $\Upsilon_i(k)$, which can be computed by Eq. (7.8). The parameters for the scenario are such that the number of groups L is set to four and the time period κ is set to 1 h. While the additive parameter of the active power AIMD is computed at each time step using Eq. (7.13), all remaining parameters of the active and reactive power AIMD are identical to the previous simulations.

Figure 7.9 depicts the active and reactive power consumption at the transformer "SubXF". Similarly to the other scenarios, the algorithm manages to mostly push the active power below the limit, while allowing the PEVs to balance a large part of the reactive power in the area.

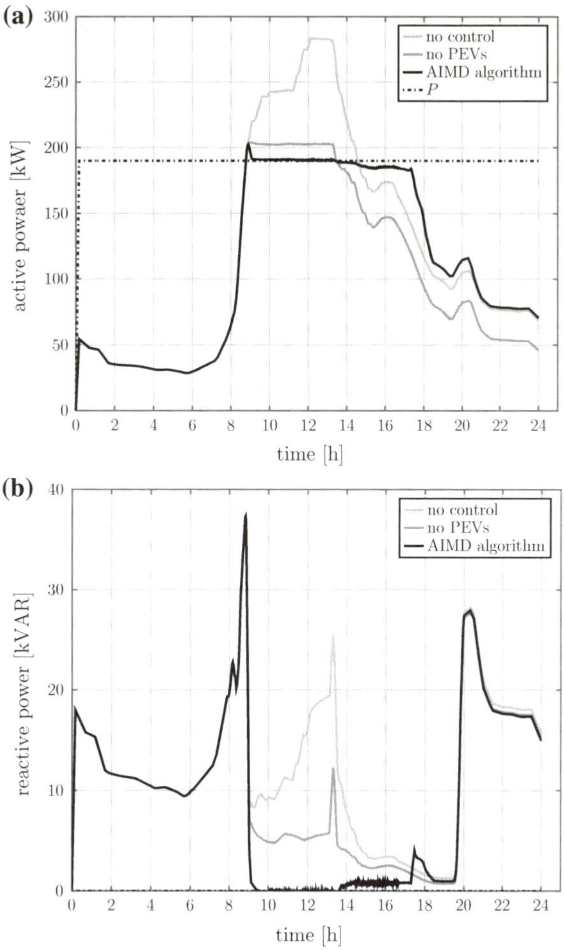

Fig. 7.7 The power consumption at the transformer "SubXF" filtered using a moving average filter with a window length of 600 time steps, while the connected PEVs apply the EF scenario. The active power consumption is depicted in (**a**), and the reactive in (**b**)

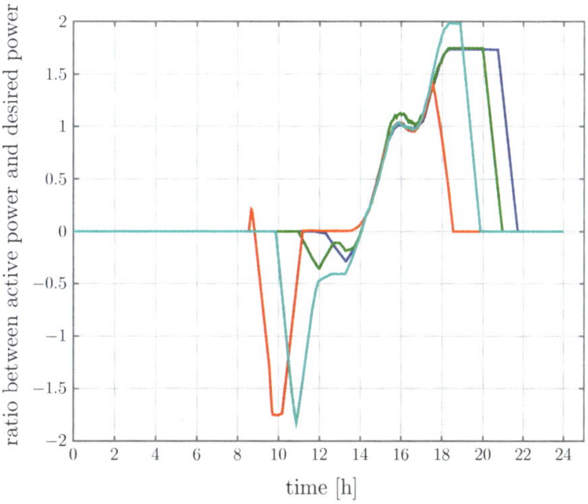

Fig. 7.8 The ratio between the active charge rates filtered using a moving average filter with a window length of 600 time steps and the desired charge rate $\tilde{p}_i(k)$ of four randomly selected PEVs applying the EF scenario

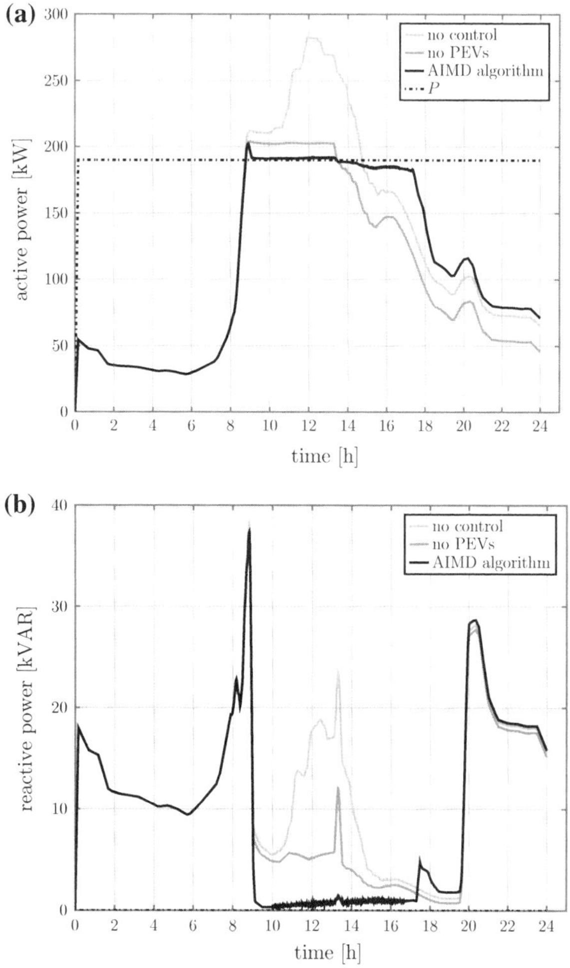

Fig. 7.9 The power consumption at the transformer "SubXF" filtered using a moving average filter with a window length of 600 time steps, while the connected PEV apply the TF scenario. The active power consumption is depicted in (**a**), and the reactive in (**b**)

Here, the PEVs power consumption or injection should be proportional to their priority. In Fig. 7.10 the power consumption of four randomly selected vehicles is depicted (dashed) and their priority (solid). The power consumption or injection is higher for PEVs that have just been connected to the grid compared to those connected for a longer period, as desired in this scenario.

Similarly to the other simulations, the power consumption is filtered using a moving average filter with a window length of 10 min.

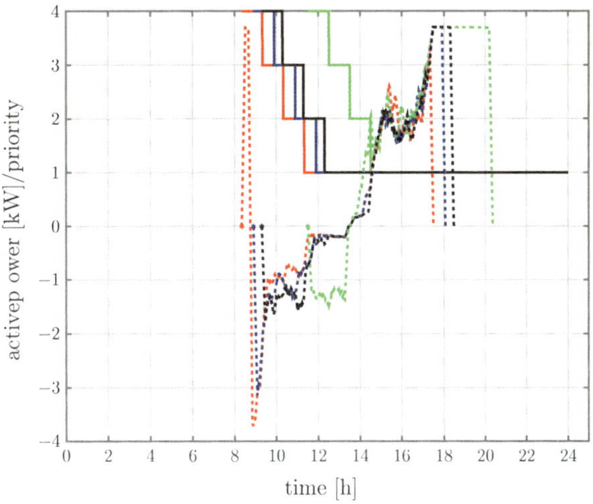

Fig. 7.10 The active charge rates of four randomly selected PEVs applying the TF scenario filtered using a moving average filter with a window length of 600 time steps (*dashed*, *star* marks connection time) and the associated priority of the PEVs (*solid*)

7.6 Future Work

We have simulated the algorithm proposed for managing the active and reactive power consumption of PEVs to support the distribution grid while PEVs are connected for charging. In this chapter a simple radial feeder model has been used to test the algorithm. In reality, however, the grid structures are more complex and may contain multiple feeders with different limitations. Those more complex cases with additional limitations have to be investigated to guarantee an efficient operation of the algorithm.

Similarly, different types of constraints such as line current or node voltage limits may be present. For example, a node voltage limit might be imposed to ensure all users have supply that meets the relevant standards. In that regard it may be important to carefully select a subset of nodes where excess voltage excursions are most likely to occur. In this case, further studies would be required to prevent unforeseen interactions among the different CE-generating sources.

Additionally, the algorithm proposed could be used to shape the demand curve that regions should follow instead of just limiting the demand in the region. Such a behavior can be achieved by intelligent adaption of the limit, P. This can be used to limit the rate of change in aggregate demand to allow generators that react slowly to compensate for the changes. The main problem in this case is how to find the optimal P to support the distribution grid while full-filling the needs of the owners of PEVs.

Similarly, extensions of this basic algorithm and intelligent adaptation of the limit P allows two more services of support. The first one is to balance the power among the three phases by controlling the phases separately. In that case, a limitation is introduced for each phase whose value depends on the demand in the three phases. However, it is not straightforward to include such an operation. For example it might be hard to define whether a phase should reduce its power consumption or the other phases should increase their power consumption. The second possible extension is to regulate currents according to the grid frequency and utilize this frequency as a signal, indicating when to locally reduce the power consumption.

Furthermore, we assumed throughout the paper that all participating PEVs are applying an identical scenario. It has to be verified whether the fairness can still be guaranteed if the connected PEVs apply different scenarios in the same distribution grid. For example in a region where a lot of PEVs are connected using the PF scenario, for example in a domestic setting, and a few vehicles connect using the EF scenario, for example at a small office building in the same area. This issue has to be studied, especially for a larger scale use of the algorithm. In the situations described above, one might try to use multiple levels of DAIMD. For instance, one DAIMD controls only the PEVs that apply the EF scenario, and a second one controls the PEVs that apply the PF scenario. The power limit of those two separate DAIMD algorithms is then controlled by a higher level DAIMD. Such multi-level DAIMD algorithms might also be used for the control of large scale distribution grids. However, as this adds higher levels of communication and control, the behavior has to be studied more carefully.

Due to the small communication overhead of a single bit, the PEVs are not able to react to future expectations of the demand by uncontrollable loads and the power generated by renewable power sources. By allowing the management unit to broadcast additional information, for example the expected increase in demand in the next hour, the PEVs may be able to react to such information and adapt their power consumption in response to predictions. What information is most useful and how the PEVs should react while maintaining a sense of fairness remains an open problem. Further, by relying on predictions it is important to consider that they normally are not exact. It needs to be verified that the algorithm can handle such prediction errors and whether the advantages are large enough to accept the higher communication requirements and the necessity of predictions.

7.7 Conclusions

This chapter presents a distributed algorithm to control the charging of PEVs and enables them to support the grid. While the algorithm manages to limit the peak demand and reduces the reactive power transported outside of an area, it also allows flexibility in how the PEVs are "fairly" controlled.

We presented three possible definitions of how the control can be interpreted as "fair". While there are many more possibilities to define "fairness" among the participants, those three scenarios illustrate the flexibility of the proposed algorithm.

Using a simple radial test feeder we simulated the behavior of the algorithm for the different scenarios and verified its usefulness.

Acknowledgments This work is supported in part by Science Foundation Ireland grant 11/PI/1177.

References

1. Acha S, Green TC, Shah N (2010) Effects of optimised plug-in hybrid vehicle charging strategies on electric distribution network losses. In: Proceedings of the IEEE PES transmission and distribution conference and exposition
2. AECOM (2009) Economic Viability of Electric Vehicles. Department of Environment and Climate Change, prepared by AECOM Australia Pty Ltd. http://www.environment.nsw.gov.au/resources/climatechange/ElectricVehiclesReport.pdf. Accessed 22 Jul 2014
3. Beil I, Hiskens I (2012) A distributed wireless testbed for plug-in hybrid electric vehicle control algorithms. In: Proceedings of North American power symposium
4. Brooks A, Lu E, Reicher D, Spirakis C, Weihl B (2010) Demand Dispatch. IEEE Power Energy Mag 8(3):20–29
5. Carradore L, Turri R (2010) Electric vehicles participation in distribution network voltage regulation. In: Proceedings of universities power engineering conference
6. Chiu D-M, Jain R (1989) Analysis of the increase and decrease algorithms for congestion avoidance in computer networks. Comput Netw ISDN Syst 17(1):1–14
7. Clement-Nyns K, Haesen E, Driesen J (2010) The impact of charging plug-in hybrid electric vehicles on a residential distribution grid. IEEE Trans Power Syst 25(1):371–380
8. Crisostomi E, Liu M, Raugi M, Shorten R (2014) Plug-and-play distributed algorithms for optimised power generation in a microgrid. IEEE Trans Smart Grid 5(4):2145–2154
9. De Hoog J, Alpcan T, Brazil M, Thomas DA, Mareels I (2014) Optimal charging of electric vehicles taking distribution network constraints into account. In: IEEE Trans on power systems, to be published
10. Deilami S, Masoum AS, Moses PS, Masoum MAS (2011) Real-time coordination of plug-in electric vehicle charging in smart grids to minimize power losses and improve voltage profile. IEEE Trans Smart Grid 2(3):456–467
11. Di Giorgio A, Liberati F, Canale S (2014) Electric vehicles charging control in a smart grid: a model predictive control. Control Eng Practice 22:147–162
12. EPRI Smart Grid Resource Centre (2014) Open DSS—IEEE test cases. http://svn.code.sf.net/p/electricdss/code/trunk/Distrib/IEEETestCases/. Accessed 22 July 2014
13. EPRI Smart Grid Resource Centre (2014) Simulation tool—OpenDSS. http://smartgrid.epri.com/SimulationTool.aspx. Accessed 22 July 2014
14. Kejun Q, Chengke Z, Allan M, Yue Y (2011) Modeling of load demand due to EV battery charging in distribution systems. IEEE Trans Power Syst 26(2):802–810
15. Kellett CM, Middleton RH, Shorten RN (2007) On AIMD congestion control in multiple bottleneck networks. IEEE Commun Lett 11(7):631–633
16. Kelly FP, Maulloo AK, Tan DKH (1998) Rate control for communication networks: shadow prices, proportional fairness and stability. J Oper Res Soc 49:237–252
17. Keshav S, Rosenberg C (2011) How internet concepts and technologies can help green and smarten the electrical grid. SIGCOMM Comput Commun Rev 41:109–114
18. Kirby B, Hirst E (1999) Load as a resource in providing ancillary services. In: Proceedings of the American power conference
19. Liu M, McLoone S (2013) Investigation of AIMD based charging strategies for EVs connected to a low-voltage distribution network. Intelligent computing for sustainable energy and environment. Springer, Berlin, pp 433–441
20. Lopes JAP, Soares FJ, Almeida PMR (2011) Integration of electric vehicles in the electric power system. Proc IEEE 99(1):168–183
21. Mets K, Verschueren T, De Turck F, Develder C (2011) Evaluation of multiple design options for smart charging algorithms. In: Proceedings of the IEEE international conference on communications
22. Palensky P, Dietrich D (2011) Demand side management: demand response, intelligent energy systems, and smart loads. IEEE Trans Ind Inform 7(3):381–388

23. Papadopoulos P, Skarvelis-Kazakos S, Grau I, Cipcigan LM, Jenkins N (2010) Predicting electric vehicle impacts on residential distribution networks with distributed generation. In: Proceedings IEEE vehicle power and propulsion conference
24. Shahidinejad S, Filizadeh S, Bibeau E (2012) Profile of charging load on the grid due to plug-in vehicles. IEEE Trans Smart Grid 3(1):131–141
25. Shorten R, Wirth F, Leith D (2006) A positive systems model of TCP-like congestion control: asymptotic results. IEEE/ACM Trans Network 14(3):616–629
26. Shorten RN, Kellett CM, Leith DJ (2007) On the dynamics of TCP's higher moments. IEEE Commun Lett 11(2):210–212
27. Srikant R (2004) The mathematics of internet congestion control. Birkhäuser, Boston
28. Stüdli S, Crisostomi E, Middleton R, Shorten R (2012) A flexible distributed framework for realising electric and plug-in hybrid vehicle charging policies. Int J Control 85(8):1130–1145
29. Stüdli S, Crisostomi E, Middleton R, Shorten R (2014) Optimal real-time distributed V2G and G2V management of electric vehicles. Int J Control 87(6):1153–1162
30. Stüdli S, Griggs W, Crisostomi E, Shorten R (2014) On optimality criteria for reverse charging of electric vehicles. IEEE Trans Intell Transp Syst 15(1):451–456
31. Su W, Rahimi-Eichi H, Zeng W, Chow M-Y (2012) A survey on the electrification of transportation in a smart grid environment. IEEE Trans Ind Inform 8(1):1–10
32. Wen C-K, Chen J-C, Teng J-H, Ting P (2012) Decentralized plug-in electric vehicle charging selection algorithm in power system. IEEE Trans Smart Grid 3(4):1779–1789
33. Wencong S, Mo-Yuen C (2012) Performance evaluation of an EDA-based large-scale plug-in hybrid electric vehicle charging algorithm. IEEE Trans Smart Grid 3(1):308–315

Chapter 8
Towards a Business Case for Vehicle-to-Grid—Maximizing Profits in Ancillary Service Markets

David Ciechanowicz, Alois Knoll, Patrick Osswald and Dominik Pelzer

Abstract Employing *plug-in electric vehicles* (PEV) as energy buffers in a smart grid could contribute to improved power grid stability and facilitate the integration of renewable energies. While the technical feasibility of this concept termed *vehicle-to-grid* (V2G) has been extensively demonstrated, economic concerns remain a crucial barrier for its implementation into practice. A common drawback of previous economic viability assessments, however, is their static approach based on average values which neglects intrinsic system dynamics. Realistically assessing the economics of V2G requires modeling an intelligent agent as a homo economicus who exploits all available information with regard to maximizing its utility. Therefore, a smart control strategy built on real-time information, prediction and more sophisticated battery models is proposed in order to optimize an agent's market participation strategy. By exploiting this information and by dynamically adapting the agent behavior at each time step, an optimal control strategy for energy dispatches of each single PEV is derived. The introduced cost-revenue model, the battery model, and the optimization model are applied in a case study building on data for Singapore. It is the aim of this work to provide a comprehensive view on the economic aspects of V2G which are essential for making it a viable business case.

Keywords Vehicle-to-grid · Plug-in electric vehicle · Economic viability · Ancillary services · Battery modeling · Optimization model

D. Ciechanowicz (✉) · P. Osswald · D. Pelzer
TUM CREATE, Singapore, Singapore
e-mail: david.ciechanowicz@tum-create.edu.sg

P. Osswald
e-mail: patrick.osswald@tum-create.edu.sg

D. Pelzer
e-mail: dominik.pelzer@tum-create.edu.sg

A. Knoll
Institute for Informatics VI, Robotics and Embedded Systems, Technische Universität München (TUM), Munich, Germany
e-mail: knoll@in.tum.de

S. Rajakaruna et al. (eds.), *Plug In Electric Vehicles in Smart Grids*,
Power Systems, DOI 10.1007/978-981-287-302-6_8

8.1 Introduction

In power systems, fluctuations of energy demand and supply cause continuous deviations from the desired frequency. Ensuring power grid stability requires an instantaneous response by the power system operator which restores the equilibrium between demand and supply. This is either achieved by power plants capable of quickly adjusting their power output or by storage facilities which buffer energy excesses or shortages. Most of these solutions are, however, either costly, entail large space or exhibit low energy efficiencies leading to the need for development of alternative approaches.

One possible solution could be the utilization of *plug-in electric vehicles* (PEV) of which batteries could be employed as short term energy storage through charging in the case of a power excess or by feeding electricity back to the grid in the opposite case. This concept termed *vehicle-to-grid* (V2G) was first mentioned in 1997 [1] and has been subject to intensive research in the last two decades. While the effectiveness of the V2G concept to improve power grid stability has been confirmed by both theoretical considerations [2–8] as well as fully functional prototypes [5, 7, 9], its economic viability is still subject to controversial discussions. This is reflected in the diverging conclusions on the profitability where some expect annual losses of several thousand dollars while others promise multiple thousand dollars of yearly income [2, 6, 9–15].

One drawback of previous economic analyses of the V2G concept is that calculations are based on average annual values for the involved parameters. In reality, however, electricity prices highly vary during the course of a day, presenting varying scenarios where V2G may yield profits in one time period but result in losses in a different one. Furthermore, individual travel itineraries impose restrictions on the temporal availability of PEVs. At the same time, factors such as battery aging typically depend non-linearly on a variety of parameters which cannot be kept constant during V2G operation. Simple averaging therefore does not yield correct cost estimations. The entity of these aspects significantly limits the explanatory power of static approaches and leaves the outcome of these methods highly sensitive to the choice of the input parameters. To correctly determine the economic viability of V2G and at the same time provide a control strategy for individual V2G agents, more dynamic approaches are required.

The purpose of this chapter is to discuss the problems of previous approaches investigating the economic viability of V2G and to identify solutions that could pave the way for making V2G an economically viable business case. The remainder of this work is structured as follows: In Sect. 8.2, the transition from a power system to a smart grid is described. In this context, the V2G concept is discussed as one possible future solution for improving power grid stability. Section 8.3 introduces an electricity market independent V2G control strategy which aims for maximizing profits in ancillary service markets. This concept includes an appropriate consideration of battery depreciation as well as an optimization methodology. The optimization model is then applied in a simple case study building on data for

Singapore in Sect. 8.4, followed by a discussion of parameter sensitivities. In Sects. 8.5 and 8.6 findings are finally discussed and an outlook on future research is given.

8.2 Power System Fundamentals

In the first part of this section, the fundamentals of the power system and the concept of ancillary services are briefly introduced. It is discussed, how the transition to a smart grid could mitigate the increasing need for balancing power demand and supply which arises from the growing share of renewable energy sources. This leads to the possible role of PEVs and the V2G concept in the future power grid, which is described in more detail in Sect. 8.2.2. One important component for the implementation of V2G is the aggregator which is finally discussed in Sect. 8.2.3.

8.2.1 Power System and Smart Grid

A *power system* is a network of power lines which connect energy producing and consuming entities with each other. Different voltage levels may distinguish the power grid into a maximum, high, medium, and low voltage grid with the first two levels forming the *transmission grid* and the latter two the *distribution grid*. The different levels are physically separated from each other by substations, switches, and transformers and are controlled by high performance computers.

In Fig. 8.1 a rough illustration of the Singapore power system as it can be derived from data on high-voltage grid, substations and consumers is exemplarily shown. The upper layer shows the transmission grid while the middle layer depicts

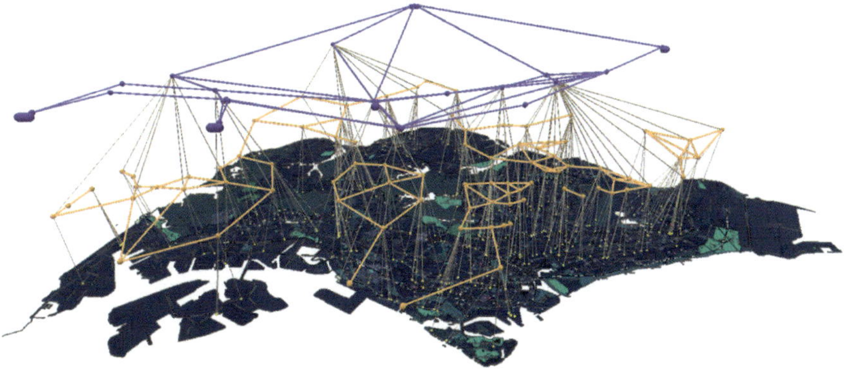

Fig. 8.1 Singapore power system

the distribution grid. The nodes in both layers represent power plants, substations, switches or transformers which in this case are not visually distinguished. The nodes at the bottom layer on the map depict a selection of consumers connected to the distribution grid.

A power grid is operated by one or more *transmission system operators* (TSO) of which its primary task is the energy transfer from the generation units to regional or local *distribution system operators* (DSO) which then deliver the energy to the consumers. One key responsibility of a TSO is to ensure power grid stability. The power grid itself only exhibits negligible energy storage capacity and is therefore an inherently unstable system. Deviations between energy demand and supply which may either be positive when supply exceeds demand or negative in the opposite case therefore lead to fluctuations of voltage and frequency which require an immediate action. This response is performed by so called *ancillary services* provided by fast responding power generators which are capable of quickly ramping up or curbing down their power output. Depending on the response time and the duration of providing ancillary services, it is distinguished between *regulation* as well as *primary*, *secondary*, and *contingency reserve*. All of the four markets usually have a ratio of around 1 % of the total annual energy generation. Providers of ancillary services receive a payment for the dispatched energy when up-regulation is required or a compensation for curbing power generation in the opposite case. These *energy payments* are usually differentiated and considerably higher for regulation than for reserve. Besides these energy payments, many national electricity markets also have a *capacity payment* which is a reward solely for holding power generation potential available instead of energy dispatch. In most markets, prices are fairly variable over time but are kept constant for a certain time period of 15 or 30 mins in most cases.

As a result of growing shares of intermittent renewable energy sources and the introduction of PEVs on a large scale, the need for ancillary services and energy storage is increasing. This is because both the availability of renewable energies and the mobility pattern of PEVs are volatile and sometimes hard to predict. To satisfy the additional demand for ancillary services, either fast reacting generators or energy storage facilities are needed. Technologies capable of providing this functionality include *pumped storage hydroelectricity* (PSH), *compressed air energy storage* (CAES), *hydrogen-driven fuel-cells*, or *supercapacitors*. These technologies are, however, often costly, energy inefficient or may entail large space leading to the need for alternative approaches.

The need for energy storage may be reduced in a smart grid which supports multi-directional energy flow instead of showing a strictly hierarchical topology. In this case, energy is not only generated at the high voltage levels but may also be provided by generators within the distribution grid. These generators could then also serve as ancillary service providers so that large power plants could keep operating at their optimal efficiency. With a communication infrastructure allowing the intelligent control of energy producers and consumers this would lead to a distributed, self-organizing grid design.

One important role in a future smart grid could be taken by PEVs which have the capability of acting as either consumers or producers by using their battery as energy buffers. PEVs are advantageous compared to classical generators in the way that they can react to demand requests virtually in real-time, have low standby and initial costs per kWh, and provide temporarily high power. With a sufficient amount of PEV participating in V2G services, capacities from conventional power sources would therefore become redundant. This V2G concept will be introduced in greater detail in the following section and will be further assessed in the remainder of this chapter.

8.2.2 The V2G Concept

The V2G concept is depicted in Fig. 8.2. Energy is generated by conventional power plants or renewable energy sources and transmitted through maximum, high, medium and low voltage lines to the consumers (e.g., households, enterprises, charging stations, etc.). The type of consumers that is of interest in this context are PEVs which may either use the energy for driving or serve as a short-term energy storage by charging their battery packs in case of power excess or feeding electricity back in the opposite case. Discharging a PEV's battery during an energy shortage and therefore providing energy to the grid is called V2G while charging the battery during an energy excess is known as *grid-to-vehicle* (G2V). The V2G concept incorporates both services so this term will be used within this chapter whenever no explicit distinction is necessary. In the following, four scenarios are introduced in which possible use cases for the energy stored in the battery packs of the PEVs are outlined.

Scenario A depicts a one-way flow of energy where a PEV is simply charged at a charging station installed in a household. Scenario B uses the same setting but in addition energy can be locally fed back to the household. This concept is termed *vehicle-to-home* (V2H). The case of allowing energy to flow back into the power grid representing the V2G concept is depicted in Scenario C. In this case, the PEVs communicate with an intelligent charging station which then dispatches or draws energy to or from the PEV. The charging station itself is controlled by an aggregator which is a unit that bundles multiple PEVs to a *virtual power plant* (VPP) [7, 16–19] in order to trade energy at the electricity market. Due to its important role, the aggregator is discussed in further detail in Sect. 8.2.3. In Scenario D, multiple PEVs are aggregated to a VPP through an operator of e.g. a car park. This operator could use the aggregated energy as described by the V2H concept, directly participate in the energy market or could again be part of a VPP of some higher level aggregator.

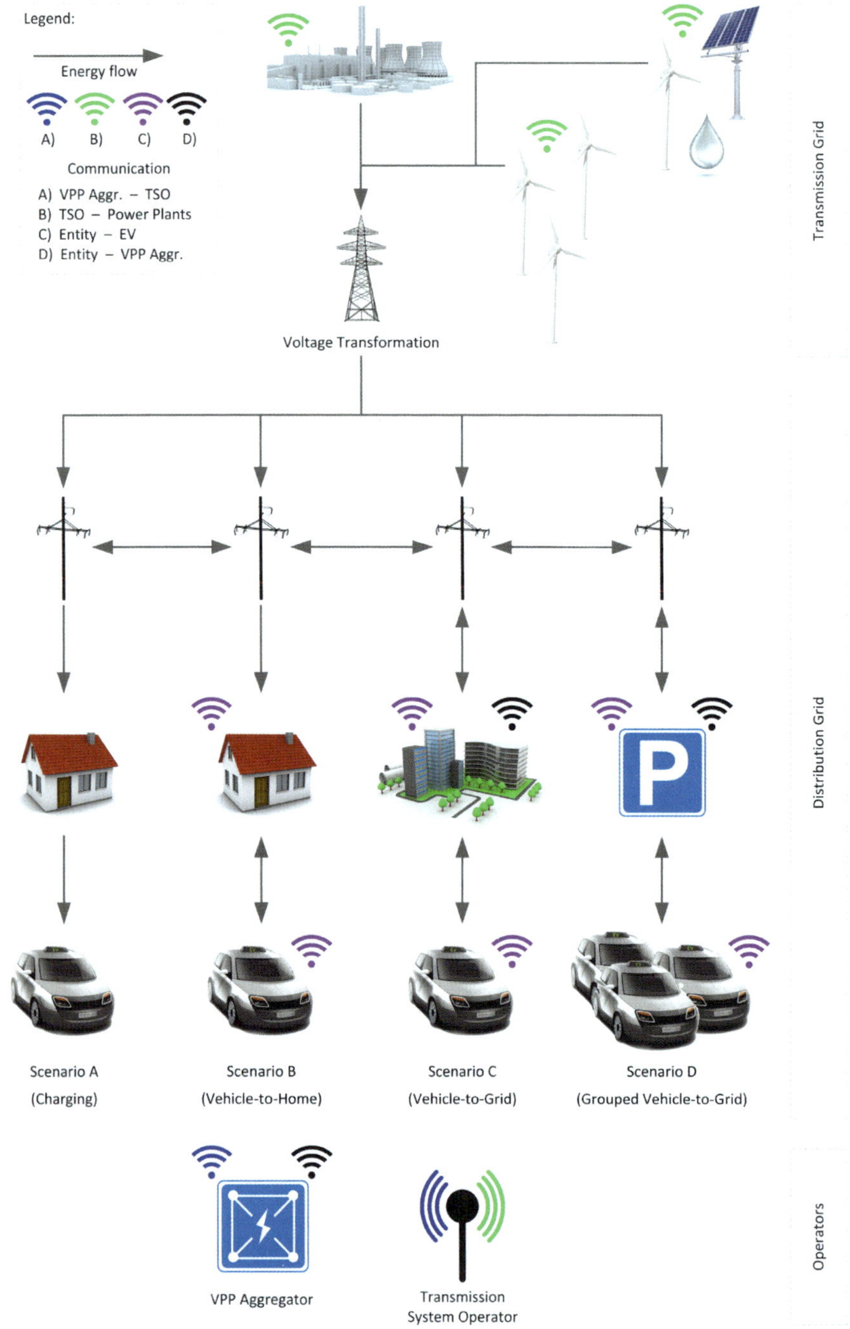

Fig. 8.2 The V2G concept

8.2.3 The Aggregator

The amount of energy and power each individual PEV can provide is too small to participate on most electricity markets (in Singapore 1 MW for half an hour is necessary). Meeting these conditions thus requires hundreds to thousands of PEVs aggregated to a VPP. This is achieved by an aggregator who serves as a mediator between the PEVs and the electricity market. The aggregator trades energy at the market and ensures that the VPP is capable of providing the contracted power at all times. An aggregator should be considered a virtual entity rather than a physical one. This means that the PEVs belonging to one aggregator do not necessarily need to be connected at neighboring locations. Instead, aggregation at the level of the same grid node or even only in the operation range of one grid operator may be sufficient in many electricity markets. In addition to PEVs, an aggregator could have access to other energy sources e.g. secondary market battery packs, conventional or renewable energy power plants or other sub-aggregators. From the TSO's point of view, the power generation capacity offered by the aggregator presents itself as a single large, fast-controllable energy source although it may originate from a variety of different sources. The relation between all involved actors is depicted in Fig. 8.3.

The challenge faced by the aggregator is to synchronize charging and discharging operations of a large number of PEVs in order for all PEVs to reach their targeted state of charge, while ensuring that the contracted ancillary services can be provided at all times. Due to the continuous fluctuation of the number of PEVs in the VPP, the heterogeneity of the carpool and the fact that both aggregator and PEV owners aim to maximize their profit, this leads to an optimization problem with a high degree of uncertainty. Since each PEV typically has its individual utility function and own constraints, a central control mechanism would quickly become

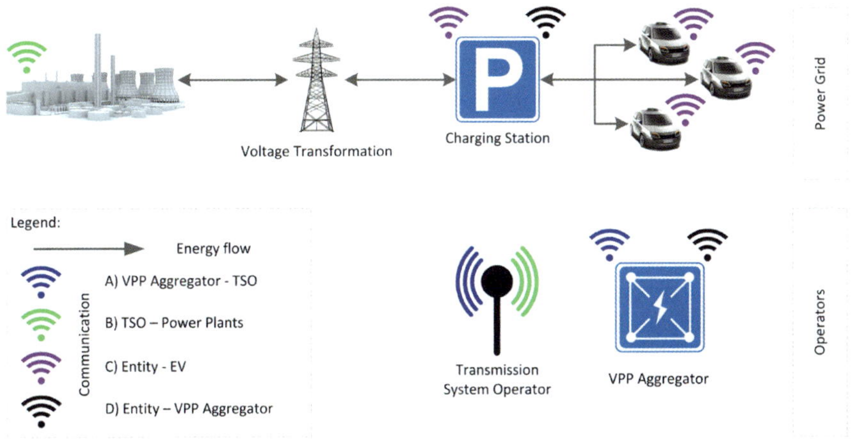

Fig. 8.3 The aggregator concept

infeasible. Therefore, a distributed approach where each PEV performs its individual optimization is discussed in Sect. 8.3.3. As such, the central remaining task of the aggregator is to achieve sound estimations on the demand and power generation capacity of its VPP and, if necessary, trigger behavior changes of the involved PEVs in case it is at risk of failing to fulfill its obligations to the TSO.

Influencing the charging and dispatching behavior of PEVs which are part of the VPP can be achieved by sending price signals which may not necessarily correspond to market prices. By decoupling prices offered by an aggregator from prices given by the electricity market, the temporal gap between a period in an electricity market in the range of minutes and the requirements for regulation on a scale of seconds can be closed. Real-time prices would also allow an aggregator to dynamically adapt the charging/dispatching power of each individual agent in real-time and not only on a period basis. Based on historical data collected by an aggregator, the algorithm would have to take an estimate of the temporal availability of each PEV as well as each agent's individual cost function and battery capacity constraints into consideration. In return, it may produce an optimal charging/dispatching schedule for each point in time optimizing its own profits by also generating (not necessarily optimal) profits for each agent.

Presuming V2G is a profitable concept, there are different types of entities that might be interested in establishing themselves as aggregators. First of all, battery pack or vehicle manufacturers have detailed knowledge about their battery inherent depreciation cost functions. The drawback of the two parties is the spatial distribution of their aggregated PEV fleet which might cause problems with feeding energy into the right section of the low voltage grid. Additionally, they may lack necessary know-how in the area of communication. Another group of interest could be mobile network operators and DSOs which both have expertise regarding communication technology and accounting systems, especially with a large amount of small-size customers. Additionally, DSOs already have a business connection with customers in the energy segment. Particularly advantageous for DSOs is their profound knowledge of power demand and supply in the grid. At last, entities who command a sufficiently large PEV fleet could promote themselves being an aggregator. Their advantage is their knowledge about the tempo-spatial availability of each PEV.

8.3 An Intelligent Agent Behavior Model

One essential criterion for making the V2G concept applicable in practice is to prove its economic viability on one hand and on the other to provide individual agents with a control strategy which maximizes their profit. For this purpose, in Sect. 8.3.1 the basic equations for an economic model for V2G are introduced. Section 8.3.2 then discusses the challenges and approaches regarding battery aging models which are a crucial factor for assessing the costs of V2G. In Sect. 8.3.3 it is then described, how the introduced equations can be utilized for a dynamic control strategy which most effectively exploits the economic potential of V2G for an individual user. Section 8.3.4 finally briefly discusses what role artificial intelligence could play for making V2G applicable in practice.

8.3.1 Economic Model

In this section, the equations which are used for investigating the economic viability of V2G valid for most electricity markets are introduced [20]. Total annual profits are calculated from the difference between revenues R and costs C

$$\prod = R - C \tag{8.1}$$

which are separately discussed in the following two sections.

8.3.1.1 Revenues

The total revenue R is the sum of the revenues made from up-regulation and the revenues attained from down-regulation services. In the event of an under-supply of power, up-regulation is necessary. In this case, the PEV acts as a generator and feeds energy into the grid. Therefore, energy is sold at the regular selling price in the respective electricity market p_E plus a compensation for providing up-regulation ancillary services $p_{\uparrow,\mathrm{Anc}}$. Depending on the energy market under consideration, $p_{\uparrow,\mathrm{Anc}}$ corresponds to either the payment for reserve or regulation. In this case, the received payment per unit of dispatched energy is

$$p_{\uparrow,E} = p_E + p_{\uparrow,\mathrm{Anc}} \tag{8.2}$$

In the opposite case where power supply exceeds demand, down-regulation is required and the PEV acts as a consumer. The owner pays the electricity tariff p_{ET} which is discounted by the down-regulation compensation $p_{\downarrow,\mathrm{Anc}}$. Since the energy purchase costs given by p_{ET} are explicitly accounted for in Sect. 8.3.1.2, the effective payment per unit of energy in this case is therefore simply

$$p_{\downarrow,E} = p_{\downarrow,\mathrm{Anc}} \tag{8.3}$$

In many national electricity markets, additional capacity payments $p_{\uparrow,\mathrm{Cap}}$ and $p_{\downarrow,\mathrm{Cap}}$ are provided for only holding power generation potential available rather than actually dispatching energy.

The total annual revenue R is the sum of the revenues resulting from the energy and the capacity payment. Each of the two payments has to be multiplied by the respective amounts of purchased and dispatched energy, or stand-by power. As previously discussed, market prices are typically time-dependent but remain constant for time periods of a certain duration Δt. As a simplification, the charging/dispatching power may also be assumed to be kept unchanged during one time period. The total annual revenue R over all time periods i can then be written as

$$R = \sum_i \left(p_{\uparrow,E,i} P_{\uparrow,E,i} + p_{\uparrow,Cap,i} P_{\uparrow,Cap,i} + p_{\downarrow,E,i} P_{\downarrow,E,i} + p_{\downarrow,Cap,i} P_{\downarrow,Cap,i} \right) \cdot \Delta t \qquad (8.4)$$

8.3.1.2 Costs

The total annual costs C_A are calculated as the variable costs $c_{var} = c_\eta + c_D$ multiplied by the total annual amount of energy cycled through the battery pack E_A, plus annual fixed costs C_{AF}:

$$C_A = E_A \left(c_\eta + c_D \right) + C_{AF} \qquad (8.5)$$

In this equation, c_η denotes the energy purchase costs which, using the charge-discharge efficiency η, can be written as

$$c_\eta = \frac{p_{ET}}{\eta} \qquad (8.6)$$

The term c_D represents the variable battery pack depreciation costs which result from the limited number of possible charge-discharge cycles. Using the purchase costs of a battery pack $C_{\text{BatterPack}}$ and the total possible energy throughput E_{Lifetime}, this turns into

$$c_D = \frac{C_{\text{BatteryPack}}}{E_{\text{Lifetime}}} \qquad (8.7)$$

The quantity of energy which can be cycled through a battery pack until it fails to meet its specific performance criteria is given by the capacity $Q_{\text{BatteryPack}}$ multiplied by the *depth of discharge* (*DOD*) and the maximum number of cycles Z possible at a certain DOD:

$$E_{\text{Lifetime}} = Z \cdot DOD \cdot Q_{\text{BatteryPack}} \qquad (8.8)$$

One cycle in this context is understood as discharging the battery from an initial *state of charge* (SOC) by a certain DOD and subsequently recharging it to the initial SOC; the charge throughput per cycle therefore depends upon the corresponding DOD. The cycle stability Z is a quantity which depends on a large number of parameters such as *charge rate* (C-rate), DOD, temperature, humidity and time and which strongly varies among different battery chemistries [21]. It is therefore not possible to reliably model the cyclic lifetime so that many studies simply assume a fixed number for Z [6, 10, 12, 13]. In Sect. 8.3.2 the challenges related to battery lifetime modeling are discussed in further detail.

The last term of (8.5) C_{AF} denotes the fixed costs which account for the investment in equipment required to make a PEV suitable for V2G. To annualize and discount the fixed costs, it can be written as

$$C_{AF} = C_C \frac{d}{1-(1+d)^{-n}} \tag{8.9}$$

with C_C being the total capital costs, d the discount rate and n the number of years until the investment is depreciated. With these considerations, the total annual costs can finally be rewritten as

$$C_A = E_A \left(\frac{p_{ET}}{\eta} + \frac{C_{\text{BatteryPack}}}{Z \cdot DOD \cdot Q_{\text{BatteryPack}}} \right) + C_C \frac{d}{1-(1+d)^{-n}} \tag{8.10}$$

8.3.2 Battery Modeling

A crucial aspect for the profitability of V2G applications is battery degradation cost. To appropriately consider the costs of battery degradation in an economic model and to account for these costs during V2G operation, an understanding of battery aging processes and their representation by a suitable battery model is required.

The performance fade of a cell can be separated into the loss of capacity (measured in Ah) as well as the increase of the cell impedance which causes energy fade (measured in Wh) and power fade (measured in W). The main effect for capacity fade is the loss of cyclable lithium, primarily caused by formation of the *solid electrolyte interface* (SEI) at the graphite anode [22, 23] as well as by lithium plating occurring at high charging currents and low temperatures. This loss of cyclable lithium in turn causes a change of the electrode balancing, preventing the battery from being fully charged and discharged at specific current rates [24]. The second contributor to capacity fade is the loss of active electrode material. When the cell is cycled at high and low SOCs, the electrodes undergo certain mechanical stress during lithium intercalation, resulting in micro cracking. These micro cracks lead to either further SEI formation or can cause a loss of contact for the active material, making them unavailable for further intercalation processes [22, 23, 25]. In addition to the capacity fade described, these mechanisms are closely correlated to the increase of the cell impedance. The ongoing SEI reformation causes a constantly growing surface layer with a low conductivity and low diffusivity, causing an increase in the charge transfer resistance [25]. The loss of active material leads to higher local currents and local SOC variations, which in turn accelerate the aging process [26].

These different aging mechanisms are triggered from the environment and the utilization mode, including the cell's temperature, the DOD, the charge and discharge current rate as well as the SOC range the cell is used in [25, 27–29]. In general, high currents as well as extremely high and low SOC conditions accelerate the aging process of the cells; high temperatures accompanying these high currents lead to an increased amount and speed of parasitic side reaction. An intelligent V2G control strategy would therefore aim at maintaining moderate SOC conditions, avoiding extreme DODs and keeping charge and discharge currents low.

Numerous studies have been performed to understand these mechanisms and to establish a quantitative relation between these aging effects and the corresponding control parameters [29, 30]. Battery aging studies considering multiple parameters are, however, complex and very time consuming, particularly at low C-rates. Hence, certain drawbacks in the accuracy of the battery aging model have to be taken into account. As a first approach, it can be assumed that the aging of the cell is dominated by the charge throughput during charge and discharge of the cell. As described earlier, the rate of damage is greater at extremely high and low SOCs which can be reflected in a DOD dependent aging parameter.

To quantitatively account for battery depreciation costs due to charging and discharging, the cost of a unit of cycled energy needs to be computed according to (8.7). A simple empirical model for battery aging which is employed in the sensitivity analysis of the case study presented in this paper was developed by Peterson and Whitcare [31]. In this model, the cyclic stability introduced in (8.8) is given as

$$Z(DOD) = \left(\frac{145.71}{DOD}\right)^{\frac{1}{0.6844}} \tag{8.11}$$

which explicitly considers the effect of the DOD on the possible number of cycles.

While the DOD can be assumed to be the most relevant parameter, the battery aging estimation can be further improved by additionally taking the non-linear behavior of the SOC-dependent aging into account. This is achieved by a model presented in [32] which was adapted to be employed in the case study in Sect. 8.4. It describes the battery capacity fade due to cyclic aging as a function of the charge throughput q according to the relation

$$CAP(q) = 1 - \beta \cdot \sqrt{q} \tag{8.12}$$

where CAP denotes the battery capacity and where β is an experimentally determined factor which was found to be

$$\beta = 7.348 \times 10^{-3} \cdot (\bar{U} - 3.667)^2 + 7.6 \times 10^{-4} + 4.081 \times 10^{-3} \cdot DOD \tag{8.13}$$

for the investigated battery type.

In this equation, $\bar{U}$ is the average voltage at which the cycling occurs which can be obtained from the open-circuit voltage of the battery cell. As $\bar{U}$ depends on the SOC, this relation implicitly accounts for the SOC as a second parameter apart from the DOD. Setting $CAP(q) = 80\,\%$ which is a common criterion for the end of life of batteries used for automotive applications then allows calculating E_{Lifetime}. This ultimately leads to the following equation for battery depreciation costs

$$c_D = C_{\text{BatteryPack}} \cdot \frac{\beta^2}{0.04 \cdot \overline{U}} \tag{8.14}$$

which consider both DOD and SOC.

8.3.3 Optimization

Cost and revenue equations similar or equivalent to the ones discussed in the previous sections have been applied in many cases to assess the economic viability of V2G. This has mostly been accomplished by using average values for prices, battery lifetime and charging and dispatching power. Most of the studies relying on realistic assumptions conclude that the PEV owner would incur monetary losses from providing V2G services. It is therefore clear that control strategies based on this averaging behavior would not lead to a valid business case. Control strategies for V2G need to be directly related to economic considerations to give the V2G concept a chance to be implemented in practice at all. This means that strategies need to account for the temporal dynamics of the market and need to reflect the behavior of intelligent agents who would attempt to maximize their profits by adapting to these fluctuations.

The resulting question therefore is how rational agents would decide on their charging and dispatching strategies presuming they have certain information on internal and external parameters. Technically, this means that a cost-benefit calculation according to the equations defined above needs to be conducted whenever any change of the relevant parameters occurs.

There have been several recent attempts in the literature which address this issue [15, 33]. A simple strategy which improves the loss-making averaging approach is to make a binary decision on when to provide V2G services, depending on whether the evaluation of the cost model yields an expected benefit or a loss. Given an additional degree of freedom where the user cannot only make a binary decision but continuously adapt the power output or input, a next step is to compute an optimal value for the charging or dispatching power for a certain point in time. This approach can be refined further by making use of predictive information. In most electricity markets, price estimates for buying and selling electricity are known a certain period of time in advance. This information may be used by an intelligent agent to decide when and at what power to charge or discharge its battery in order to achieve the greatest possible profit. The agent may then even accept losses in some periods to attain higher profits in the following ones. Technically, this can be formulated as a mathematical optimization problem with various constraints.

The function to be maximized is the profit Π which can be attained during a certain number of time periods T. Since electricity prices fluctuate over time, revenues and costs according to (8.4) and (8.10) can be expected to be different in each time interval so that the profit has to be written in the form

$$\Pi = \sum_{i=1}^{T} (R_i(P_i) - C_i(P_i)) \tag{8.15}$$

In a simple scenario, the power P_i in a certain time interval is limited by the maximum C-rate defined by the battery specifications and by an SOC constraint which determines how much energy can be charged into the battery or dispatched to the grid. The optimization problem can then be written in the form

$$\begin{aligned} &\underset{P_i}{\text{maximize}}\ \Pi \\ &\text{subject to } P_{\min} \leq P_i \leq P_{\max} \\ &\text{and } 0 \leq SOC_i \leq 1 \end{aligned} \tag{8.16}$$

The SOC change between two time steps is simply calculated according to the relation

$$SOC_i = SOC_{i-1} + \frac{P_i \cdot \Delta t}{Q} \tag{8.17}$$

A more sophisticated control strategy should also account for time periods at which the PEV is expected to be in use. This leads to additional constraints which ensure that the battery contains enough energy to complete the next trip. Given a battery capacity Q and energy consumption e, a trip starting at time interval m with an expected driving distance d implies the following condition for the SOC at time interval $m - 1$:

$$SOC_{m-1} \geq \frac{e \cdot d}{Q} \tag{8.18}$$

During the trip from time interval m to n no grid connection can be established so that

$$P_i = 0\ \forall i : m \leq i \leq n \tag{8.19}$$

The SOC change between the start and the end of the trip is then calculated by

$$SOC_n = SOC_m - \frac{e \cdot d}{Q} \tag{8.20}$$

In general, due to the non-linearity of realistic battery aging models, this problem has to be treated as a non-linear optimization problem. It may therefore either be addressed by a non-linear solver, or be piecewisely linearized and then be solved using a linear solver.

8.3.4 Artificial Intelligence

Given the dynamics of the system and complexity of the problem, the decision on whether to charge or discharge the battery needs to be automatized by an intelligent control unit in the PEV [19]. This control system should not only be capable of performing the mathematical optimization but should also be able to autonomously define the optimization constraints. One example for these constraints is the battery SOC required for driving. A user cannot be expected to be willing to manually specify the time, duration and expected energy consumption of the next trip. Instead, the system needs to be able to make appropriate predictions which ensure that the user does not run out of energy at any point in time. The better the prediction quality, the higher the expected profits because safety buffers can be kept small resulting in more battery capacity being available for V2G. This prediction, however, needs to be tailored to every individual user. Different users have different driving patterns, different driving styles and corresponding differences in energy consumption. Some may exhibit very regular commuting patterns while others might have highly varying itineraries. Different agents would therefore require different V2G strategies. In order to facilitate V2G, intelligent mechanisms are thus required to keep the user free of these concerns.

Artificial intelligence may also be beneficial in the context of price prediction. While 24 h predictions of electricity prices are available in a day ahead market, these are generally subject to an error which grows with the number of lookahead periods. Using these predictions may therefore compromise optimization efforts. With an increasing number of individual market participants, the market can be expected to gain additional dynamics that may further increase this error. For best optimization results it would therefore be crucial for an intelligent system to provide error estimations for certain times and locations. Also aggregators may require machine learning mechanisms in order to optimize their bids at the electricity markets.

8.4 Case Study

In this section, the cost and revenue model is applied to the electricity market data of Singapore using the optimization model from Sect. 8.3.3. The purpose of this case study is to demonstrate how different models and parameters lead to highly different conclusions on the economic viability of the V2G concept and to show how previous studies relate to a model which accounts for the dynamics of the problem. This case study should not be considered a thorough economic viability analysis of the V2G concept. Instead, its purpose is to create a sense for the influence of certain parameters and therefore demonstrate the importance of choosing models and parameters with care.

Due to the non-linearity of the battery model and the existence of integer variables, the optimization problem is treated as a *mixed integer non-linear program* (MINLP). The problem was implemented in the *general algebraic modeling system* (GAMS) and the COUENNE solver was used for optimization.

General parameters that are used for the optimization model are described in Sect. 8.4.1. Section 8.4.2 then introduces the specific electricity market data of Singapore. Findings of this case study are discussed in different scenarios in Sect. 8.4.3. Since several of the mentioned parameters broadly disperse in reality and are expected to change over time, the general parameters are varied as part of the sensitivity analysis presented in Sect. 8.4.4.

8.4.1 General Parameters

In all calculations of the case study, a battery pack capacity of 20 kWh is assumed. This is in accordance with the battery dimensions of the Nissan Leaf (24 kWh), the Mitsubishi i-MiEV (16 kWh), or the BMW i3 (18.8 kWh). The battery pack replacement costs are set to S$ 770[1] per kWh which reflects present prices according to [34, 35]. Additional equipment that enables PEVs to provide V2G services is expected to yield fixed costs of at most a few hundred S$. These costs are negligibly low when prorated over the whole lifetime of the battery pack and are therefore not considered in this case study.

As described in Sect. 8.3.3, power is treated as a continuous variable in the model and is kept in the range between −40 and +40 kW. This ensures a maximum C-rate of 2C meaning that the battery pack can be fully charged or discharged within half an hour. The energy efficiency of a charge-discharge cycle is determined by the efficiency of charging and discharging electronics as well as the efficiency of the battery pack. In the given C-rate range, the efficiency can be considered the same for charge and for discharge processes [36, 37]. In accordance with values from this literature, the total energy efficiency of a charge-discharge process is set to $\eta = 0.80$. The cycle stability model used for the assessment of the V2G concept in this case study was already described in Sect. 8.3.2. Prices for the different energy markets are described in Sect. 8.4.2.

[1] S$ 1 equals 0.80 USD (November 6, 2014).

8.4.2 Market Data

In Singapore, energy is traded at the *national electricity market singapore* (NEMS) which is controlled by the *energy market authority* (EMA) [38]. As already discussed in Sect. 8.2.1, this case study focuses on the ancillary service market. Depending upon the response time and the duration of providing ancillary services, the market distinguishes between *regulation* as well as *primary*, *secondary*, and *contingency reserve* [39]. When buying electricity from a generator an entity has to pay the *uniform singapore energy price* (USEP). This is considered the energy payment an entity receives when offering up-regulation [see (8.2)]. For the opposite case of down-regulation, no energy payment is provided [see (8.3)]. In addition to the energy payment, there is a compensation for holding power generation or remission potential available. This capacity payment is called *market regulation price* (MFP) and *market reserve price* (MRP) for regulation and reserve, respectively. While there is only one MFP, a distinct MRP is associated with each of the three classes of reserve. Due to the lack of concrete data it is assumed that offered energy will be entirely dispatched. This is to ensure that participants in the NEMS only earn money if they actually dispatch energy.

The electricity market price data used in this study cover the USEP, MFP and all classes of MRP for the entire year 2012 [40]. At the NEMS, all of these prices are adjusted on a half-hourly basis so that all presented calculations build on time series with a 30-min resolution, dividing one day in 48 periods. These prices are known 24 h in advance with an increasing average deviation, depending on the lookahead time. Calculations in this case study are based on a lookahead of 2 periods having a mean uncertainty of slightly above 1 ‰. Additionally, the end-consumer price for electricity, called *electricity tariff* (ET), is used. It mainly consists of energy costs (82 %) as well as transmission costs (17 %) and is subject to quarterly adaptation. To provide a rough overview of these prices and their temporal variance, their average values as well as standard deviations are given in Table 8.1.

8.4.3 Results

In the simplest possible scenario, a PEV is grid-connected 24 h per day, 365 days a year. It can therefore be considered a stationary energy buffer with a service level

Table 8.1 Key figures of the NEMS price data in 2012

	ET [S$/MWh]	USEP [S$/MWh]	MFP [S$/MWh]	MRP [S$/MWh]		
				Primary	Secondary	Contingency
Average	279.3	222.49	91.53	0.33	1.37	11.40
Standard deviation	5.69	112.92	40.35	2.26	4.48	64.86

agreement on availability of 100 %. Although this assumption is fairly unrealistic, it allows an upper bound estimate on the economic attractiveness of the different ancillary service types introduced in Sect. 8.2.1.

An illustration of the functioning of the method can be found in Fig. 8.4a–d which exemplarily shows the optimization result for a period of two days. Figure 8.4a depicts the MFP and USEP. In this figure it can roughly be distinguished between four different regions, the first exhibiting high prices, the second showing a period of lower prices and another high price period followed by a region of again lower prices. In Fig. 8.4b, the calculated optimal power is shown. It can be seen that during high price periods high charging and discharging power is applied while power remains low or even zero in the low price regions. The alternation between charging and discharging is due to the SOC constraint and ensures that the cycling occurs at moderate SOC levels. Naturally, as shown in Fig. 8.4c, profits in each time period follow the power curve. Oscillations into the negative direction are, however, fairly moderate since the compensation for down-regulation is credited. Figure 8.4d shows the accumulation of profits over time. The oscillations occur because losses are accepted in one period in order to make even higher profits in another. With increasing lookahead, these oscillations can be expected to become less regular since the algorithm has a higher degree of freedom for optimizing profits.

The outcome of the analysis for the different energy markets of Singapore can be found in Table 8.2. Results for all of the three reserve markets show that annual profits in the range from S$ 177 to S$ 912 can be gained. A large fraction of those

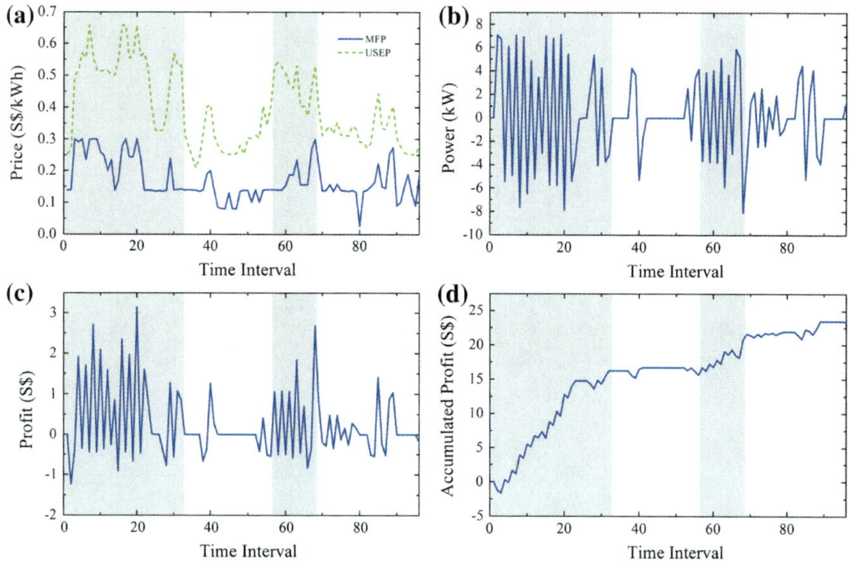

Fig. 8.4 Exemplary illustration of the optimization result for a time period of two days regarding **a** prices, **b** power, **c** profit and **d** accumulated profit

Table 8.2 Profits in different electricity markets regarding the simple scenario

Market	Profits [S$/year]	Profits, adjusted[a] [S$/year]
Reserve, primary	177	0
Reserve, secondary	183	0
Reserve, contingency	912	36
Regulation	394	109

[a] Extraordinarily profitable periods are left out

profits, however, results from just a dozen extraordinarily profitable periods which are most likely the outcome of a disruption in the power system. Leaving out those periods would result in annual profits of only up to S$ 36 for the contingency reserve and no profit at all for primary and secondary reserve. The reserve market is therefore not of interest for an economic application of the V2G concept and can be neglected in further analyses. For the regulation market the situation is more beneficial so that up to S$ 394 can be gained per year. Again, by neglecting the highest-price periods annual profits decrease to S$ 109.

To illustrate the fluctuations of achievable profits, Fig. 8.5 shows revenues and variable costs for one exemplary week in March 2012 for a fixed charging/dispatching power of 2 kW. It can be observed that revenues are highly variable over time. Some of these fluctuations have a considerable impact on annual income which leads to the discrepancy between profits and adjusted profits shown in Table 8.2.

A more realistic scenario assumes typical commuting habits of the population of the area of investigation. Therefore, mobility patterns of Singapore residents representing about 90 % of the population are used [20]. These patterns describe the trips various groups of people undertake on different days of the week. In particular, the data specify the start and end time of a trip as well as the type of destination categorized by *home*, *work* and *leisure*. This reveals information on the time windows at which PEVs can be connected to the grid depending upon the availability of charging stations at the various types of destinations.

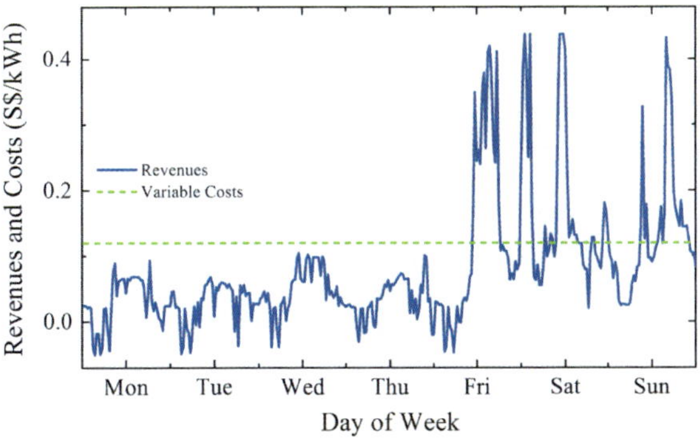

Fig. 8.5 Exemplary illustration of revenues and variable costs at fixed charging/dispatching power for a week in March 2012

Table 8.3 Profits in different electricity markets for the mobility pattern-based scenario

Market	Profits [S$/year]	Profits, adjusted[a] [S$/year]
Reserve, primary	60–178	0
Reserve, secondary	128–171	0
Reserve, contingency	597–863	21–25
Regulation	262–357	71–90

[a] Extraordinarily profitable periods are left out

Findings for different electricity markets regarding the mobility pattern-based scenario are presented in Table 8.3. Results for all of the three reserve markets show an annual profit in the range from S$ 60 to S$ 863, depending on the market and the applied mobility pattern. In the regulation market, annual profits lie in the range from S$ 262 to S$ 357. In the mobility pattern based approach, highest-price periods are sometimes left out anyway and therefore do not contribute as much to the resulting profits as in the simple scenario. Nevertheless, by completely leaving out these periods, annual profits decrease to S$ 0 to S$ 25 for reserve and S$ 71 to S$ 90 for regulation, respectively, again depending on the applied mobility pattern. Concluding from the results, profits in this configuration might not be high enough to practically apply the V2G concept.

8.4.4 Sensitivity Analysis

As discussed in Sect. 8.3, V2G profits strongly depend on multiple parameters. Above all are the battery inherent variable depreciation costs, the energy efficiency, and the electricity market prices whose influence will be discussed in this section. All investigations in this section are based on the simple scenario introduced in Sect. 8.4.3. Analyses are done ceteris paribus, meaning that each section discusses the variation of only one specific parameter.

8.4.4.1 Battery Model

The simplest view on battery lifetime which has been broadly employed in V2G literature is to assume a fixed number of possible cycles. Using this approach, battery lifetimes between 1,000 and 6,000 cycles would yield annual profits in the range between S$ 343 and S$ 2,992 in the presented case. While any of these cycle stabilities may be theoretically achievable under specific conditions, fixing the number of possible cycles to one particular value is a completely arbitrary decision because it neglects the dynamic processes within the battery. To realistically assess profits, a proper battery aging model is of utmost importance. A battery aging model provides the battery inherent variable depreciation costs from charging and

Table 8.4 Profits depending on the battery model

Battery model	Cyclic lifetime/Depending variables	Profits [S$/year]
Static	1,000	343
	2,000	785
	3,000	1,238
	4,000	1,736
	5,000	2,322
	6,000	2,992
Non-linear	DOD	653
	DOD, SOC	394

dispatching and is the main factor influencing the magnitude of annual profits. As described in Sect. 8.3.2 the cycle stability highly depends on a large number of parameters and varies among different cell chemistries.

In [20], a battery model for V2G profit calculation is presented that accounts for the important dependency of battery lifetime from the DOD. This model provides a significantly more realistic representation of battery aging costs than assuming a constant cyclic lifetime, however, it still lacks the SOC as the second most important parameter. Section 8.3.2 refers to a refined battery model that incorporates both, DOD and SOC, as parameters for the cyclic lifetime of a battery. This model more realistically assesses battery depreciation costs and was used for all calculations in this case study except stated otherwise. Results obtained by using the simpler model from [20] indicate annual profits of about S$ 650 while incorporating the SOC dependency reduces annual profits to roughly S$ 400. The significant discrepancies to the static approach and even between the fairly realistic models thus demonstrate that for any estimation of V2G profitability, a careful choice for a proper battery model needs to be made. Annual profits achievable depending on different battery models are presented in Table 8.4 (Fig. 8.6).

Varying the initial price of the battery pack has the same effect on the variable costs as proportionally changing the cyclic lifetime. Cutting fixed battery pack costs in half thus results in the same profits as doubling the cyclic lifetime. PEV manufacturers are already pre-selling battery packs to be delivered in 8 years at a price four times lower than the current one.[2] Besides the proper choice for the battery model, the initial battery pack costs are therefore a crucial factor when re-investigating V2G in the future.

8.4.4.2 Efficiency Factor

With the increasing maturity of battery technology or the integration of super capacitors into PEVs, the charging/discharging efficiency is expected to undergo

2 http://www.teslamotors.com/blog/2013-model-s-price-increase.

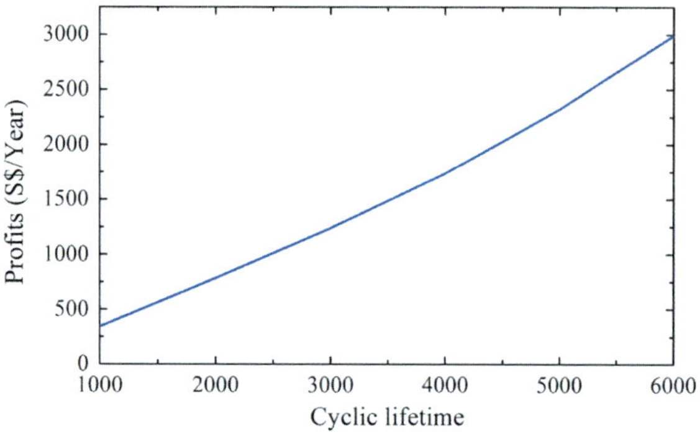

Fig. 8.6 Profits as a function of the cyclic lifetime

further improvements. The reduced energy dissipation will therefore result in decreased variable costs and ultimately in an increase in profits per period. Analysis show that on an absolute value in the range from 0.80 to 0.90, annual profits increase by roughly S$ 8 with a 1 % increase in efficiency. This relation results in a 15 % increase of annual profits when increasing the efficiency in the given range by 10 %. This relation is presented in Table 8.5 and Fig. 8.7 respectively. The impact of the efficiency factor is therefore slightly higher than proportional but cannot be considered a game changing parameter with regard to V2G profits.

8.4.4.3 Market Prices

Although prices are fixed by the electricity market, it might be useful to understand their influence on the profit. For this purpose, the end-consumer price for electricity (ET), the energy payment (USEP), as well as the capacity payment (MFP) were altered. As shown in Table 8.6, an increase/decrease in the ET by a factor will result in a decrease/increase of profits by less/more than this factor. For the USEP and the MFP it is the opposite case, meaning an increase/decrease in the USEP or MFP by a factor will result in an increase/decrease of profits by more/less than this factor. The ratio of

Table 8.5 Profits depending on the efficiency factor

Efficiency (one-way)	Efficiency (two-way)	Profits [S$/year]
0.7	0.49	207
0.75	0.56	248
0.8	0.64	293
0.85	0.72	343
0.9	0.81	396
0.95	0.9	468
1	1	565
0.89	0.80	394

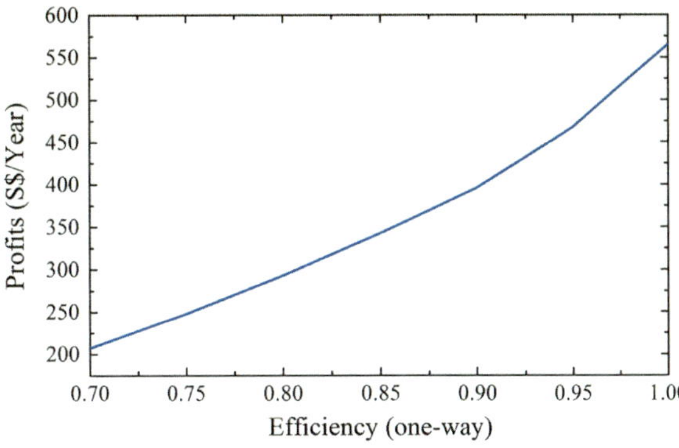

Fig. 8.7 Profits as a function of the efficiency factor

Table 8.6 Profits when varying the ET, USEP and MFP

Price	Price factor	Profits [S$/year]
ET	0.5	1,062
	0.75	565
	1	394
	1.25	324
	1.5	279
MFP	0.5	121
	0.75	233
	1	394
	1.25	639
	1.5	1,025
USEP	0.5	248
	0.75	310
	1	394
	1.25	542
	1.5	858

those factors favors the USEP and the MFP over the ET. An increase of the ET may therefore be more than compensated by a corresponding increase of the energy or capacity payment (Fig 8.8).

In Fig. 8.9 the distribution of USEP and MFP for the whole year 2012 is shown. It can be observed that the upper 3 % of prices (Region 3) exhibit a high variance with maximum values of up to S$ 4,000 in case of the USEP. This domain is followed by a broad plateau which consists of about 83 % of all time periods (Region 2). Finally, 14 % of the time intervals exhibit low prices with again higher fluctuations (Region 1). As already pointed out in Sect. 8.4.3, periods of Region 3 represent the extraordinarily profitable periods. In case of a low value of a parameter with a positive influence on profits (e.g., efficiency), only the periods of Region 3 are profitable. By increasing the value of this parameter, the intervals

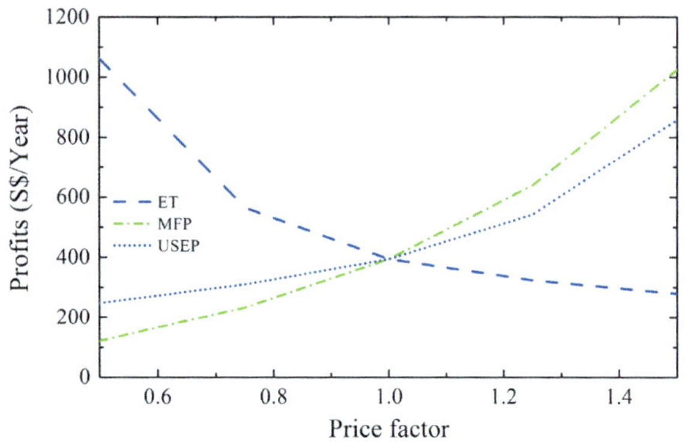

Fig. 8.8 Profits as a function of the ET, MFP and USEP

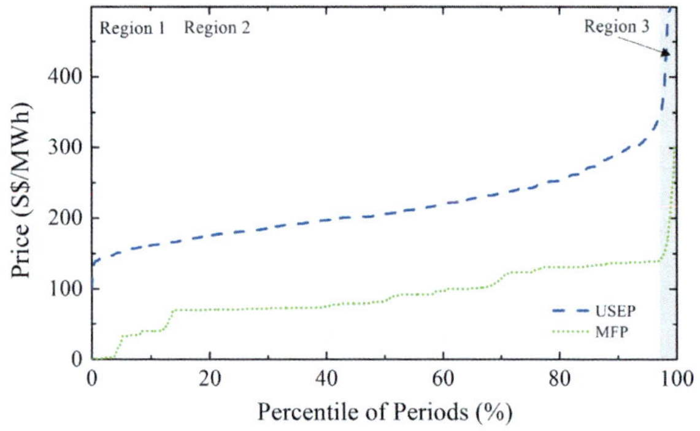

Fig. 8.9 Distribution of USEP and MFP

belonging to the plateau of Region 2 also become economically viable. Once this area is reached, a slight increase in this parameter significantly raises the number of profitable time periods leading to a considerable profit increase.

8.5 Discussion

This section summarizes and evaluates the findings of applying the presented intelligent agent behavior model to the electricity market data of Singapore. To show the relevance of these results for other countries, a qualitative discussion of the characteristics of the Singaporean market compared to other national markets is given. Furthermore, the benefits of the proposed strategy on power grid stability as well as the limitations of the applied cost and revenue model are examined. Finally, advantageous conditions for the practical implementation of the V2G concept are discussed.

8.5.1 Profitability of V2G in Singapore

The results presented in Sect. 8.4 show that given present market conditions of Singapore and realistic technical parameters, a maximum annual profit of S$ 110 could be achieved at the regulation market. This value, however, is only valid when the PEV is continuously grid-connected at all times except for the dozen outlier periods. When investigating the more realistic mobility pattern based scenario with a lower PEV availability profits decrease to S$ 80 per year. Since these numbers are the outcome of an optimization specifically targeted on profits, it is clear that alternative approaches would yield lower annual incomes. This shows that an economically motivated optimization strategy is a necessary condition for making V2G profitable in practice.

Nevertheless, these values indicate that under the given conditions V2G is yet unlikely to be an attractive concept for PEV owners in Singapore. There are, however, three main factors which could increase profits and thereby create conditions under which V2G could become a profitable business case. The first aspect relates to the battery where further development may either lead to reduced investment costs or where advancements in cell chemistry could yield higher cyclic lifetimes. This is, however, unlikely to happen on very short time scales. A second more realistic scenario to increase profits is therefore to utilize more information on future prices in the optimization process. Since longer temporal lookaheads also come with a higher uncertainty, this would, however, require improvements of the optimization approach to efficiently deal with uncertain information. A third factor could be to give a higher weight to capacity payments. In this study it was assumed that all energy offered to the ancillary service market is also being dispatched. In practice it could, however, also be possible to only receive a capacity payment without actually delivering energy. In this case, no depreciation costs occur which would ultimately have a positive effect on profits. Summing up all chances of realistically increasing profits, PEV owners may then be able to achieve an additional income of a few hundred S$ per year.

8.5.2 Applicability to Other National Markets

The cost-revenue model, the battery model, and the optimization method are generic under the described conditions and can be equally applied to other markets. In contrast, the results of the case study are specific to the conditions of Singapore and would not necessarily be identical in other markets. A quantitative conclusion for V2G in other countries would therefore require other case studies building on the presented models. Hence, in this section only a qualitative discussion putting the results obtained for Singapore into a greater context can be provided.

In the NEMS, as in many other markets, there is an energy and a capacity payment. Especially the latter has been identified as a major source of profits [2, 10, 13].

Economic analyses have to show whether or not the abolishment of this revenue stream in certain countries like Germany can be compensated by a higher energy payment. To investigate these questions, no changes in the cost-revenue model and the presented optimization method are required.

In Singapore, the fiscal framework appears to be advantageous in regard to the economic viability of V2G. The end-consumer electricity price almost entirely consists of generation and transmission costs without much taxes added. This is in contrast to other markets where a PEV owner would have to pay consumption taxes on the electricity price even if the energy may just be bought for the purpose of feeding it back into the grid. A different taxation policy could therefore yield higher profits in these countries while there is little potential for improvements in Singapore.

8.5.3 Model Limitations

A remaining weakness of the optimization approach is the inability to deal with uncertain price information and to make improved predictions based on knowledge from the past. This limits the temporal lookahead and therefore leaves parts of the optimization potential unutilized. This deficiency can, however, be addressed by further elaborating the optimization algorithm to incorporate these aspects.

Another issue is the difficulty of determining battery depreciation costs. As discussed above, this work employs empirically validated battery models which are believed to give a good estimation of battery aging costs. Nevertheless, conducting measurements regarding cell aging is time consuming and results in battery models that are always one generation behind the cells implemented in newest PEVs. The discussion in Sect. 8.3.2 also shows that the aging process depends on a large variety of parameters and may significantly differ among various cell chemistries. It therefore needs to be considered that for application purposes, battery aging models specifically developed for the corresponding battery type need to be employed.

8.6 Conclusion and Outlook

In this chapter, it is argued that static approaches for assessing the economic viability of V2G are of only limited informative value because market dynamics are neglected and the optimization potential individual agents have remains unexploited. In contrast, models which take the dynamics of market prices into account to optimize the charging/dispatching strategy for each individual PEV are considered more suitable for showing the economic potential of the V2G concept. Using dynamic approaches based on real-time information, a PEV may autonomously decide on its individual charging/dispatching strategy. This is achieved by the introduced optimization model which dynamically adapts charging/dispatching power of a PEV in each time period depending on its internal cost function and externally given market prices.

From the discussion of the results, it becomes clear that a control strategy motivated by an economic optimization approach is a necessary condition for the realization of the V2G concept in practice. Above all, this requires that V2G participants gain access to dynamic market prices for electricity instead of being bound to fixed tariffs. This is also necessary to trigger the behavior of V2G providers in order to achieve an effective load curve flattening. Current profits are assumed to be at the lower range of what PEV owners would accept for providing their batteries for ancillary services. An increase of prices for regulation and reserve energy which may follow growing shares of renewable energy sources or the introduction of premium tariff rates for V2G power would therefore be beneficial for the introduction of the V2G concept. On the cost side, battery depreciation is a crucial factor for the economic viability of V2G. While only moderate improvements of cyclic stabilities are expected in the near future, battery purchase costs are assumed to undergo a more rapid decrease, which in turn would lead to a significant cost reduction of V2G. A game changing innovation could additionally be the introduction of supercapacitors which exhibit significantly higher cyclic stabilities than batteries.

A possible soft factor obstructing the acceptance of the V2G concept is the reluctance of PEV owners to assign control of their vehicle battery to a third party. The possibility to manually take over control of the charging process by means of a smart mobile device could therefore be helpful for creating appropriate framework conditions for the practical employment of V2G.

A next step towards the implementation of V2G is to extend the optimization algorithms to address the issue of uncertain price information. By also incorporating machine learning mechanisms to improve price prediction, this can boost further exploitation of the optimization potential. Another important aspect of a feasible system in practice is the implementation of an aggregator. The challenge faced by this entity is to synchronize charging and discharging operations of a large number of PEVs under individual constraints. For this purpose, an algorithm has to be developed which estimates a VPPs power generation capacity and triggers behavior changes of the involved PEVs in case the power dispatch obligations cannot be fulfilled. Building on the entity of individual optimization algorithms combined with an aggregator mechanism would allow simulating the entire system in a nanoscopic simulation environment to quantify the overall impact on power grid stability. Together with temporally resolved data on the required quantity of regulation and reserve energy, this could then also yield a sound estimation of the V2G market size.

Acknowledgments This work was financially supported by the Singaporean National Research Foundation (NRF) under its Campus for Research Excellence and Technological Enterprise (CREATE).

References

1. Kempton W, Letendre SE (1997) Electric vehicles as a new power source for electric utilities. Transp Res Part D Transp Environ 2(3):157–175
2. Kempton W, Tomic J (2005) Vehicle-to-grid power fundamentals: calculating capacity and net revenue. J Power Sources 144(1):268–279
3. Clement-Nyns K, Haesen E, Driesen J (2011) The impact of vehicle-to-grid on the distribution grid. Electr Power Syst Res 81(1):185–192
4. Tomic J, Kempton W (2007) Using fleets of electric-drive vehicles for grid support. J Power Sources 168(2):459–468
5. Brooks AN (2002) Vehicle-to-grid demonstration project: grid regulation ancillary service with a battery electric vehicle. California Environmental Protection Agency, Air Resources Board, Research Division
6. Han S, Han S (2013) Economic feasibility of V2G frequency regulation in consideration of battery wear. Energies 6(2):748–765
7. Binding C, Gantenbein D, Jansen B et al (2010) Electric vehicle fleet integration in the Danish Edison project—a virtual power plant on the Island of Bornholm. In: IEEE power and energy society general meeting
8. Morais H, Sousa T, Vale Z et al (2014) Evaluation of the electric vehicle impact in the power demand curve in a smart grid environment. Energ Convers Manage 82:268–282
9. Peterson SB, Whitacre JF, Apt J (2010) The economics of using plug-in hybrid electric vehicle battery packs for grid storage. J Power Sources 195(8):2377–2384
10. Ciechanowicz D, Leucker M, Sachenbacher M (2012) Ökonomische Bewertung von Vehicle-to-Grid in Deutschland. In: Proceedings of the Multikonferenz Wirtschaftsinformatik (MKWI), Braunschweig, Feb, 2012
11. Hartmann N, Özdemir ED (2011) Impact of different utilization scenarios of electric vehicles on the German grid in 2030. J Power Sources 196(4):2311–2318
12. Dallinger D, Krampe D, Wietschel M (2011) Vehicle-to-grid regulation reserves based on a dynamic simulation of mobility behavior. IEEE Trans Smart Grid 2(2):302–313
13. Andersson SL, Elofsson AK, Galus MD et al (2010) Plug-in hybrid electric vehicles as regulating power providers: case studies of Sweden and Germany. Energ policy 38 (6):275–2762
14. Mullan J, Harries D, Bräunl T et al (2012) The technical, economic and commercial viability of the vehicle-to-grid concept. Energ Policy 48:394–406
15. Schuller A, Rieger F (2013) Assessing the economic potential of electric vehicles to provide ancillary services: the case of Germany. Zeitschrift für Energiewirtschaft 37(3):177–194
16. Jansen B, Binding C, Sundstrom O (2010) Architecture and communication of an electric vehicle virtual power plant. In: Proceedings of the first IEEE international conference on smart grid communications (SmartGridComm), Gaithersburg, Oct 2010
17. Musio M, Damiano A (2012) A virtual power plant management model based on electric vehicle charging infrastructure distribution. In: 3rd IEEE PES international conference and exhibition on innovative smart grid technologies (ISGT Europe), Berlin, Oct 2012
18. Sousa T, Morais H, Soares J et al (2012) Day-ahead resource scheduling in smart grids considering vehicle-to-grid and network constraints. Appl Energ 96:183–193
19. Ramchurn SD, Vytelingum P, Rogers A et al (2012) Putting the 'smarts' into the smart grid: a grand challenge for artificial intelligence. Commun ACM 55(4):86–97
20. Pelzer D, Ciechanowicz D, Aydt H et al (2014) A price-responsive dispatching strategy for vehicle-to-grid: an economic evaluation applied to the case of Singapore. J Power Sources 256:345–353
21. Wang J, Liu P, Hicks-Garner J et al (2011) Cycle-life model for graphite-$LiFePO_4$ cells. J Power Sources 196(8):3942–3948
22. Fu LJ, Endo K, Sekine K et al (2006) Studies on capacity fading mechanism of graphite anode for Li-ion battery. J Power Sources 162(1):663–666

23. Vetter J, Novák P, Wagner MR et al (2005) Ageing mechanisms in lithium-ion batteries. J Power Sources 147(1–2):269–281
24. Dubarry M, Svoboda V, Hwu R et al (2007) Capacity loss in rechargeable lithium cells during cycle life testing: the importance of determining state-of-charge. J Power Sources 174 (2):1121–1125
25. Abraham DP, Knuth JL, Dees DW et al (2007) Performance degradation of high-power lithium-ion cells—electrochemistry of harvested electrodes. J Power Sources 170(2):465–475
26. Zhang G, Shaffer CE, Wang CY et al (2013) In-situ measurement of current distribution in a Li-Ion cell. J Electrochem Soc 160(4):A610–A615
27. Abraham DP, Reynolds EM, Sammann E et al (2005) Aging characteristics of high-power lithium-ion cells with LiNi0.8Co0.15Al0.05O2 and Li4/3Ti5/3O4 electrodes. Electrochim Acta 51(3):502–510
28. Broussely M, Herreyre S, Biensan P et al (2001) Aging mechanism in Li ion cells and calendar life predictions. J Power Sources 98:0–8
29. Ramadass P, Haran B, White R et al (2002) Capacity fade of Sony 18650 cells cycled at elevated temperatures Part I. Cycling performance 112:606–613
30. Dubarry M, Truchot C, Liaw BY et al (2011) Evaluation of commercial lithium-ion cells based on composite positive electrode for plug-in hybrid electric vehicle applications. Part II. Degradation mechanism under 2C cycle aging. J Power Sources 196(23):10336–10343
31. Peterson SB, Apt J, Whitacre JF (2010) Lithium-ion battery cell degradation resulting from realistic vehicle and vehicle-to-grid utilization. J Power Sources 195(8):2385–2392
32. Schmalstieg J, Käbitz S, Ecker M et al (2014) A holistic aging model for $Li(NiMnCo)O_2$ based 18650 lithium-ion batteries. J Power Sources 257:325–334
33. De Los Rios A, Göntzel J, Nordstrom KE et al (2012) Economic analysis of vehicle-to-grid (V2G)-enabled fleets participating in the regulation service market. In: IEEE PES transactions on innovative smart grid technologies (ISGT), Tianjin, China, May 2012
34. Hensley R, Newman J, Rogers M (2012) Battery technology charges ahead. Technical report
35. Nelson PA, Bloom KG, Dees DW (2011) Modeling the performance and cost of lithium-ion batteries for electric-drive vehicles. Technical report, Argonne National Laboratory (ANL), Argonne
36. Schmidt H (2011) Technical data and start-up, NLG513, BRUSA Electronic AG. http://media3.ev-tv.me/BrusaManual2012.pdf. Accessed 17 Jun, 2014
37. Krieger EM (2013) Effects of variability and rate on battery charge storage and lifespan. Dissertation, Princeton University
38. Energy Market Authority (EMA) (2009) Introduction to the national electricity market of Singapore. Technical report
39. Energy Market Company (EMC) (2013) Glossary. https://www.emcsg.com/glossary. Accessed 13 May, 2014
40. Energy Market Company (EMC) (2012) Market data. https://www.emcsg.com/marketdata/priceinformation. Accessed 13 May, 2014

Chapter 9
Integration of PEVs into Power Markets: A Bidding Strategy for a Fleet Aggregator

Marina González Vayá, Luis Baringo and Göran Andersson

Abstract With a large-scale introduction of plug-in electric vehicles (PEVs), a new entity, the PEV fleet aggregator, is expected to be responsible for managing the charging of, and for purchasing electricity for, the vehicles. This book chapter deals with the problem of an aggregator bidding into the day-ahead electricity market with the objective of minimizing charging costs while satisfying the PEVs' flexible demand. The aggregator is assumed to potentially influence market prices, in contrast to what is commonly found in the literature. Specifically, the bidding strategy of the aggregator is formulated as a bi-level problem, which is implemented as a mixed-integer linear program. The upper-level problem represents the charging cost minimization of the aggregator, whereas the lower-level problem represents the market clearing. An aggregated representation of the PEV end-use requirements as a virtual battery, with time varying power and energy constraints, is proposed. This aggregated representation is derived from individual driving patterns. Since the bids of other market participants are not known to the aggregator ex ante, a stochastic approach is proposed, using scenarios based on historical data to describe such uncertain bids. The output of the proposed approach is a set of bidding curves, one for each hour of the day. Results show that by using PEV demand flexibility, the aggregator significantly reduces the charging cost. Additionally, the aggregator's bidding strategy has an important impact on market prices.

Keywords Electric vehicles · Aggregator · Smart-charging · Demand side management · Strategic bidding · Bilevel optimization · MPEC

M. González Vayá (✉) · L. Baringo · G. Andersson
EEH—Power Systems Laboratory, ETH Zurich, Zurich, Switzerland
e-mail: vayam@ethz.ch

S. Rajakaruna et al. (eds.), *Plug In Electric Vehicles in Smart Grids*,
Power Systems, DOI 10.1007/978-981-287-302-6_9

9.1 Introduction

9.1.1 Motivation

The introduction of plug-in electric vehicles (PEVs) in the vehicle fleet contributes to the reduction of greenhouse gas emissions and fossil fuel demand. However, a large-scale introduction of PEVs may cause problems in the electricity network. Most drivers have similar driving patterns. For example, on a typical day, most drivers depart from home to work in the morning, then vehicles are parked for several hours at the work location. After that, drivers may perform other activities such as shopping, and finally they drive back home, where vehicles remain parked until the next day. Without charging control or incentives, PEV owners would start charging their PEVs as soon as they park and until the battery is completely filled or they depart for the next trip. This means that most PEVs would be charging at the same time, which may cause voltage and congestion problems in the grid. Therefore, it becomes important to manage the charging of PEVs to avoid such problems [1, 2].

PEVs usually remain parked more time than what is needed to completely fill their batteries. In such a situation, it may be optimal to postpone the charging until a more suitable time, instead of starting charging as soon as PEVs are parked. This charging flexibility might be used, for example, to shift charging to hours of low electricity prices, which would reduce the charging costs, or to avoid voltage and congestion problems in the grid that may arise if a very large number of PEVs are simultaneously charged. Moreover, flexible charging may also alleviate problems caused by intermittent generators, such as photovoltaic and wind.

However, although it is potentially possible to achieve some improvements by letting each individual PEV owner manage the charging of its own PEV based on local information, such as voltage and frequency, or an exogenous time-of-use tariff, the optimal use of the flexibility potential can only be achieved when charging decisions are coordinated. Thus, a new entity, the so-called aggregator, is envisaged to be in charge of managing a PEV fleet. PEV owners communicate their driving requirements to the aggregator, which is responsible for purchasing electricity on behalf of the PEV owners.

In this context, we develop a strategic bidding algorithm for an aggregator in charge of a set of PEVs. The aggregator participates in the day-ahead market with the aim to minimize the charging costs of the PEV fleet, while at the same time satisfying the driving requirements of PEV owners. To do so, the aggregator decides both the bid price and the bid power level to be submitted to the day-ahead market.

On one hand, the bidding strategy is formulated using a bi-level model in which we explicitly model the working of the day-ahead market [3, 4]. On the other hand, the aggregation of PEVs is modeled using a virtual battery model with time-varying power and energy constraints [5].

Finally, note that the aggregator decides its bidding strategy one day in advance. At this point in time the aggregator faces a number of uncertainties, e.g., uncertainty in the market in which the aggregator participates and uncertainty in the driving requirements of the PEV fleet. Thus, it is important to model these uncertainties in order to obtain informed decisions. To do so, we consider a stochastic bi-level model in which uncertainties are efficiently modeled using a set of scenarios.

The focus of this chapter is on how to establish an optimal bidding strategy for a PEV aggregator, which results in an aggregated charging schedule. We do not describe how this aggregated profile is broken down into individual charging schedules for single PEVs. This latter problem can be treated separately in a consecutive step [6]. Distribution network constraints can be considered at this later stage. For example, distributed optimization approaches [7], as well as market based control approaches [6] could be used to determine individual charging schedules.

9.1.2 Literature Review and Contributions

First, we briefly review the literature on the integration of PEVs into power systems through an aggregation agent [8–13]. The aggregator is seen as an intermediary agent between PEVs and other system's entities, such as transmission system operators (TSOs), distribution system operators (DSOs), energy suppliers, balance group managers, and electricity markets [10–12]. For example, the aggregator could sell ancillary services to the TSO [14, 15], or negotiate demand side management contracts with the DSO, e.g. to avoid the overloading of lines and distribution transformers [16, 17]. Moreover, the aggregator could purchase the charging energy directly on electricity markets, such as the day-ahead market, and use the available flexibility to do so at the lowest cost possible. This is the subject addressed in this chapter. Different possible business models for the aggregator are discussed in [9].

Several publications have proposed optimization algorithms for the participation of PEV aggregators in electricity markets [4, 18–22]. In [18] an approach to determine the optimal charging and discharging profiles for a fleet, by forecasting the electricity prices in the Nordic day-ahead power exchange Nord Pool, is introduced. The charging and discharging flexibility is modeled through a set of aggregated representative PEV demand types, obtained by clustering individual driving patterns. This paper proposes both a model for the case where the aggregator is assumed to be a price-taker, and where it acts a price-maker, which leads to linear and quadratic programming techniques, respectively. An optimization framework for the aggregator to participate both in the day-ahead market and in the secondary reserve market simultaneously is presented in [19], and a case study for the Iberian market is analyzed. The fleet flexibility is modeled through constraints on the aggregated charging power and on the amount of aggregated energy to be purchased by a given time. The same authors extend the framework in [19] by introducing a model where PEVs are modeled individually [20], as opposed to the aggregated model in [19]. In [21] a method for the participation of a PEV aggregator in day-ahead and regulation markets is introduced, where a method for clustering driving patterns is proposed. The stochastic nature of driving patterns and market prices is taken into account but, instead of solving a stochastic optimization, a sequence of deterministic optimizations are performed, based on point estimates. As [19, 21], [22] proposes a method for bidding in the day-ahead and regulation markets, with a case study with data corresponding to PJM markets. It proposes a stochastic optimization framework where the uncertainty in both the fleet model and the market environment is taken into account.

Compared with the existing approaches, the framework described in this chapter, which extends the work in [4], comprises the following contributions:

- The aggregator is modeled as a *price-maker*. Apart from the approach in [18], all other models in the literature assume exogenous prices. Instead of estimating the relationship between demand and prices through a regression as in [18], the market clearing process is explicitly modeled, taking into account estimates of the bids of other market participants. Therefore the proposed framework not only provides the optimal bid *volumes* to be entered into the market, but also the optimal bid *prices*.
- A stochastic optimization framework is proposed to take into account market uncertainties, which are only explicitly modeled in [22], and partially in [21]. Since prices are not an exogenous parameter in our model, uncertainty here comes from bid prices and volumes in the market. Therefore, by considering several scenarios of market bid configurations, a *bid curve*, instead of a single bid, is computed for each time step.

As aforementioned, using a bi-level model we explicitly represent the working of the electricity pool, which allows us to consider endogenous market prices within the model. Bi-level models have been recently used in a wide range of energy-market topics [22]. Regarding bidding strategies, bi-level models have been applied for conventional producers [3], renewable generating units [23, 24], and PEV aggregators [4].

Bi-level models such as the one described in this chapter can be transformed into mathematical programs with equilibrium constraints (MPECs) [25, 26]. These MPECs are generally non-convex. However, several techniques have been proposed in the literature to efficiently solve them, e.g., mixed-integer linear programming (MILP) reformulations [27], relaxation methods [28], and interior methods [29].

9.1.3 Chapter Organization

The remaining of this chapter is organized as follows. Section 9.2 describes the main features of the proposed problem. Section 9.3 outlines the bi-level model used to derive the optimal bidding strategy of the aggregator, as well as its transformation into a MILP problem. Section 9.4 analyzes a clarifying example to illustrate the working of the model. Finally, Sect. 9.5 summarizes the chapter and provides some conclusions.

9.2 Problem Description

9.2.1 Notation

The main notation used in this chapter is stated below for quick reference. Other symbols are defined as needed throughout the text.

Index:

d	Index for demand bids $(d = d_1, \ldots, d_D)$
s	Index for supply bids $(s = s_1, \ldots, s_S)$
t	Index for time periods $(t = t_0, t_1, \ldots, t_T)$
v	Index for vehicles $(v = v_1, \ldots, v_V)$
ϖ	Index for market scenarios $(\varpi = \varpi_1, \ldots, \varpi_\Omega)$

Constants:

$b^{D}_{d\varpi t}$	Bid price of the dth demand bid at time period t and scenario ϖ [€/MWh]
$c^{S}_{s\varpi t}$	Bid price of the sth supply bid at time period t and scenario ϖ [€/MWh]
E^{ARR}_{t}	Energy contribution of vehicles arriving at time period t [MWh]
E^{DEP}_{t}	Energy drop of vehicles departing at time period t [MWh]
$E^{A,min}_{t}$	Lower bound of the energy content of the PEV aggregation at time period t [MWh]
$E^{A,max}_{t}$	Upper bound of the energy content of the PEV aggregation at time period t [MWh]
$E^{V,min}_{vt}$	Lower bound of the energy content of the battery of vehicle v at time period t [MWh]
$E^{V,max}_{vt}$	Upper bound of the energy content of the battery of vehicle v at time period t [MWh]
$P^{A,min}_{t}$	Lower bound of the aggregated charging power at time period t [MW]
$P^{A,max}_{t}$	Upper bound of the aggregated charging power at time period t [MW]
$P^{V,min}_{vt}$	Lower bound of the charging power of vehicle v at time period t [MW]
$P^{V,max}_{vt}$	Upper bound of the charging power of vehicle v at time period t [MW]
$\bar{P}^{D}_{d\varpi t}$	Volume of the dth demand bid at time period t and scenario ϖ [MW]
$\bar{P}^{S}_{s\varpi t}$	Volume of the sth supply bid at time period t and scenario ϖ [MW]
η	Charging efficiency of the PEV aggregation [pu]
π_{ϖ}	Weight of market scenario ϖ [pu]
Δt	Duration of time periods [h]

Variables:

$b^{A}_{\varpi t}$ Bid price of aggregator at time period t and scenario ϖ [€/MWh]

$E^{A}_{\varpi t}$ Energy content of the PEV aggregation at time period t and scenario ϖ [MWh]

$P^{A}_{\varpi t}$ Accepted volume of aggregator's demand bid at time period t and scenario ϖ [MW]

$P^{D}_{d\varpi t}$ Accepted volume of the dth demand bid at time period t and scenario ϖ [MW]

$P^{S}_{s\varpi t}$ Accepted volume of the sth supply bid at time period t and scenario ϖ [MW]

$\beta_{\varpi t}$ Marginal market price at time period t and scenario ϖ [€/MWh]

9.2.2 Aggregation Model

The aggregator needs a representation of the fleet's demand, to be incorporated as a set of constraints in its cost minimization problem. Since this demand is flexible, i. e., not fixed, it is not sufficient to forecast a demand profile, but a model that represents the set of feasible demand profiles is needed. For this purpose, the fleet can be modeled as a virtual battery, with a set of constraints on the aggregation's charging power and on the energy state of the virtual battery. The energy constraints introduce a link between the different time-steps of the problem, and therefore make the bidding strategy problem an inter-temporal problem. To derive the parameters of the aggregated virtual battery, a bottom-up approach is adopted, based on the driving patterns and characteristics of individual PEVs.

Starting at the individual PEV level, it is possible to define a lower and upper energy bound, $E^{V,\min}_{vt}$ and $E^{V,\max}_{vt}$, for the energy in the battery of vehicle v at time step t. The lower bound is calculated assuming that charging is deferred as much as possible, while the upper bound is calculated assuming charging takes place as soon as the vehicle plugs in. A minimum state of charge (SOC) is also considered to avoid excessive battery degradation.

Figure 9.1 shows an example of the computation of the energy bounds for a given vehicle. Before a trip, the energy content must reach a certain minimum level, so that the battery can provide enough energy for that trip. In fact, this minimum level is calculated with foresight, e.g. in the case of a short parking break before a long trip, this is taken into account before the previous trip. After a trip, the energy content can be built up again at the maximum charging rate until the battery is full. Moreover, it is assumed that batteries are fully charged at some point in time, and therefore the energy content of the battery cannot drop by more than the total daily energy consumption from this reference.

Similarly, lower and upper bounds can be derived for the charging power of a vehicle at a given time step $P^{V,\min}_{vt}$ and $P^{V,\max}_{vt}$. When a vehicle is connected, the upper bound of the charging power is equivalent to the lowest of the following

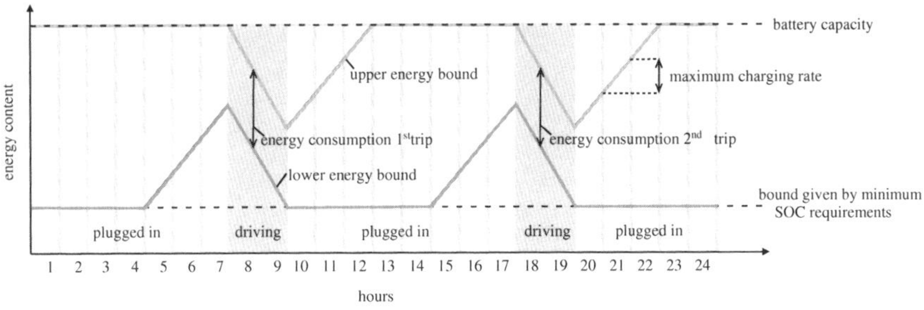

Fig. 9.1 Example of the energy bound computation for an individual vehicle

(i) the maximum charging power of the charging station or (ii) the maximum charging power of the PEV's battery. The lower bound of the charging power when the vehicle is connected is either zero or whatever is necessary to fulfill the trip energy requirements (inflexible charging). This inflexible charging can be derived by comparing the upper and lower energy bounds at subsequent time steps: If $E_{v(t-1)}^{V,max} < E_{vt}^{V,min}$, this implies that some charging is required at time step t. When the vehicle is disconnected, both the lower and upper power bounds are equal to zero.

Based on these descriptions at the individual vehicle level, an aggregated model of a virtual battery can be derived. The following equations describe the aggregated virtual battery model:

$$E_t^{A} = E_{(t-1)}^{A} + P_t^{A}\eta\Delta t + E_t^{ARR} - E_t^{DEP} \quad \forall t \tag{9.1}$$

$$P_t^{A,min} \le P_t^{A} \le P_t^{A,max} \quad \forall t \tag{9.2}$$

$$E_t^{A,min} \le E_t^{A} \le E_t^{A,max} \quad \forall t \tag{9.3}$$

The energy content of the virtual battery E_t^{A} stands for the aggregation of energy contents of all PEVs plugged in at a given time step. The dynamics of this variable are defined by Eq. (9.1) and given by the aggregated charging power at a given time step P_t^{A}, and by the positive/negative energy contributions of arriving/departing vehicles, E_t^{ARR}/E_t^{DEP}. The power and energy of the virtual battery should be within certain bounds as defined by Eqs. (9.2) and (9.3), respectively.

The aggregated parameters E_t^{ARR}, E_t^{DEP}, $E_t^{A,min}$, $E_t^{A,max}$, $P_t^{A,min}$ and $P_t^{A,max}$ are determined out of the individual PEV bounds. We denote u_{vt} a binary variable specifying the connection status of vehicle v at time step t, i.e. $u_{vt} = 1$ when the vehicle is connected and $u_{vt} = 0$ otherwise. Then the aggregated parameters are calculated as follows:

$$E_t^{\mathrm{A,max}} = \sum_v u_{vt} E_{vt}^{\mathrm{V,max}};\ E_t^{\mathrm{A,min}} = \sum_v u_{vt} E_{vt}^{\mathrm{V,min}} \tag{9.4}$$

$$P_t^{\mathrm{A,max}} = \sum_v u_{vt} P_{vt}^{\mathrm{V,max}};\ P_t^{\mathrm{A,min}} = \sum_v u_{vt} P_{vt}^{\mathrm{V,min}} \tag{9.5}$$

$$E_t^{\mathrm{DEP}} = \sum_v u_{v(t-1)} (u_{v(t-1)} - u_{vt}) E_{v(t-1)}^{\mathrm{V,max}};\ E_t^{\mathrm{ARR}} = \sum_v u_{vt} (u_{vt} - u_{v(t-1)}) E_{v(t-1)}^{\mathrm{V,max}} \tag{9.6}$$

The power and energy bounds of the individual vehicles contribute to the aggregated bounds when these vehicles are connected, as defined by Eqs. (9.4) and (9.5). The departure and arrival energy is estimated using the upper energy bound at the time of departure or arrival, respectively, see Eq. (9.6).

Figures 9.2 and 9.3 show examples for the aggregated energy and power bounds of a PEV fleet, respectively. These are the bounds used in our case study (more details on fleet composition are given later in Sect. 9.4). It can be seen that two dips in the upper bounds occur in the morning and in the evening, which correspond to typical commuting patterns.

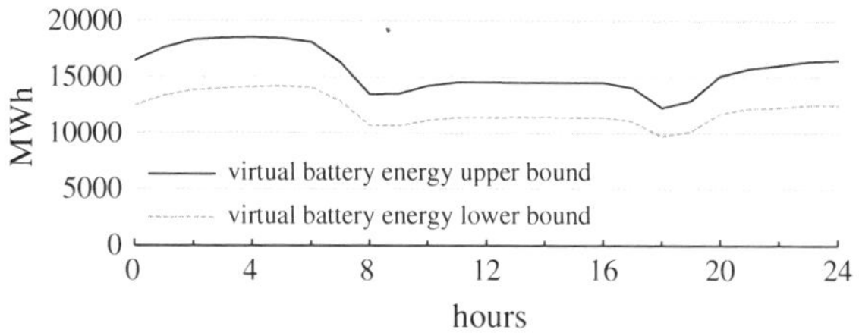

Fig. 9.2 Aggregated energy bounds

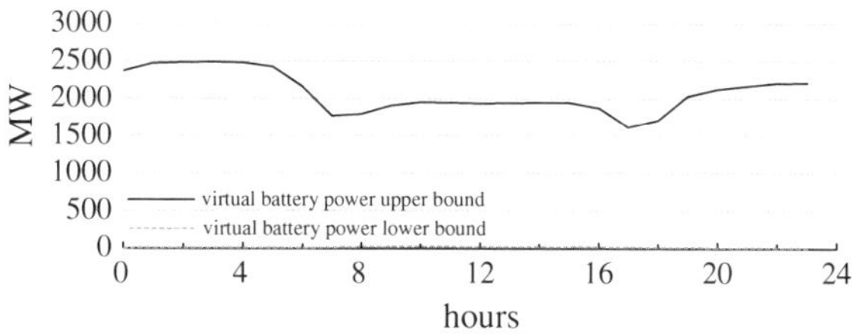

Fig. 9.3 Aggregated power bounds

9.2.3 Uncertainty Characterization

The aggregator decides on its bidding strategy in the day-ahead market one day in advance. At this point in time, (i) the aggregator does not know the actual driving requirements of the PEV owners in the PEV fleet that the aggregator manages, and (ii) the aggregator does not know the bidding decisions of the remaining market participants. That is, the bidding decisions are made in an uncertain environment that should be modeled correspondingly in order to obtain accurate and informed bidding decisions.

On one hand, the aggregator faces the uncertainty in the driving requirements of the PEV owners. However, note that we do not model each PEV individually but an aggregation of PEVs. While the driving requirements of an individual PEV are highly uncertain, the uncertainty in the driving requirements of a PEV fleet as a whole decreases as the number of PEVs in the fleet increases [4, 30]. For the sake of simplicity, we consider that the number of PEVs managed by the aggregator is large enough, and thus, we assume that the driving requirements of the PEV fleet are perfectly forecasted throughout this book chapter. From previous tests, fleets composed of more than 100,000 vehicles exhibited a small variance in the aggregated model parameters to be forecasted.

On the other hand, the aggregator faces the uncertainty in the bidding decisions of other market participants. Note that if these bidding decisions are not precisely described, the aggregator's bidding decisions may be suboptimal and, as a consequence, the aggregator may be scheduled in the day-ahead market an amount of energy lower/higher than that needed by the PEV fleet. Therefore, it is important to model the uncertainty in such bidding decisions of other market participants. To do this, we consider a set of market scenarios indexed by ϖ. These scenarios represent different realizations of the bidding decisions of other market participants for each time period of the considered planning horizon and are generated based on historical observations in the day-ahead market under study.

9.2.4 Bi-Level Structure

As previously explained, the aggregator is responsible for the charging of a PEV fleet. To do so, the aggregator participates in the day-ahead electricity market in which it submits a bid power volume and a bid price for each hour of the following day. This bid price is the maximum price the aggregator is willing to pay for its submitted bid power volume. Most markets allow market participants to submit bidding curves, i.e., instead of submitting just one bid power volume with its corresponding bid price, participants can submit different bid power volumes with the different prices they are willing to pay for each of them.

The aggregator derives its bidding strategy with the aim to minimize the charging costs of the PEV fleet, while at the same time taking into account the driving requirements of the PEV fleet. Additionally, the aggregator should incorporate the clearing of the market in which it participates into its decision framework. However, this market clearing is itself an optimization problem. Thus, we have an optimization problem (minimizing charging costs) subject to a set of constraints (driving requirements), as well as to another optimization problem (market clearing). This type of problems can be formulated as a bi-level model, also known as complementarity or hierarchical models [25].

Figure 9.4 schematically depicts the structure of the considered bi-level model:

1. There is an upper-level (UL) problem where the aggregator makes its bidding decisions, i.e. the bid prices to be submitted to the market so that the expected charging costs are minimized and the driving requirements of the PEV fleet are satisfied.
2. There is a set of lower-level (LL) problems representing the clearing of the day-ahead market under different market scenario realizations. In this market clearing problem, the total surplus of the market participants is maximized, given the bid volumes and prices submitted to the market. From the market clearing outputs, the aggregator is informed about its scheduled volume $\left(P^{A}_{\varpi t}\right)$ and the price $(\beta_{\varpi t})$ to pay for the charging power. We consider a common marginal market clearing price to pay/be paid by each market participant, which is the common situation of most day-ahead markets.

Both the UL and the LL problems are interconnected. On one hand, the aggregator decides its bidding strategy in the UL problem, which influences the market clearing through the aggregator's bid prices. On the other hand, from the market clearing problem represented in the LL problems, the aggregator obtains its

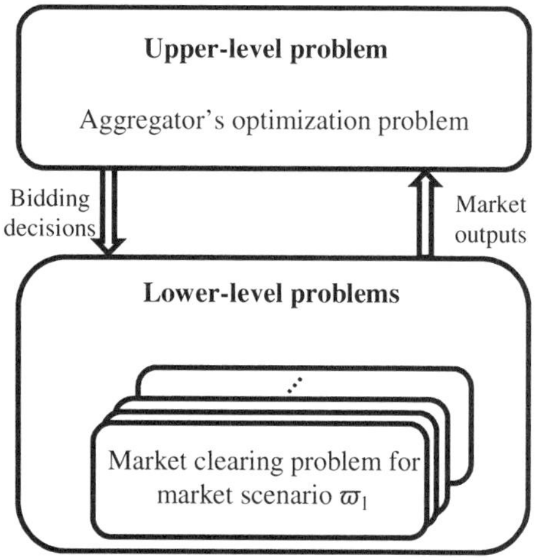

Fig. 9.4 Bi-level structure of the bidding problem

scheduled charging power, as well as the price it has to pay for this charging power, which in turn influence its costs, and therefore its bidding strategy. Thus, the UL and LL problems must be jointly solved. This is further explained in Sect. 9.3.

9.3 Model Formulation

9.3.1 Bi-Level Model

The optimal bidding strategy of the PEV aggregator can be formulated using the bi-level model below:

$$\text{Minimize}_{\Psi_{\varpi t}^{\text{UL}} \bigcup \Psi_{\varpi t}^{\text{LL,P}}} \sum_{\varpi} \pi_{\varpi} \sum_{t} \beta_{\varpi t} P_{\varpi t}^{\text{A}} \Delta t \tag{9.7}$$

subject to

$$E_{\varpi t}^{\text{A}} = E_{\varpi(t-1)}^{\text{A}} + P_{\varpi t}^{\text{A}} \eta \Delta t + E_{t}^{\text{ARR}} - E_{t}^{\text{DEP}} \quad \forall \varpi, \forall t \tag{9.8}$$

$$E_{t}^{\text{A,min}} \le E_{\varpi t}^{\text{A}} \le E_{t}^{\text{A,max}} \quad \forall \varpi, \forall t \tag{9.9}$$

$$P_{t}^{\text{A,min}} \le P_{\varpi t}^{\text{A}} \le P_{t}^{\text{A,max}} \quad \forall \varpi, \forall t \tag{9.10}$$

$$E_{\varpi t_F}^{\text{A}} = E_{\varpi t_0}^{\text{A}} = E_{\varpi' t_0}^{\text{A}} \quad \forall \varpi, \forall \varpi' \tag{9.11}$$

$$\beta_{\varpi t} = \lambda_{\varpi t} \quad \forall \varpi, \forall t \tag{9.12}$$

where $\lambda_{\varpi t}, P_{\varpi t}^{\text{A}} \in \arg\{$

$$\text{Minimize}_{\Psi_{\varpi t}^{\text{LL,P}}} \sum_{s} c_{s\varpi t}^{\text{S}} P_{s\varpi t}^{\text{S}} - \sum_{d} b_{d\varpi t}^{\text{D}} P_{d\varpi t}^{\text{D}} - b_{\varpi t}^{\text{A}} P_{\varpi t}^{\text{A}} \tag{9.13}$$

subjecto

$$\sum_{s} P_{s\varpi t}^{\text{S}} = \sum_{d} P_{d\varpi t}^{\text{D}} + P_{\varpi t}^{\text{A}} : \lambda_{\varpi t} \tag{9.14}$$

$$0 \le P_{s\varpi t}^{\text{S}} \le \bar{P}_{s\varpi t}^{\text{S}} : \underline{\varphi}_{s\varpi t}^{\text{S}}, \bar{\varphi}_{s\varpi t}^{\text{S}} \quad \forall s \tag{9.15}$$

$$0 \le P_{d\varpi t}^{\text{D}} \le \bar{P}_{d\varpi t}^{\text{D}} : \underline{\varphi}_{d\varpi t}^{\text{D}}, \bar{\varphi}_{d\varpi t}^{\text{D}} \quad \forall d \tag{9.16}$$

$$0 \le P_{\varpi t}^{A} \le P_{t}^{\text{A,max}} : \underline{\varphi}_{\varpi t}^{\text{A}}, \bar{\varphi}_{\varpi t}^{\text{A}} \tag{9.17}$$

$\}, \forall \varpi, \forall t$, where

$$\Psi^{\text{UL}}_{\varpi t} = \left\{ b^{\text{A}}_{\varpi t}, E^{\text{A}}_{\varpi t}, E^{\text{A}}_{\varpi t_0} \right\} \forall \varpi, \forall t, \tag{9.18}$$

$$\Psi^{\text{LL,P}}_{\varpi t} = \left\{ P^{\text{S}}_{s\varpi t}, \forall s; P^{\text{D}}_{d\varpi t}, \forall d; P^{\text{A}}_{\varpi t} \right\} \forall \varpi, \forall t, \tag{9.19}$$

and

$$\Psi^{\text{LL,D}}_{\varpi t} = \left\{ \lambda_{\varpi t}; \underline{\varphi}^{\text{S}}_{s\varpi t}, \bar{\varphi}^{\text{S}}_{s\varpi t}, \forall s; \underline{\varphi}^{\text{D}}_{d\varpi t}, \bar{\varphi}^{\text{D}}_{d\varpi t}, \forall d; \underline{\varphi}^{\text{A}}_{\varpi t}, \bar{\varphi}^{\text{A}}_{\varpi t} \right\} \forall \varpi, \forall t. \tag{9.20}$$

Bi-level problem (9.7)–(9.17) comprises UL problem (9.7)–(9.12) and a collection of LL problems (9.13)–(9.17), one for each market scenario and time period. The dual variables associated with the constraints of the LL problems are indicated following a colon.

On one hand, the UL problem (9.7)–(9.12) represents the aggregator's optimization problem. The aggregator aims to decide the bidding strategy that minimizes its expected charging costs (9.7) and that satisfies the driving requirements of the PEV fleet. Constraints (9.8) represent the evolution of the energy content of the virtual battery at each time period and for each scenario. Constraints (9.9) and (9.10) impose bounds on the energy content and the charging power of the virtual battery, respectively, as explained in Sect. 9.2.2. Constraints (9.11) impose that the energy content of the virtual battery at the end of the planning horizon must be equal to the energy content of the virtual battery at the beginning of the planning horizon, which must have the same value across scenarios. Otherwise, and due to the cost minimization nature of the problem, the aggregator would tend to deplete the energy of the virtual battery over time. Constraints (9.12) state that the price paid for charging is the market clearing price, which is obtained from the day-ahead market clearing problem.

On the other hand, the LL problems (9.13)–(9.17) represent the clearing of the day-ahead market under different market scenarios and time periods. In these day-ahead market problem, the minus surplus of the market is minimized (9.13). Constraints (9.14) define the supply-demand balance, while constraints (9.15)–(9.17) impose bounds on the accepted volumes of supply, demand and aggregator bids, respectively. The market clearing prices are defined as the sensitivity associated with the supply-demand balance constraints (9.14). Here we represent a simple market clearing process, where supply and demand bids are cleared together independent of their physical locations, i.e. network constraints are not taken into account. This type of market clearing is typical for spot power markets in many European countries. However, it is also possible to represent the lower level problem as a DC-Optimal Power Flow (DC-OPF), which also leads to a MILP problem [3].

9.3.2 MPEC

In order to obtain the solution of bi-level model (9.7)–(9.17), the UL and LL problems need to be jointly solved as explained below.

Note that each LL problem is continuous and linear, and thus, convex. Hence, the Karush-Kuhn-Tucker conditions are necessary and sufficient optimality conditions [31]. Therefore, we replace each LL problem by its KKT optimality conditions rendering an MPEC, which is provided below:

$$\text{Minimize}_{\Psi^{\text{UL}}_{\varpi t} \bigcup \Psi^{\text{LL,P}}_{\varpi t} \bigcup \Psi^{\text{LL,D}}_{\varpi t}} \quad (9.7) \tag{9.21}$$

subject to

$$\text{Constraints (9.8)–(9.12)} \tag{9.22}$$

$$\{\text{Constraint (9.14)} \tag{9.23}$$

$$c^{\text{S}}_{s\varpi t} - \lambda_{\varpi t} + \bar{\varphi}^{\text{S}}_{s\varpi t} - \underline{\varphi}^{\text{S}}_{s\varpi t} = 0 \quad \forall s \tag{9.24}$$

$$-b^{\text{D}}_{\varpi t} + \lambda_{\varpi t} + \bar{\varphi}^{\text{D}}_{d\varpi t} - \underline{\varphi}^{\text{D}}_{d\varpi t} = 0 \quad \forall d \tag{9.25}$$

$$-b^{\text{A}}_{\varpi t} + \lambda_{\varpi t} + \bar{\varphi}^{\text{A}}_{\varpi t} - \underline{\varphi}^{\text{A}}_{\varpi t} = 0 \tag{9.26}$$

$$0 \le P^{\text{S}}_{s\varpi t} \perp \underline{\varphi}^{\text{S}}_{s\varpi t} \ge 0 \quad \forall s \tag{9.27}$$

$$0 \le \bar{P}^{\text{S}}_{s\varpi t} - P^{\text{S}}_{s\varpi t} \perp \bar{\varphi}^{\text{S}}_{s\varpi t} \ge 0 \quad \forall s \tag{9.28}$$

$$0 \le P^{\text{D}}_{d\varpi t} \perp \underline{\varphi}^{\text{D}}_{d\varpi t} \ge 0 \quad \forall d \tag{9.29}$$

$$0 \le \bar{P}^{\text{D}}_{d\varpi t} - P^{\text{D}}_{d\varpi t} \perp \bar{\varphi}^{\text{D}}_{d\varpi t} \ge 0 \quad \forall d \tag{9.30}$$

$$0 \le P^{\text{D}}_{\varpi t} \perp \underline{\varphi}^{\text{D}}_{\varpi t} \ge 0 \tag{9.31}$$

$$0 \le P^{\text{A,max}}_{t} - P^{\text{A}}_{\varpi t} \perp \bar{\varphi}^{\text{A}}_{\varpi t} \ge 0 \tag{9.32}$$

$\}, \forall \varpi, \forall t.$

MPEC (9.21)–(9.32) comprises the objective function of the UL problem (9.21), constraints of the UL problem (9.22), equality constraints of the LL problems (9.23), equality constraints (9.24)–(9.26) obtained by differentiating the Lagrangian function of each LL problem with respect to the primal variables included in sets $\Psi^{\text{LL,P}}_{\varpi t}$, and complementarity constraints (9.27–9.32).

9.3.3 Linearization

MPEC (9.21)–(9.32) includes two types of nonlinear terms, namely:

1. The objective function (9.21).
2. The complementarity constraints (9.27)–(9.32).

However, these nonlinear terms can be replaced by exact equivalent mixed-integer linear expressions as explained in the Appendix. Using such mixed-integer linear terms, the problem of identifying the optimal bidding strategy of a PEV aggregator can be finally formulated as the following MILP problem:

$$\text{Minimize}_{\Psi^{\text{UL}}_{\varpi t} \bigcup \Psi^{\text{LL,P}}_{\varpi t} \bigcup \Psi^{\text{LL,D}}_{\varpi t} \bigcup \Psi^{\text{AUX}}_{\varpi t}} \sum_{\varpi} \pi_{\varpi} \sum_{t} \Upsilon_{\varpi t} \tag{9.33}$$

subject to

$$\text{Constraints (9.2)–(9.7)} \tag{9.34}$$

$$\text{Constraints (9.23)–(9.32)} \tag{9.35}$$

$$\text{Constraints (9.67)–(9.81)} \tag{9.36}$$

$\}, \forall \varpi, \forall t$, where

$$\Psi^{\text{LL,AUX}}_{\varpi t} = \{\underline{u}^{S}_{s\varpi t}, \bar{u}^{S}_{s\varpi t}, \forall s; \underline{u}^{D}_{d\varpi t}, \bar{u}^{D}_{d\varpi t}, \forall d; \underline{u}^{A}_{\varpi t}, \bar{u}^{A}_{\varpi t}\} \forall \varpi, \forall t. \tag{9.37}$$

9.3.4 Building Bidding Curves

From the optimal solution of the MILP problem (9.33)–(9.37), the aggregator obtains, for each market scenario realization ϖ, a value for the bid price $b^{\text{A}}_{\varpi t}$ and a value for the scheduled power volume $P^{\text{A}}_{\varpi t}$. Based on the resulting scheduled charging power in the day-ahead market and the computed bid prices for each scenario, the aggregator builds its bid blocks, as schematically depicted in Fig. 9.5.

The resulting MILP problem (9.33)–(9.37) allows the aggregator to decide its optimal bidding strategy in the market. For each market scenario, the aggregator obtains a value for the bid price and a value for the bid volume to be submitted to the market. However, demand bid curves need to be decreasing in price. With the formulation provided in Sect. 9.3.1, this is not ensured since bid prices and volumes for different market scenarios are not linked. In order to ensure that the resulting bid blocks are decreasing in price, it is necessary to include an additional set of constraints in MILP problem (9.33)–(9.37):

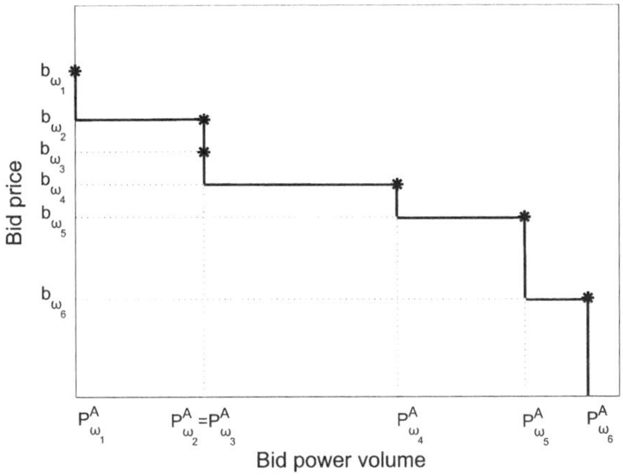

Fig. 9.5 Illustrative example of bid blocks

$$b^{\mathrm{A}}_{\varpi t} - b^{\mathrm{A}}_{\varpi' t} \leq x_{\varpi\varpi' t} M^{b} \quad \forall \varpi, \forall \varpi' > \varpi, \forall t \tag{9.38}$$

$$b^{\mathrm{A}}_{\varpi t} - b^{\mathrm{A}}_{\varpi' t} \geq (x_{\varpi\varpi' t} - 1) M^{b} \quad \forall \varpi, \forall \varpi' > \varpi, \forall t \tag{9.39}$$

$$P^{\mathrm{A}}_{\varpi' t} - P^{\mathrm{A}}_{\varpi t} \leq y_{\varpi\varpi' t} M^{\mathrm{P}} \quad \forall \varpi, \forall \varpi' > \varpi, \forall t \tag{9.40}$$

$$P^{\mathrm{A}}_{\varpi' t} - P^{\mathrm{A}}_{\varpi t} \geq (y_{\varpi\varpi' t} - 1) M^{\mathrm{P}} \quad \forall \varpi, \forall \varpi' > \varpi, \forall t \tag{9.41}$$

$$x_{\varpi\varpi' t} + y_{\varpi\varpi' t} = 2 z_{\varpi\varpi' t} \quad \forall \varpi, \forall \varpi' > \varpi, \forall t \tag{9.42}$$

$$x_{\varpi\varpi' t}, y_{\varpi\varpi' t}, z_{\varpi\varpi' t} \in \{0, 1\} \quad \forall \varpi, \forall \varpi' > \varpi, \forall t \tag{9.43}$$

where M^{b} and M^{P} are large enough positive constants.

The working of constraints (9.38)–(9.43) is illustrated in Fig. 9.6. Let us consider two market scenarios, namely ϖ and ϖ'. We obtain, for each of them, a value for the bid price $(b^{\mathrm{A}}_{\varpi t}, b^{\mathrm{A}}_{\varpi' t})$ and a value for the bid power volume $(P^{\mathrm{A}}_{\varpi t}, P^{\mathrm{A}}_{\varpi' t})$. Figure 9.6 depicts the four different possibilities for these pairs of values. However, only cases (a) and (b) are feasible bidding curves since case (c) and (d) have a positive slope, and thus, cannot be used as demand curves.

In case (a), we observe that $b^{\mathrm{A}}_{\varpi} - b^{\mathrm{A}}_{\varpi'} \geq 0$ and $P^{\mathrm{A}}_{\varpi' t} - P^{\mathrm{A}}_{\varpi t} \geq 0$. This is ensured by considering $x_{\varpi\varpi'} = 1$ and $y_{\varpi\varpi'} = 1$ as explained below:

1. We obtain that $0 \leq b^{\mathrm{A}}_{\varpi} - b^{\mathrm{A}}_{\varpi'} \leq M^{b}$ by Eqs. (9.38) and (9.39). Provided that M^{b} is a large enough positive constant, this ensures $b^{\mathrm{A}}_{\varpi} - b^{\mathrm{A}}_{\varpi'} \geq 0$.
2. We obtain that $0 \leq P^{\mathrm{A}}_{\varpi' t} - P^{\mathrm{A}}_{\varpi t} \leq M^{\mathrm{P}}$ by Eqs. (9.40) and (9.41). Provided that M^{P} is a large enough positive constant, this ensures $P^{\mathrm{A}}_{\varpi' t} - P^{\mathrm{A}}_{\varpi t} \geq 0$.

Note also that $x_{\varpi\varpi'} = 1$ and $y_{\varpi\varpi'} = 1$ satisfy Eq. (9.42), with $z_{\varpi\varpi'} = 1$.

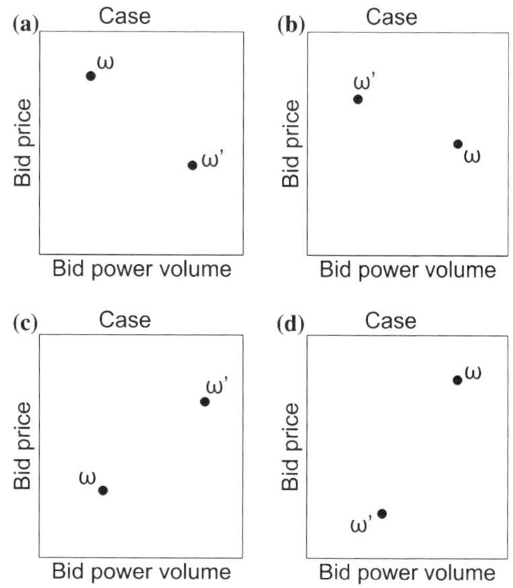

Fig. 9.6 Different alternatives for bidding curves

Similarly, in case (b), we observe that $b^{\mathrm{A}}_{\varpi} - b^{\mathrm{A}}_{\varpi'} \leq 0$ and $P^{\mathrm{A}}_{\varpi' t} - P^{\mathrm{A}}_{\varpi t} \leq 0$, which is ensured by considering $x_{\varpi\varpi'} = 0$ and $y_{\varpi\varpi'} = 0$. These values of $x_{\varpi\varpi'}$ and $y_{\varpi\varpi'}$ also satisfy Eq. (9.42), with $z_{\varpi\varpi'} = 0$.

On the other hand, in cases (c) and (d) we observe $b^{\mathrm{A}}_{\varpi} - b^{\mathrm{A}}_{\varpi'} \leq 0$, $P^{\mathrm{A}}_{\varpi' t} - P^{\mathrm{A}}_{\varpi t} \geq 0$ and $b^{\mathrm{A}}_{\varpi} - b^{\mathrm{A}}_{\varpi'} \geq 0$, $P^{\mathrm{A}}_{\varpi' t} - P^{\mathrm{A}}_{\varpi t} \leq 0$, respectively. To achieve this, we need to consider $x_{\varpi\varpi'} = 0$, $y_{\varpi\varpi'} = 1$ in case (c) and $x_{\varpi\varpi'} = 1$, $y_{\varpi\varpi'} = 0$ in case (d). However, these two cases violate constraint (9.42), i.e. there is no corresponding feasible value for $z_{\varpi\varpi'}$. By including constraints (9.38)–(9.43) in the bi-level model described in Sect. 9.3.1 we ensure non-increasing demand curves.

Finally, note that constraints (9.38)–(9.43) allow us to represent those cases with equal power volumes and different bid prices. These cases represent, for example, situations where the aggregator needs a specific power volume to satisfy the driving requirements regardless of the price to be paid.

9.3.5 Central Dispatch Model

As a benchmark for the proposed bidding strategy we consider a central dispatch model [4] whose formulation is provided below:

$$\text{Minimize}_{\Psi^{\mathrm{CD}}_t} \sum_s c^{\mathrm{S}}_{st} P^{\mathrm{S}}_{st} - \sum_d b^{\mathrm{D}}_{dt} P^{\mathrm{D}}_{dt} \tag{9.44}$$

subject to

$$E_t^{\mathrm{A}} = E_{(t-1)}^{\mathrm{A}} + P_t^{\mathrm{A}}\eta\Delta t + E_t^{\mathrm{ARR}} - E_t^{\mathrm{DEP}} \quad \forall t \tag{9.45}$$

$$E_t^{\mathrm{A,min}} \leq E_t^{\mathrm{A}} \leq E_t^{\mathrm{A,max}} \quad \forall t \tag{9.46}$$

$$P_t^{\mathrm{A,min}} \leq P_t^{\mathrm{A}} \leq P_t^{\mathrm{A,max}} \quad \forall t \tag{9.47}$$

$$E_{t_F}^{\mathrm{A}} = E_{t_0}^{\mathrm{A}} \tag{9.48}$$

$$\sum_s P_{st}^{\mathrm{S}} = \sum_d P_{dt}^{\mathrm{D}} + P_t^{\mathrm{A}} \tag{9.49}$$

$$0 \leq P_{st}^{\mathrm{S}} \leq \bar{P}_{st}^{\mathrm{S}} \tag{9.50}$$

$$0 \leq P_{dt}^{\mathrm{D}} \leq \bar{P}_{dt}^{\mathrm{D}} \tag{9.51}$$

where

$$\Psi_t^{\mathrm{CD}} = \left\{P_{st}^{\mathrm{S}}, \forall s; P_{dt}^{\mathrm{D}}, \forall d; P_t^{\mathrm{A}}; E_t^{\mathrm{A}}; E_{t_0}^{\mathrm{A}}\right\}, \forall t, \tag{9.52}$$

In this central dispatch model, the constraints of the virtual battery used to model the PEV aggregation (9.45)–(9.48) are included as additional constraints of the market clearing problem (9.13)–(9.17). Note that in this problem there is no uncertainty in the market since the market operator knows the bids of the participants. This benchmark represents a theoretical welfare-maximizing dispatch of the PEV fleet together with supply and demand bids by a central agent. It can therefore be used to assess the market power potential of the aggregator.

9.4 Illustrative Example

9.4.1 Data

The proposed approach is tested with price curves of the electricity spot market for the bidding area of Germany/Austria, obtained from the European Energy Exchange (EEX). The market clearing volumes and prices of this day-ahead market are public and reported on the EEX website [32]. However, the price curves containing detailed information on demand and supply bids, which are used here, are only commercially available. The bid prices in this market have to lie between −3000 € and 3000 €. The simulation period spans the period from February to November 2013. Each day is divided into hourly time steps, i.e. Δt corresponds to

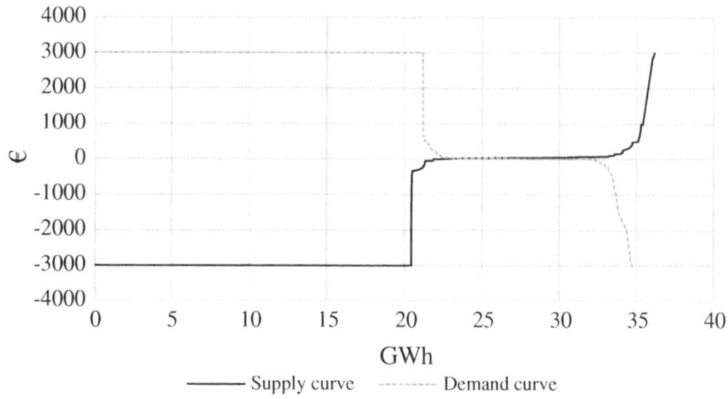

Fig. 9.7 Supply and demand curves of the day-ahead market of the Germany/Austria area on March 8, 2013, hour 24

one hour. Figure 9.7 shows the supply and demand curves, without aggregator bids, for a given day and hour.

The scenarios for supply and demand bids are derived from past data. For the sake of simplicity, we consider Ω equally weighted scenarios, which are defined as the supply and demand bids of the Ω previous weeks to the week under study. For example, for a given Monday at noon, the bids of the previous Ω Mondays at noon are used as scenarios. This simple method is used here for demonstration purposes, but more advanced estimation methods are possible. Note that although there is a rich literature in price estimation techniques, bid estimation has received little attention.

We assume a 2 % penetration of PEVs in Germany and Austria, which corresponds to almost one million PEVs [33, 34]. The value of 2 % is chosen because it is a number low enough to be realistic in the near future, but already high enough to affect market prices, as will be shown later. In Germany, the goal is to have 1 million PEVs by 2020 [35].

The driving patterns used in the case study are obtained from the transportation simulation tool MATSim [36]. MATSim is an agent-based simulation where each agent has a set of activities to be performed (e.g. go to work, go shopping, etc.) and the optimization selects the driving patterns that maximize agents' utility, taking into account factors such as the induced traffic and the available means of transportation. This simulation is specific to Switzerland and represents typical mobility behavior on weekdays. No equivalent simulations exist for Germany and Austria and for weekend mobility. For this reason we use the Swiss data for the numerical example, although it does not allow an exact quantitative analysis. Nevertheless, it is still possible to qualitatively test the performance of the described bidding strategy. The assumed maximum charging power for vehicles is 3.7 kW and the charging efficiency 90 %. The energy consumption is derived from trip distances with the factor 0.2 kWh/km. Battery capacities of 16 and 24 kWh are used, each for half of the fleet. A minimum SOC of 20 % is enforced. Table 9.1 summarizes the

Table 9.1 PEV parameters

Charging power	Charging efficiency	Minimum SOC	Battery capacity	Average consumption
3.7 kW	90 %	20 %	16/24 kWh	0.2 kWh/km

main PEV parameters. It is assumed that PEVs can potentially charge whenever they are parked for 1 h. Therefore we indirectly assume that charging infrastructure is available at every location. Our results therefore represent a lower bound on the costs of charging. In practice, charging flexibility would be partially reduced because of the unavailability of charging infrastructure at some locations.

The simulations were run with MATLAB R2012a on a 2.67 GHz Intel Xeon processor. The optimization problems (MILP problems for the bidding strategy and LP problems for the central dispatch) were solved using CPLEX v12.5.1.

9.4.2 Results

Table 9.2 gives an overview of the results of both the central dispatch and the bidding strategy. For the bidding strategy, results are computed under perfect information and with uncertainty in market bids. In the latter case, different numbers of scenarios are considered. Results under perfect information are computed with a single scenario, which corresponds to a perfect forecast of the market's supply and demand bids. The central dispatcher, since it decides on the charging of

Table 9.2 Results overview

	Central dispatch	Bidding strategy			
		Perfect information	One scenario	Two scenarios	Three scenarios
Average cost [€/MWh]	28.52	27.88	26.53	28.43	28.98
Average cost including additional purchases/sales [€/MWh]	28.52	27.88	33.64	31.92	31.33
Average energy purchased [MWh]	6,092	6,092	2,615	4,026	4,571
Average energy shortfall [MWh]	–	–	3,477	2,188	1,671
Average energy overshoot [MWh]	–	–	0	121	150
Average problem size (continuous/integer)	5,180/0	15,487/10,292		30,954/ 20,611	46,350/ 30,971
Average solving time [s]	0.041	28		319	3,612

the PEV fleet simultaneously with the market clearing, does not face any uncertainty in market bids.

The costs reported in Table 9.2 are the total costs of purchasing electricity divided by the total purchased electricity for the considered period. Since the aggregator's bidding strategy can lead to purchasing insufficient or, more rarely, excess energy in the day-ahead market, these costs are not directly comparable. To really be able to compare the strategies, it would be necessary to simulate a second step, after the day-ahead market clearing, where the aggregator would purchase the missing energy in a different market, or through bilateral contracts. This is outside of the scope of this chapter. Instead, an additional cost computation is performed, estimating the potential costs of purchasing/selling the mismatch between the required energy and the assigned volume in the day-ahead market. For this purpose, the additional purchases/sales were priced at the average price for each of the days.

It can be seen that if the aggregator had perfect information of market bids, it would face slightly lower costs than under a central dispatch. This means that the aggregator could potentially exercise market power. With the bidding strategy under uncertainty, the costs are also around 28 €/MWh in the case where several scenarios are considered, and 26.53 €/MWh for a single scenario. The costs of the bidding strategy using a single scenario are lower than under perfect information, but in this case the aggregator does not even manage to purchase 50 % of the required energy, which is not a desirable situation. Although the costs are the highest in the three-scenario case, the energy shortfall is significantly lower than in the other cases. Taking the additional purchases/sales into account in the cost computation, the three-scenario case performs best. However, the difference between the two- and three-scenario cases is not very large. The size of the problem grows approximately linearly with the number of scenarios, whereas the computation time grows approximately exponentially. For larger number of scenarios partitioning decomposition techniques could be applied to reduce the computation time [37, 38]. We also computed the costs that the aggregator would incur if demand were left uncontrolled, i.e. each PEV would start charging as soon as it is parked. In this case the aggregator has to purchase this predefined demand in the market, independent of the price. This is equivalent to a bid with the maximum allowed bid price and the aggregated charging demand at each time step as the bid volume. The costs for charging in this case are 43.07 €/MWh in average, i.e. a much higher value than the costs reported in Table 9.2. This means that it is in principle economically attractive to exploit the charging flexibility.

Figure 9.8 shows profiles for market clearing volumes and prices, as well as accepted aggregator bid volumes (charge profiles) for a particular day. Results are shown for both the bidding strategy under perfect information, as well as under uncertainty, using three scenarios. Although the aggregator's purchased volume corresponds to a small percentage of the total market volume only, the impact on prices is clearly visible. The aggregator tends to shift most demand to the night hours, when prices are lowest, and also part of demand to the time between the daily peaks. The latter cannot be avoided since demand is not flexible enough to move all charging to the night. It can be seen that the charge profiles with perfect

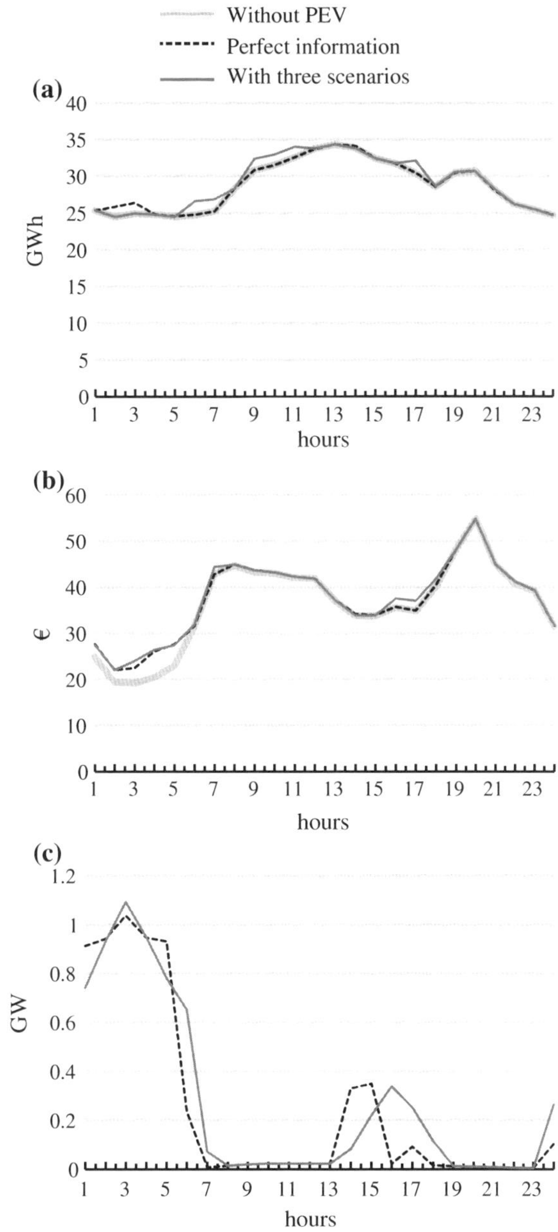

Fig. 9.8 Profiles for Friday March 8, 2013 for the following cases: (i) without PEV—reference market clearing without the participation of an aggregator, (ii) perfect information—bidding strategy with perfect information on market bids, (iii) with three scenarios—bidding strategy using three scenarios. **a** Market volumes, **b** market prices, **c** charge profile

information and with uncertainty are relatively similar, since the aggregator is capable of predicting the general price trends by using several scenarios. However, it is not capable of perfectly optimizing it strategy without full information on market bids.

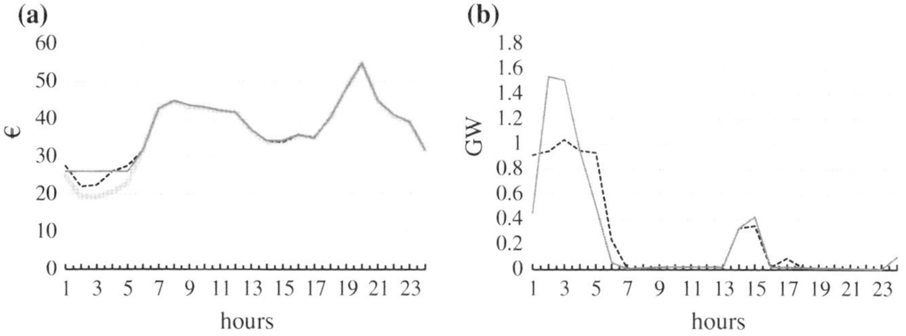

Fig. 9.9 Profiles for Friday March 8, 2013 for the following cases: (i) without PEV—reference market clearing without the participation of an aggregator, (ii) perfect information—bidding strategy with perfect information on market bids, (iii) central dispatch. **a** Market prices, **b** charge profile

Figure 9.9 focuses on the differences between the central dispatch and the aggregator's bidding strategy under perfect information, i.e. the aggregator has the same information (market bids) as the central dispatcher. It can be seen that whereas the central dispatcher tries to generate a price curve that is as flat as possible, the aggregator tries to increase prices as little as possible, which leads to a flatter charge profile. This explains the small differences in charging costs between the two approaches, as reported in Table 9.2.

Figure 9.10 shows two exemplary bid curves resulting from the simulation considering three scenarios. It can be seen that one of these bid curves (hour 7) consists of three bid blocks, each corresponding to a different price/volume

Fig. 9.10 Examples of aggregator bid curves considering thee scenarios, Thursday April 18, 2013

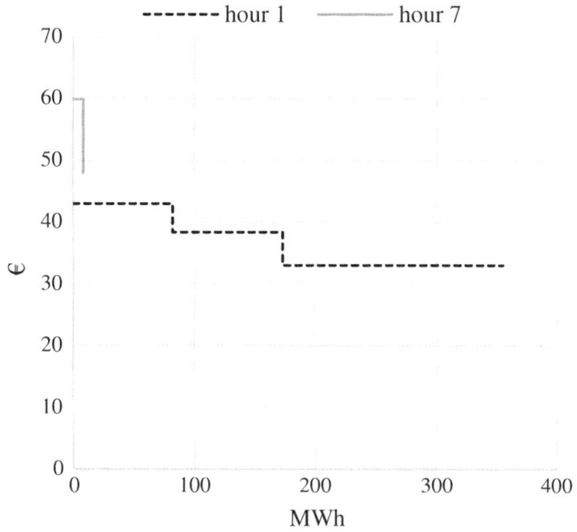

combination in each of the three scenarios. The other bid curve consists of a single block, which happens when for all the prices in all three scenarios the same volume is chosen. This situation actually happens in 74 % of the hours simulated. This can be interpreted as the aggregator choosing an optimal aggregated charge profile independently of the prices, as long as these are not higher than a given value. This maximum willingness to pay at a particular hour is defined by the highest price chosen for that hour across the scenarios.

9.5 Summary and Conclusions

This chapter describes a mathematical tool to assist an aggregator responsible for charging a PEV fleet in deciding its optimal bidding strategy in the day-ahead electricity market. The aim of the aggregator is to minimize the charging costs while at the same time satisfying the driving requirements of PEV drivers.

The problem is formulated as a bi-level model comprising an UL problem, which represents the charging optimization problem, and a set of LL problems, which represent the clearing of the day-ahead market. This bi-level model is recast as a MILP problem that can be solved using conventional branch-and-cut solvers.

PEVs are represented in an aggregated way. In this sense, PEVs are modeled as a virtual battery.

Uncertainty in the bids of the market participants is efficiently modeled using a set of scenarios.

Given the framework described in this chapter and the results of the illustrative example carried out in Sect. 9.4, the conclusions below are in order:

1. The aggregator uses the flexibility of PEV demand to significantly reduce charging costs.
2. Even at a low penetration of PEVs (2 %), the aggregator significantly impacts market prices through its bidding strategy.
3. An accurate modeling of the uncertainty in the market bids of market participants is important to avoid energy shortfalls.
4. The resulting MILP problem is computationally tractable provided that the number of scenarios is small enough.

Appendix

This appendix explains the procedure used to linearize the non-linear terms that appear in problem (9.21)–(9.32) provided in Sect. 9.3.2.

Linearization of Objective Function (9.21)

Objective function (9.21) includes the terms $\sum_t \beta_{\varpi t} P^{\mathrm{A}}_{\varpi t}$. These terms are nonlinear since they include the product of two decision variables, namely $\beta_{\varpi t}$ and $P^{\mathrm{A}}_{\varpi t}$. However, these nonlinear terms can be replaced by exact equivalent linear expressions as explained below [3].

The strong duality theorem states that if a problem is convex, the objective functions of the primal and dual problems have the same value at the optimum [31]. This is the case of LL problems (9.13)–(9.17), and thus:

$$\begin{aligned}\sum_s c^{\mathrm{S}}_{s\varpi t} P^{\mathrm{S}}_{s\varpi t} - \sum_d b^{\mathrm{D}}_{\varpi t} P^{\mathrm{D}}_{d\varpi t} - b^{\mathrm{A}}_{\varpi t} P^{\mathrm{A}}_{\varpi t} &= -\sum_s \bar{\varphi}^{\mathrm{S}}_{s\varpi t} \bar{P}^{\mathrm{S}}_{s\varpi t} \\ &\quad - \sum_d \bar{\varphi}^{\mathrm{D}}_{d\varpi t} \bar{P}^{\mathrm{D}}_{d\varpi t} - \bar{\varphi}^{\mathrm{A}}_{\varpi t} P^{\mathrm{A,max}}_t \quad \forall \varpi, \forall t.\end{aligned} \tag{9.53}$$

On the other hand, from Eq. (9.26) we obtain

$$b^{\mathrm{A}}_{\varpi t} = \lambda_{\varpi t} + \bar{\varphi}^{\mathrm{A}}_{\varpi t} - \underline{\varphi}^{\mathrm{A}}_{\varpi t} \quad \forall \varpi, \forall t, \tag{9.54}$$

and thus,

$$b^{\mathrm{A}}_{\varpi t} P^{\mathrm{A}}_{\varpi t} = \left(\lambda_{\varpi t} + \bar{\varphi}^{\mathrm{A}}_{\varpi t} - \underline{\varphi}^{\mathrm{A}}_{\varpi t}\right) P^{\mathrm{A}}_{\varpi t} \quad \forall \varpi, \forall t, \tag{9.55}$$

In a similar way, from constraints (9.26) and (9.27) we obtain

$$P^{\mathrm{A}}_{\varpi t} \underline{\varphi}^{\mathrm{A}}_{\varpi t} = 0 \quad \forall \varpi, \forall t, \tag{9.56}$$

and

$$\bar{\varphi}^{\mathrm{A}}_{\varpi t} P^{\mathrm{A,max}}_t = \bar{\varphi}^{\mathrm{A}}_{\varpi t} P^{\mathrm{A}}_{\varpi t} \quad \forall \varpi, \forall t, \tag{9.57}$$

respectively.

Next, we use (9.56) and (9.57) to simplify (9.55) and we obtain:

$$b^{\mathrm{A}}_{\varpi t} P^{\mathrm{A}}_{\varpi t} = \lambda_{\varpi t} P^{\mathrm{A}}_{\varpi t} + \bar{\varphi}^{\mathrm{A}}_{\varpi t} P^{\mathrm{A,max}}_t \quad \forall \varpi, \forall t, \tag{9.58}$$

Finally, we replace term $b^{A}_{\varpi t} P^{A}_{\varpi t}$ in Eq. (9.53) by its equivalent expression defined in (9.58) and we obtain:

$$\begin{aligned}&\sum_s c^{\mathrm{S}}_{s\varpi t} P^{\mathrm{S}}_{s\varpi t} - \sum_d b^{\mathrm{D}}_{\varpi t} P^{\mathrm{D}}_{d\varpi t} - \lambda_{\varpi t} P^{\mathrm{A}}_{\varpi t} - \bar{\varphi}^{\mathrm{A}}_{\varpi t} P^{\mathrm{A,max}}_t \\ &\quad = -\sum_s \bar{\varphi}^{\mathrm{S}}_{s\varpi t} \bar{P}^{\mathrm{S}}_{s\varpi t} - \sum_d \bar{\varphi}^{\mathrm{D}}_{d\varpi t} \bar{P}^{\mathrm{D}}_{d\varpi t} - \bar{\varphi}^{\mathrm{A}}_{\varpi t} P^{\mathrm{A,max}}_t \quad \forall \varpi, \forall t.\end{aligned} \tag{9.59}$$

Since $\beta_{\varpi t} = \lambda_{\varpi t}, \forall \varpi, \forall t$, as defined in (9.6), Eq. (9.59) allows us to calculate nonlinear terms $\beta_{\varpi t} P^{A}_{\varpi t}$ as a function of exclusively linear terms:

$$\frac{\Upsilon_{\varpi t}}{\Delta t} = \beta_{\varpi t} P^{\mathrm{A}}_{\varpi t} = \sum_{s} c^{\mathrm{S}}_{s\varpi t} P^{\mathrm{S}}_{s\varpi t} - \sum_{d} b^{\mathrm{D}}_{\varpi t} P^{\mathrm{D}}_{d\varpi t} + \sum_{s} \bar{\varphi}^{\mathrm{S}}_{s\varpi t} \bar{P}^{\mathrm{S}}_{s\varpi t} + \sum_{d} \bar{\varphi}^{\mathrm{D}}_{d\varpi t} \bar{P}^{\mathrm{D}}_{d\varpi t} \quad \forall \varpi, \forall t. \quad (9.60)$$

Linearization of Complementarity Constraints

Complementarity constraints (9.27)–(9.32) have the form $0 \leq a \perp e \geq 0$, which is equivalent to the following set of equations:

$$a,\ e \geq 0 \quad (9.61)$$

$$a \cdot e = 0 \quad (9.62)$$

Equations (9.61)–(9.62) can be replaced by a set of exact equivalent mixed-integer linear equations as explained in [27]:

$$a, e \geq 0 \quad (9.63)$$

$$a \leq Mu \quad (9.64)$$

$$e \leq M(1 - u) \quad (9.65)$$

$$u = \{0, 1\} \quad (9.66)$$

where M is a large enough positive constant.

Using (9.63)–(9.64) to linearize complementarity constraints (9.22)–(9.27) we obtain:

$$\{0 \leq P^{\mathrm{S}}_{s\varpi t} \leq M\underline{u}^{\mathrm{S}}_{s\varpi t} \quad \forall s \quad (9.67)$$

$$0 \leq \underline{\varphi}^{\mathrm{S}}_{s\varpi t} \leq M\left(1 - \underline{u}^{\mathrm{S}}_{s\varpi t}\right) \quad \forall s \quad (9.68)$$

$$0 \leq \bar{P}^{\mathrm{S}}_{s\varpi t} - P^{\mathrm{S}}_{s\varpi t} \leq M\bar{u}^{\mathrm{S}}_{s\varpi t} \quad \forall s \quad (9.69)$$

$$0 \leq \bar{\varphi}^{\mathrm{S}}_{s\varpi t} \leq M\left(1 - \bar{u}^{\mathrm{S}}_{s\varpi t}\right) \quad \forall s \quad (9.70)$$

$$\underline{u}^{\mathrm{S}}_{s\varpi t}, \bar{u}^{\mathrm{S}}_{s\varpi t} = \{0, 1\} \quad \forall s \quad (9.71)$$

$$0 \leq P^{\mathrm{D}}_{d\varpi t} \leq M\underline{u}^{\mathrm{D}}_{d\varpi t} \quad \forall d \quad (9.72)$$

$$0 \leq \underline{\varphi}^{\mathrm{D}}_{d\varpi t} \leq M\left(1 - \underline{u}^{\mathrm{D}}_{d\varpi t}\right) \quad \forall d \tag{9.73}$$

$$0 \leq \bar{P}^{\mathrm{D}}_{d\varpi t} - P^{\mathrm{D}}_{s\varpi t} \leq M\bar{u}^{\mathrm{D}}_{d\varpi t} \quad \forall d \tag{9.74}$$

$$0 \leq \bar{\varphi}^{\mathrm{D}}_{d\varpi t} \leq M\left(1 - \bar{u}^{\mathrm{D}}_{d\varpi t}\right) \quad \forall d \tag{9.75}$$

$$\underline{u}^{\mathrm{D}}_{d\varpi t}, \bar{u}^{\mathrm{D}}_{d\varpi t} = \{0, 1\} \quad \forall d \tag{9.76}$$

$$0 \leq P^{\mathrm{A}}_{\varpi t} \leq M\underline{u}^{\mathrm{A}}_{\varpi t} \tag{9.77}$$

$$0 \leq \underline{\varphi}^{\mathrm{A}}_{\varpi t} \leq M\left(1 - \underline{u}^{\mathrm{A}}_{\varpi t}\right) \tag{9.78}$$

$$0 \leq P^{\mathrm{A,max}}_{t} - P^{\mathrm{A}}_{\varpi t} \leq M\bar{u}^{\mathrm{A}}_{\varpi t} \tag{9.79}$$

$$0 \leq \bar{\varphi}^{\mathrm{A}}_{\varpi t} \leq M\left(1 - \bar{u}^{\mathrm{A}}_{\varpi t}\right) \quad \forall d \tag{9.80}$$

$$\underline{u}^{\mathrm{A}}_{\varpi t}, \bar{u}^{\mathrm{A}}_{\varpi t} = \{0, 1\} \tag{9.81}$$

$\}, \forall \varpi, \forall t$, where M is a large enough positive constant.

For the sake of clarity in the notation, a single M has been defined. However, note that appropriate and different values of this large constant should be defined for each constraint (9.67)–(9.81) as explained in [27].

References

1. Galus MD, Waraich RA, Andersson G (2011) Predictive, distributed, hierarchical charging control of PHEVs in the distribution system of a large urban area incorporating a multi agent transportation simulation. In: Proceedings of the 17th power systems computation conference, Stockholm, Aug 22–26, 2011
2. González Vayá M, Galus MD, Waraich RA, Andersson G (2012) On the interdependence of intelligent charging approaches for plug-in electric vehicles in transmission and distribution networks. In: Proceedings of the IEEE PES international conference and exhibition on innovative smart grid technologies (ISGT Europe), Berlin, Oct 14–17, 2012
3. Ruiz C, Conejo AJ (2009) Pool strategy of a producer with endogenous formation of locational marginal prices. IEEE Trans Power Syst 24(4):1855–1866
4. González Vayá M, Andersson G (2013) Optimal bidding strategy of a plug-in electric vehicle aggregator in day-ahead electricity markets under uncertainty. IEEE Trans Power Syst 99:1–11
5. González Vayá M, Andersson G (2012) Centralized, decentralized approaches to smart charging of plug-in vehicles. In: Proceedings of IEEE power and energy society general meeting, San Diego, July 22–26, 2012
6. Vandael S, Claessens B, Hommelberg M, Holvoet T, Deconinck G (2013) A scalable three-step approach for demand side management of plug-in hybrid vehicles. IEEE Trans Smart Grid 4(2):720–728

7. González Vayá M, Andersson G, Boyd S (2014) Decentralized control of plug-in electric vehicles under driving uncertainty. In: Proceedings of the IEEE PES conference on innovative smart grid technologies (ISGT Europe), Istanbul, Oct 12–15, 2014
8. Bessa RJ, Matos MA (2012) Economic and technical management of an aggregation agent for electric vehicles: a literature survey. European Trans Electr Power 22(3):334–350
9. Gómez San Román T, Momber I, Rivier Abbad M, Sánchez Miralles A (2011) Regulatory framework and business models for charging plug-in electric vehicles: infrastructure, agents, and commercial relationships. Energy Policy 39(10):6360–6375
10. Pecas Lopes JA, Soares FJ, Almeida PMR (2011) Integration of electric vehicles in the electric power system. Proc IEEE 99(1):168–183
11. Galus MD, Zima M, Andersson G (2010) On the integration of plug-in hybrid electric vehicles into existing power system structures. Energy Policy 38(11):6736–6745
12. Galus MD, González Vayá M, Krause T, Andersson G (2013) The role of electric vehicles in smart grids. Wiley Interdisc Rev Energy Environ 2(4):384–400
13. Guille C, Gross G (2008) Design of a conceptual framework for the V2G implementation. In: Proceedings of IEEE energy 2030 conference, Atlanta, Nov 17–18, 2008
14. González Vayá M, Andersson G (2013) Combined smart-charging and frequency regulation for fleets of plug-in electric vehicles. In: Proceedings of IEEE power and energy society general meeting, Vancouver, July 21–25, 2013
15. Tomic J, Kempton W (2007) Using fleets of electric-drive vehicles for grid support. J Power Sources 168(2):459–468
16. Sundström O, Binding C (2012) Flexible charging optimization for electric vehicles considering distribution grid constraints. IEEE Trans Smart Grid 3(1):26–37
17. Hilshey AD, Hines PDH, Rezaei P, Dowds JR (2013) Estimating the impact of electric vehicle smart charging on distribution transformer aging. IEEE Trans Smart Grid 4(2):905–913
18. Kristoffersen TK, Capion K, Meibom P (2011) Optimal charging of electric vehicles in a market environment. Appl Energy 88(5):1940–1948
19. Bessa RJ, Matos MA, Soares F, Pecas Lopes J (2012) Optimized bidding of a EV aggregation agent in the electricity market. IEEE Trans Smart Grid 3(1):443–452
20. Bessa RJ, Matos MA (2013) Global against divided optimization for the participation of an EV aggregator in the day-ahead electricity market. Part I: theory. Electr Power Syst Res 95:309–318
21. Pantos M (2012) Exploitation of electric-drive vehicles in electricity markets. IEEE Trans Power Syst 27(2):682–694
22. Vagropoulos SI, Bakirtzis AG (2013) Optimal bidding strategy for electric vehicle aggregators in electricity markets. IEEE Trans Power Syst 28(4):4031–4041
23. Baringo L, Conejo AJ (2013) Strategic offering for a wind power producer. IEEE Trans Power Syst 28(4):4645–4654
24. Zugno M, Morales JM, Pinson P, Madsen H (2013) Pool strategy of a price-maker wind power producer. IEEE Trans Power Syst 28(3):3440–3450
25. Gabriel SA, Conejo AJ, Fuller JD, Hobbs BF, Ruiz C (2012) Complementarity modeling in energy markets. Springer, New York
26. Luo ZQ, Pang JS, Ralph D (1996) Mathematical programs with equilibrium constraints. Cambridge University Press, Cambridge
27. Fortuny-Amat J, McCarl B (1981) A representation and economic interpretation of a two-level programming problem. J Oper Res Soc 32:783–792
28. de Miguel A, Friedlander MP, Nogales FJ, Scholtes S (2006) A two-sided relaxation scheme for mathematical programs with equilibrium constraints. SIAM J Optim 16(2):587–609
29. Leyffer S, López-Calva G, Nocedal J (2007) Interior methods for mathematical programs with complementarity constraints. SIAM J Optim 17(1):52–77
30. González Vayá, M.; Andersson, G., "Optimal Bidding Strategy of a Plug-In Electric Vehicle Aggregator in Day-AheadElectricity Markets Under Uncertainty,' Power Systems, IEEE Transactions on , vol.PP, no.99, pp.1,11doi: 10.1109/TPWRS.2014.2363159
31. Luenberger DG, Ye Y (2008) Linear and nonlinear programming. Springer, New York

32. European energy exchange. http://www.eex.com/en/market-data/power/spot-market/. Accessed 3 Jun 2014
33. KBA. (2013). http://www.kba.de/DE/Statistik/Fahrzeuge/Bestand/2013/2013_bestand_node.html. Accessed 3 Jun 2014
34. Statistik Austria. (2014). http://www.statistik.gv.at/web_de/statistiken/verkehr/strasse/kraftfahrzeuge_-_bestand/index.html. Accessed 3 Jun 2014
35. Deutsche Bundesregierung (2009) Nationaler Entwicklungsplan Elektromobilität der Bundesregierung
36. Balmer M, Axhausen K, Nagel K (2006) Agent-based demand-modeling framework for large-scale microsimulations. Transp Res Rec J Transp Res Board 1985:125–134
37. Baringo L, Conejo AJ (2012) Wind power investment: a Benders decomposition approach. IEEE Trans Power Syst 27(1):433–441
38. Kazempour S, Conejo AJ (2012) Strategic generation investment under uncertainty via Benders decomposition. IEEE Trans Power Syst 27(1):424–432

Chapter 10
Optimal Control of Plug-in Vehicles Fleets in Microgrids

Guido Carpinelli, Fabio Mottola and Daniela Proto

Abstract This chapter focuses on the optimal operation of plug-in vehicle fleets in a microgrid characterized by the presence of other distributed resources, such as distributed generation units. The possible services that vehicle aggregators can provide in the microgrid are discussed and the control actions to be performed in order to obtain such services while integrating the vehicle aggregator actions with those of the other distributed resources of the grid are outlined. The problems with optimal operation are formulated as single-objective and multi-objective optimization problems, specifying, in both cases, objective functions, equality and inequality constraints. The differences and criticisms of both approaches are extensively analyzed in the chapter, where the two approaches are also implemented and solved with reference to the practical cases.

Keywords Microgrids · Distributed generation · Electric vehicles · Single-objective optimization models · Multi-objective optimization models · Ancillary services

G. Carpinelli · F. Mottola
Dipartimento di Ingegneria Elettrica e Tecnologie dell'Informazione,
Università degli Studi di Napoli Federico II, Napoli, Italy
e-mail: guido.carpinelli@unina.it

F. Mottola
e-mail: fmottola@unina.it

D. Proto (✉)
Dipartimento di Ingegneria Industriale e dell'Informazione,
Seconda Università degli Studi di Napoli, Aversa (Caserta), Italy
e-mail: danproto@unina.it

S. Rajakaruna et al. (eds.), *Plug In Electric Vehicles in Smart Grids*,
Power Systems, DOI 10.1007/978-981-287-302-6_10

10.1 Introduction

In the last few years, the electrical power system has been witnessing an increasing penetration of plug-in vehicles (PEVs). The number of PEVs on the road is expected to increase even more in the near future, as they represent a key element for global emissions reduction. In addition, their double role of being a load and energy source makes them strategic resources for grid operation, as they are able to provide several services that can improve the operation of distribution networks. This is possible because most vehicles are typically only driven a few hours per day and are parked the rest of the time (during the night or while the owner is working) [1].

In the case of a vehicle fleet, this capability is even greater since, by grouping together a large number of vehicles, an effective and significant contribution can be obtained on the grid service provision. Theoretically, a certain number of parked PEVs, managed through an *aggregator*, can provide several important ancillary services to the grid, such as regulation, peak power, and spinning reserve [1–3]. These services can be furnished by the vehicle fleets requiring two-way or one-way energy flows to the grid. A typical service that requires a one-way energy flow from the distribution network to the vehicle, as an example, is the so called *Smart Charging*. Smart Charging is the coordination of vehicle charging aimed at preventing congestion in the grid and providing frequency regulation [2].

In addition, when included in a microgrid (μG), the PEV fleets can be integrated with other energy resources. All of these PEV fleets can provide, in an optimized fashion, some or all of the previously mentioned services to the same μG or to the distribution grid to which the μG is linked. In this case, the distributed energy resources (DERs) can be coordinated by a control system to provide the usual energy service, as well as a number of ancillary services. This configuration is even more effective when the μG is supposed to operate in both grid connected and islanded modes. This last operating mode is particularly critical and requires the contribution of all μG resources to the regulation, as well as the provision of other power quality services.

In order to perform the optimal control of vehicle fleets, an optimization strategy is required. This optimization strategy is aimed at operating the grid efficiently. The strategy must take into account all technical and operating limitations of the grid resources to be integrated, as well as those regarding the grid itself. The optimization can be formulated as a single-objective (SO) or multi-objective (MO) problem. The choice of the approach to be used depends on several aspects as, for instance, the perspective of the stakeholder who is performing the control. The first formulation can be particularly practical from the distribution system operator point-of-view, who can directly and easily identify the best solution for system operation. Multi-objective approaches could be more effective, as they consider, at the same time, several, and sometimes conflicting, objectives.

In this chapter, a low voltage (LV) μG derived from an interconnection bus is considered which is characterized by the presence of DG units and PEVs fleets (Fig. 10.1).

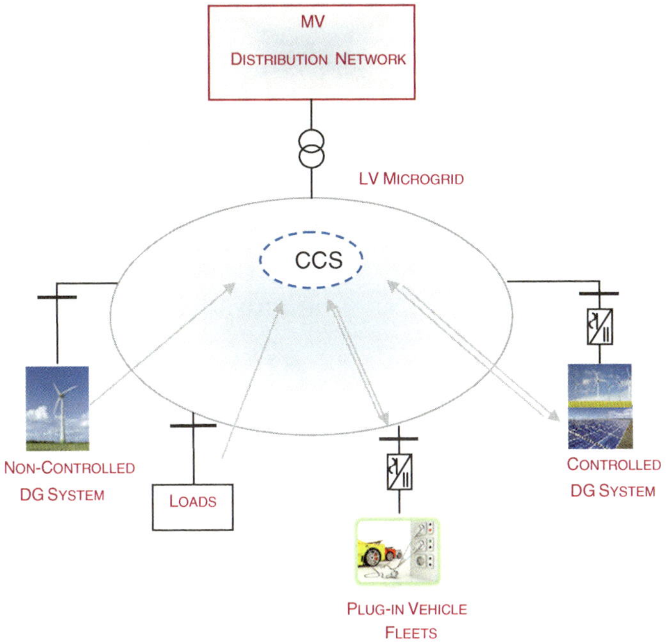

Fig. 10.1 Schematic diagram of the μG

The DG includes both controlled and non-controlled units; all DG units are undispatchable renewable power plants. The term 'controlled' refers to the presence of an interfacing converter which is able to regulate some typical quantities such as the reactive power, so giving the possibility to provide services to the grid where the DG unit is connected. Non-controlled DG units are systems directly connected to the grid without any interfacing device. The vehicle fleets are all connected through power electronic converters to the μG, that is supposed to be controlled and monitored by a centralized control system (CCS). The CCS is responsible for the optimal management of the distributed resources and is able to coordinate the grid resources by sending appropriate command signals. The converters receive reference signals of active/reactive power from the CCS. The signals are obtained by solving appropriate SO and MO optimization models that are able to guarantee specified μG internal and external services[1] while meeting operational and technical constraints.

It has to be outlined that in this chapter the μG operator is not directly responsible of the charging requirements of the single vehicles since aggregators and μG operator are supposed to be independent. In particular, the management of

[1] Internal services are those required for the correct or efficient operation of the μG, such as loss minimization and voltage regulation. External services are those provided to the upstream electrical system to which the μG is interconnected for its correct operation.

single vehicles is assumed to be effected by the aggregators who, at this purpose, apply appropriate procedures [4, 5].

In the next sections, the services that can be performed by the PEVs, their configurations and operating modes, are outlined (Sect. 10.2). The SO and MO optimization problems are then formulated (Sects. 10.3, 10.4). Finally, the results of the numerical applications are presented to verify the feasibility of the proposed approaches (Sect. 10.5).

10.2 Plug-in Vehicle Services, Configurations and Operating Modes

There are several services that PEVs or aggregations of PEVs could provide to the grid, such as regulation, peak power, spinning reserve and other ancillary services. In practice, only some of them represent attractive market opportunities.

In order to provide services, two operating modes are foreseen for PEVs. The first is PEVs that are connected to the grid and require one-way energy flows to the distribution network, in which only charging is considered. The second is two-way energy flows in which PEVs provide also power to the grid while parked to allow a bi-directional energy stream between the vehicles and the grid. In the charging mode, the PEV battery charging can be conducted either without the application of any particular control strategy (e.g., the charging of the PEV automatically starts when the vehicle is plugged in and ends when the storage device is fully charged) or with a local control strategy aimed at maximizing the advantages of PEV owners in terms of costs (e.g., PEV automatically charges during low energy price periods) [6]. Interesting applications include charging strategies based on the possibility of vehicles communicating with the grid in real time and charging according to the grid needs to provide services to the grid. The applications requiring one-way energy flows are usually referred to as V1G. The bi-directional operation mode is usually referred to as V2G. In the V2G operating mode, the possibility of bidirectional energy flows allows for the performing of all of the services mentioned previously [1, 7, 8].

Regarding the use of PEVs as generation units during peak load periods (*peak power* service), the required duration of energy to be supplied by PEVs could reach 3–5 h. As such, this service is impractical, due to the on board storage limitations. This limitation could be overcome by a series of vehicles drawing power sequentially (e.g., in car parking managed by aggregators) [7, 9].

Two interesting market opportunities include *spinning reserve* and *regulation*. *Spinning reserve* is an additional power capacity that can be requested in the case of an outage. It represents an attractive market opportunity for PEVs, since it is paid for by the amount of time the generators are available and ready (capacity price), even if no energy is produced; an extra amount is paid for the energy that is actually furnished, if needed (energy price) [8]. Compared to spinning reserve, regulation requires availability several times per day. It also requires a faster response, but only for a short duration of generation. These reasons make PEVs good candidates for the provision of such a service. *Regulation* consists of the use of controllable generation units or aggregated loads to regulate frequency and voltage [10]. There are two possible regulation modes: 'regulation up' is the ability to increase power generation (or decrease loads) from a baseline level and 'regulation down' is aimed at decreasing power generation (or increasing loads) from a baseline. PEVs are suited for this market, because they can respond very quickly. Moreover, the provision of both regulation up and down can imply only a little net discharge of batteries. PEVs provide regulation service through the so called *smart charging* service that involves the control of the charging of each vehicle to meet both the needs of the vehicle owner (i.e., to have the vehicle charged at a certain time) and the needs of the grid (e.g., providing regulation).

In conclusion, the services for which PEVs are more suitable are those characterized by a high power value and fast response. PEVs cannot provide a base load power at a competitive price, whereas peak power, spinning reserve and regulation are services that could be provided, because of the short duration of supply [1, 2].

As mentioned previously, two primary PEVs configurations can be considered: the first refers to individual vehicles, while the second refers to fleets. In the first case, the individual vehicle gives its contribution as a service provider, whereas in the second case, the grid interacts with a fleet (a significant number of vehicles plugged simultaneously into the same connection point). In this case, PEV fleets are managed by the so-called "aggregators" who contract with the grid operator through the day-ahead and hour-ahead markets to provide the required services [4]. In this way, all vehicles controlled by the aggregator would appear to the grid as a unique storage device. The fleets, compared to the individual vehicles directly plugged into the grid, represent a large source of rapidly-controllable generations or loads and are more convenient to manage when services to the grid have to be provided. In fact, when the grid has to face individual vehicles, the control procedure usually needs tighter constraints [11]. Regarding the typology of the aggregators, car parking can be either "residential", where cars are usually parked at night, or "office", where cars are parked during the morning or afternoon when the owner is working. This will impact the cost and availability of energy throughout the day. Swapping stations are automated stations where the dry batteries are changed with charged ones [12]. Besides satisfying this primary duty, they seem particularly suitable for providing grid services [2, 6].

10.3 Optimal Control of Vehicle Fleets: Single Objective Approach

As mentioned in the introduction, the converter control systems for all of the controlled DERs included in the considered LV μG receive reference signals of active and/or reactive power obtained by solving either the SO or MO optimization problems.

In the case of a control strategy formulated in terms of an SO optimization problem, a unique objective is required to be fixed, while the other operation requirements are considered constraints. In the most general case, the problem can be formulated as the following non-linear constrained optimization problem:

$$\min f_{obj}(\mathbf{X}, \mathbf{C}) \tag{10.1}$$

Subject to

$$\psi(\mathbf{X}, \mathbf{C}) = 0 \tag{10.2}$$

$$\eta(\mathbf{X}, \mathbf{C}) \leq 0 \tag{10.3}$$

where f_{obj} is the objective function to be minimized, ψ, η the constraints to be satisfied, $\mathbf{X}$ is the system state vector (e.g., the bus voltages) and $\mathbf{C}$ is a control vector which includes the variables to be controlled in the optimization (e.g., the PEV aggregators' active and reactive power). The choice of the objective function depends on the strategy that has to be applied and the services to be provided. Constraints are typically related to the technical limitations of the grid and the DERs. Moreover, constraints could also be related to services to be provided. Irrespective of the strategy adopted, the inputs of the problem are the requirements of both the grid and DERs. Examples of grid requirements are the satisfaction of limits imposed on the node voltages, which must fall into admissible ranges, and line currents, which must be lower than the line ampacity. DERs' requirements can refer to the DGs active power production and energy required by the aggregator for charging the PEVs. The outputs are the values of the control variables that are, for example, the unknown active and/or reactive powers of the controlled DERs.

In what follows, an example of an optimization strategy is reported that deals with the economical aspects of the μG management and is aimed at minimizing the cost of the energy consumption of the whole μG. Other examples of SO strategies for PEV operations and their comparisons are reported in [10, 13].

The proposed example of a control strategy is based on an SO optimization and refers to a LV μG which includes several DERs (Fig. 10.1). It is assumed that the μG is an 'active cell' able to bid and offer energy as a whole.

The active/reactive powers of the aggregators, as well as the reactive powers of the controlled DG units, are optimally scheduled in order to minimize the daily energy costs and, contemporaneously, satisfy the technical constraints on μG bus voltages and line currents. The peak-shaving service is performed by imposing an upper boundary to the power imported from the grid to which the μG is interconnected. A further boundary can be imposed also to the power exported to the grid.

The proposed scheduling strategy is based on the solution of the SO optimization problems (10.1)–(10.3), where the vector of the decision variables $\mathbf{C}$ includes the schedule of active powers to be provided for by the aggregators and of the reactive powers that the DG and aggregator converters must supply to perform the desired services. The vector of the state variables **X** includes, as an example, the voltages at the grid nodes during the day.

The model input variables are the forecasted grid load demand, DG active power production, day-ahead market price and aggregator requirements. The model output variables are the daily profile of the: (1) active power imported/exported from/to the upstream grid, (2) active and reactive powers of the aggregators, and (3) reactive power of the DG units.

The objective function (10.1) to be minimized is:

$$f_{obj} = \sum_{t=1}^{N_T} EC_t \cdot P_{1,t} \cdot \Delta T_t, \tag{10.4}$$

where t is the time-slot code, EC_t is the forecasted value of the day-ahead market energy cost, $P_{1,t}$ is the active power at the interconnection bus, which is positive when it is absorbed from the upstream grid or negative when it is supplied to the upstream grid, and ΔT_t is the duration of the time-slot. For the sake of simplicity, the price for the energy supplied to the upstream grid is assumed equal to the price for that absorbed from the upstream grid. If the μG is not allowed to sell energy to the grid, a constraint could be imposed on the power at the interconnection node which must be always positive.

The equality/inequality constraints to be satisfied are listed separately for each bus and for the entire μG. For the sake of clarity, in what follows, it is assumed that each bus is considered to be of one type (e.g., DG bus, load bus, etc.) and if there were "mixed" cases these could be resolved easily.

- **DG bus**

The constraints for the controlled DG bus are related to the power flow equations and the rating of the interfacing converter:

$$DP_{i,t}^{sp} = V_{i,t} \sum_{k=1}^{N} V_{k,t} \left[G_{i,k} \cos\left(\delta_{i,t} - \delta_{k,t}\right) + B_{i,k} \sin\left(\delta_{i,t} - \delta_{k,t}\right) \right] \tag{10.5}$$

$$DQ_{i,t} = V_{i,t} \sum_{k=1}^{N} V_{k,t} \left[G_{i,k} \sin\left(\delta_{i,t} - \delta_{k,t}\right) - B_{i,k} \cos\left(\delta_{i,t} - \delta_{k,t}\right) \right] \tag{10.6}$$

$$\left[\left(DP_{i,t}^{sp} \right)^2 + \left(DQ_{i,t} \right)^2 \right]^{1/2} \le DS_i^{\max} \tag{10.7}$$

$$t = 1, 2, \ldots, N_T, i \in \Omega_D$$

where i is the bus code, N is the number of grid nodes, $DP_{i,t}^{sp}$ is the DG's specified active power, $DQ_{i,t}$ is the DG's reactive power, $V_{i,t}$ $(\delta_{i,t})$ is the magnitude (argument) of the bus voltage, $G_{i,k}(B_{i,k})$ is the (i, k)-term of the system's conductance (susceptance) matrix, DS_i^{max} is the size of the interfacing converter and Ω_D is the set of DG buses. In (10.5)–(10.7), $DQ_{i,t}$ is an optimization variable, and the value of $DP_{i,t}^{sp}$ is a predicted value obtained as the result of a deterministic forecast. It should be noted that forecasts, of course, are typically characterized by errors whose entity depends on the time horizon of their application.

It also is important to note that, if the DG has the capability of controlling voltages, a specified value of the DG node voltage ($V_{i,t}^{sp}$) also must be assigned [14], then the following constraint has to be imposed:

$$V_{i,t} = V_{i,t}^{sp} \tag{10.8}$$

Other types of DG units can be also considered as, for example, wind turbines equipped with asynchronous generators directly connected to the grid [15]. The problem formulation in case this last category of DG units is considered, requires to regard both active and reactive powers as specified values [14] obtained by means of deterministic or probabilistic forecasting procedures, i.e. based on neural networks or Bayesian Inference [16–18]. Thus, in this case, in Eq. 10.6 the reactive power is not an optimization variable and Eq. 10.8 is not assigned.

- **Load bus**

The constraints at the load bus refer only to the power flow equations:

$$LP_{i,t}^{sp} = V_{i,t} \sum_{k=1}^{N} V_{k,t}\left[G_{i,k}\cos\left(\delta_{i,t} - \delta_{k,t}\right) + B_{i,k}\sin\left(\delta_{i,t} - \delta_{k,t}\right)\right] \tag{10.9}$$

$$LQ_{i,t}^{sp} = V_{i,t} \sum_{k=1}^{N} V_{k,t}\left[G_{i,k}\sin\left(\delta_{i,t} - \delta_{k,t}\right) - B_{i,k}\cos(\delta_{i,t} - \delta_{k,t})\right] \tag{10.10}$$

$$t = 1, 2, \ldots, N_T, i \in \Omega_{LD}$$

where $LP_{i,t}^{sp}(LQ_{i,t}^{sp})$ is the load-specified active (reactive) power, and Ω_{LD} is the set of load buses. Also load powers can be predicted by means of deterministic or probabilistic forecasting procedures.

- **PEV aggregator bus**

The PEV aggregator was assumed to be characterized by forecasted data related to its charging/discharging potential throughout the day [4, 19], depending on the forecasted requirements of the vehicles. To avoid critical situations in case the actual data are different from the forecasts, a backup battery bank could be used [5]. The forecasted data related to the aggregator's charging/discharging potential are given in terms of both deliverable power and energy. With reference to energy, the aggregators are allowed to charge during N_c pre-assigned time intervals; each of the N_c intervals (lasting several hours) is characterized by a specified value of contracted energy that the aggregator has to absorb from the grid for charging the fleet of vehicles [13, 19]. The N_c time interval duration depends on the fleets features (e.g., for a home park, it is forecasted that the vehicles stay plugged in for a long time during the night, so that a time interval of a few hours can be chosen). With reference to the power, admissible ranges of charging/discharging powers that can be potentially exchanged with the grid are specified for each time interval. This information is based on available forecasts on the PEVs requirements and is supposed to be treated and provided by the aggregator.

The constraints at the aggregator busbars referring to the power flow equations and the rating of the interfacing converters are:

$$PEVP_{i,t} = V_{i,t} \sum_{k=1}^{N} V_{k,t} \left[G_{i,k} \cos\left(\delta_{i,t} - \delta_{k,t}\right) + B_{i,k} \sin(\delta_{i,t} - \delta_{k,t}) \right] \tag{10.11}$$

$$PEVQ_{i,t} = V_{i,t} \sum_{k=1}^{N} V_{k,t} \left[G_{i,k} \sin\left(\delta_{i,t} - \delta_{k,t}\right) - B_{i,k} \cos(\delta_{i,t} - \delta_{k,t}) \right] \tag{10.12}$$

$$\left[\left(PEVP_{i,t}\right)^2 + \left(PEVQ_{i,t}\right)^2 \right]^{1/2} \le PEVS_i^{\max} \tag{10.13}$$

$$t = 1, 2, \ldots, N_T, i \in \Omega_{PEV}$$

while the following additional constraints are included to take into account previously-mentioned allowable aggregator energy/powers:

$$PEVP_i^{\min} \le PEVP_{i,t} \le PEVP_i^{\max} \tag{10.14}$$

$$\sum_{t=N_j^{start}}^{N_j^{end}} PEVP_{i,t} \Delta T_t = -PEVE_{i,j}^{sp} \tag{10.15}$$

$$t = 1, 2, \ldots, N_T, \; j = 1, 2, \ldots, N_C, \; i \in \Omega_{PEV}$$

In (10.11)–(10.15), $PEVP_{i,t}(PEVQ_{i,t})$ is the active (reactive) aggregator power, $PEVE_{i,j}^{sp}$ is the aggregator contracted energy, $PEVP_{i,t}^{\min}$ ($PEVP_{i,t}^{\max}$) is the minimum (maximum) aggregator active power, $PEVS_i^{\max}$ is the maximum apparent power of the aggregator, N_j^{start} (N_j^{end}) is the first (last) slot of the jth time interval of the ith aggregator, j is the aggregator's time interval index code and Ω_{PEV} is the set of aggregator buses.

- **Whole micro grid**

The constraints at the bus that interconnects the μG and the upstream MV grid refer to the power flow equations, the peak-shaving, if required, and the rating of the interfacing MV/LV transformer:

$$P_{1,t} = V_{1,t}^{sp} \sum_{k=1}^{N} V_{k,t} \left[G_{1,k} \cos\left(\delta_{1,t}^{sp} - \delta_{k,t}\right) + B_{1,k} \sin(\delta_{1,t}^{sp} - \delta_{k,t}) \right] \quad (10.16)$$

$$Q_{1,t} = V_{1,t}^{sp} \sum_{k=1}^{N} V_{k,t} \left[G_{1,k} \sin\left(\delta_{1,t}^{sp} - \delta_{k,t}\right) - B_{1,k} \cos(\delta_{1,t}^{sp} - \delta_{k,t}) \right] \quad (10.17)$$

$$P_{1,t} \leq P_{MG}^{\max} \quad (10.18)$$

$$\left[(P_{1,t})^2 + (Q_{1,t})^2 \right]^{1/2} \leq S_{MG}^{\max} \quad (10.19)$$

with $t = 1, 2, \ldots, N_T$, where $P_{1,t}(Q_{1,t})$ is the active (reactive) power at the interconnection bus, $P_{MG}^{\max}(S_{MG}^{\max})$ is the maximum allowed active (apparent) power and $\delta_{1,t}^{sp}$ is the reference for the phase angle. The value of $S_{MG}^{\max}$ imposes also a limitation on the maximum power that can be injected into the upstream grid. This limitation which regards the transformer rating could be applied also in compliance with other technical or economical aspects regarding the grids' interconnection. A constraint on a specific maximum value of active power injected into the grid can also be considered.

It must be noted that the value of $P_{MG}^{\max}$ can be a stringent constraint as it depends on the contractual conditions for the peak-shaving service. In the case of a lack of resources, the provision of the service to the upstream grid would be quite small or even absent.

Constraints related to the μG's bus voltages and line currents are also imposed to improve the μG's behavior:

$$V^{\min} \leq V_{i,t} \leq V^{\max} \quad (10.20)$$

$$I_{l,t} \leq I_l^r \quad (10.21)$$

$$t = 1, 2, \ldots, N_T, \; i = 1, \ldots, N, \; l \in \Omega_{lines}$$

where $V^{\min}(V^{\max})$ is the minimum (maximum) magnitude of the bus voltage, $I_{l,t}$ is the line current expressed as a function of the state bus voltages, I_l^r is the maximum value of the current (lth line rating), and Ω_{lines} is the set of μG lines.

10.4 Optimal Control of Vehicle Fleets: Multi Objective Approach

In some cases the strategy requires the satisfaction of many objectives that may be in conflict with each other. In this case, the optimization model to be solved is particularly complex because the different objectives must be optimized simultaneously while several constraints must be met. The most adequate framework to formulate this problem is multi-objective optimization. In fact, an MO approach permits the processing of many objectives, which are sometimes conflicting, and several constraints to determine a compromised solution that takes into consideration different perspectives.

As is well known, an MO optimization problem can be formulated as follows [20]:

$$\min F(\mathbf{X},\mathbf{C}) = \left[f_{obj1}(\mathbf{X},\mathbf{C}), f_{obj2}(\mathbf{X},\mathbf{C}), \ldots, f_{obj\,m}(\mathbf{X},\mathbf{C})\right] \quad (10.22)$$

subject to the equality and inequality constraints such as (10.2) and (10.3).

In (10.22), m is the number of objective functions to be optimized, **X** and **C** have been already defined in (10.1)–(10.3). In particular, in the case of an MO problem, they are defined as the feasible design space (often called the feasible decision space or constraint set) [21].

Again, the choice of the objective functions depend on the strategy that has to be applied (i.e., the services to be provided). Constraints are typically related to the technical limitations of the grid and DERs [22, 23]. Examples of inputs of the problem are the requirements of both the grid and DERs. Outputs are the values of the control variables. However, in this case, unlike the single objective approach, an MO problem does not have a unique solution. That being said, a set of points can be available, all of which fit a predetermined definition for an optimum.

Pareto optimality is the predominant concept in defining an optimal point [20]. A solution is Pareto optimal if there exists no feasible solution that can decrease some objective functions without causing at least one objective function increase [21]. Theoretically, there are an infinite number of Pareto optimal solutions (Pareto frontier), so it often is necessary to incorporate the decision-maker's (DM's) preferences to obtain a single, suitable solution, where the term preference is intended to mean the relative importance of the different objective functions. In this framework, there are several criteria to articulate preferences, each defining an MO problem solution method. Methods with a priori articulation of preferences, with a posteriori articulation of preferences and with no articulation of preferences can be used. A classification, as well as a discussion, of the best-performing methods in terms of programming and computational efforts is described in [20].

In what follows, an example of a possible strategy based on an MO optimization is described to operate the LV μG described in Fig. 10.1. In this strategy, the DERs are optimally controlled to provide both services internal to the μG and external to it. It is assumed that there is more than one μG internal service to be satisfied that is the minimization of power losses, optimization of the voltage profile and maximization of the security margin of the currents. The external service is the smart charging [2, 24]. It has to be noted that the proposed strategy is aimed at enabling both PEV aggregators and μG operator to provide smart charging, which is a remunerated service, and at the same time guaranteeing an efficient, secure and high quality operation of the grid.

To perform the smart charging service, it is assumed that the CCS receives a load dispatch command from the grid operator. Consequently, the optimization model must provide for sharing the load dispatch command signal among the aggregators of the fleets that are connected to different μG busbars. The load dispatch command has to fall into a range compatible with the expected aggregators energy availability, thus avoiding the possibility that the command signal is not complied with (otherwise, as outlined in Sect. 10.3, to avoid critical situations, a backup battery bank could be used [5, 22]). Each aggregator provides information to the CCS on its expected energy availability.

In the problem under study, the vector of decision variables, **C**, includes the active powers to be provided by the aggregators and the reactive powers that the DG and aggregator converters must supply to perform the desired services. The vector of the state variables **X** includes, as an example, the voltages at the grid nodes.

Unlike the example of SO optimization reported in Sect. 10.3, the control interval slot *t*, in this case, is a few seconds, as smart charging is performed as an external service. In addition, the MO model is solved at each dispatch command sent from the grid operator for every control interval *t*.

The objective functions to be minimized in (10.22) are the power losses, the squared voltage deviation along all the μG busbars, and the security margin related to the line currents:

$$f_{obj1} = P_{loss,t}, \tag{10.23}$$

$$f_{obj2} = \frac{1}{N}\sum_{i=1}^{N}\left(V_{i,t} - V_{i,t}^{sp}\right)^2, \tag{10.24}$$

$$f_{obj3} = 1 - \min_{l\in\Omega_l}\left|\frac{I_l^r - I_{l,t}}{I_l^r}\right|, \tag{10.25}$$

where t is the control interval code, $P_{loss,t}$ is the value of power losses expressed as a function of the state variables, $V_{i,t}$ is the voltage amplitude at node i, $V_{i,t}^{sp}$ is its specified value, N is the total number of nodes, $I_{l,t}$ is the current of the lth line expressed as a function of the state variables, I_l^r is its rating, and Ω_l is the set of all the μG lines.

The use of the objective function (10.23) refers to an energy saving objective, whereas the use of the voltage objective function (10.24) is related to a quality improvement that is nowadays an essential requirement in the context of SGs where the power quality problem is a crucial issue. Regarding the objective function (10.25), the security margin, which consists of the amount by which power transfers can change before a security violation is encountered, its inclusion among the objective functions makes it possible for the system to better support unexpected loads and DG variations. This aspect can be a critical issue in the context of μGs, where aggregators are present; in fact, excessively high charging/discharging rates of aggregators can be avoided.

The equality/inequality constraints that must be satisfied are (10.5)–(10.14) and (10.16)–(10.21) which in this case refer only to each time slot of 4 s. Unlike the case of the scheduling procedure of Sect. 10.3, in this case the constraints in the time slot have to be verified independently of the other time slots. As a consequence, constraint (10.15) which refers to a number of consecutive time slots, is not included here. The energy required by PEVs on larger time intervals, in fact, is managed directly by the aggregator, which performs an optimal dispatch by determining how and when each vehicle is to be charged. To perform the smart charging service, the power absorbed by each aggregator has to also be compatible with the following constraint:

$$\sum_{i=1}^{v} PEVP_{i,t} = P^{sp}_{load\ dispatch}, \qquad (10.26)$$

where $P^{sp}_{load\ dispatch}$ is the specified load dispatch command sent by the grid operator to the CCS, $PEVP_{i,t}$ is the active power at the ith aggregator, and v is the number of aggregators.

Since, for the solution of the previously mentioned problem, the control interval is a few seconds (due to the need to perform the smart charging service), the optimization model must be quickly solved. As such, the CPU time is an important and constraining issue. Among the possible solution methods, the *weighted sum approach* and the *objective sum method* seem to be highly attractive, since they require low CPU time efforts. In addition, these methods are characterized by less programming complexity [20]. They are all scalar methods that provide sufficient conditions for Pareto optimality. The first approach is characterized by a priori articulation of preferences, while the second is with no articulation of preferences. Methods characterized by a posteriori articulation of preferences cannot be taken into account, because this category is less efficient in terms of CPU time when only one solution has to be selected. More details on these methods can be found in [20, 25]. In Appendix 1, the *objective sum method*, the *weighted sum approach* and the problem of setting the preferences are summarized. Moreover, two rank methods are reported and discussed, which were used in the numerical application: the *rank order* and the *rank centroid*.

10.5 Verification and Simulation Studies

Numerical applications were effected to verify the SO and MO optimization models shown in Sects. 10.3 and 10.4, respectively. The results are reported in the following sub-sections. Both the control strategies were tested on the 30-busbar μG, balanced, 3-phase, low-voltage distribution system shown in Fig. 10.2.

The parameters of the test system are reported in Appendix 2. The LV network was connected to an MV grid through a 20/0.4-kV, 250-kVA transformer

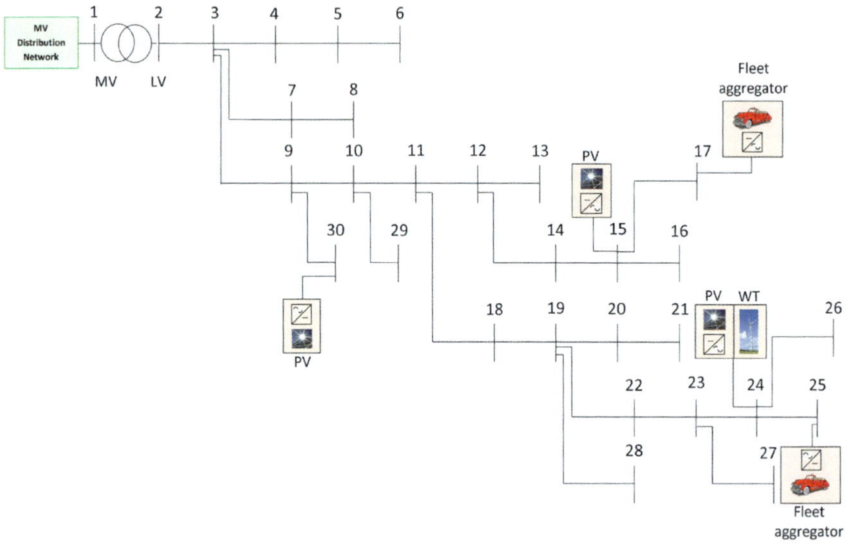

Fig. 10.2 Low-voltage distribution test system

with $v_{cc\%}$ = 4.2 %. The system included three photovoltaic (PV) units, one wind turbine (WT) unit, and two PEV aggregators. The PV units (20-kW peak power units) were located at busbars #15, #24, and #30 and connected through DC/AC converters having size of 20 kVA. The WT unit, equipped with a 7.5-kW asynchronous generator, was located at busbar #24 and was connected directly to the distribution system. The fleet's aggregators (40 kW peak power) were connected at busbars #17 and #25 through DC/AC converters having size of 40 kVA.

As examples of the input data, Fig. 10.3a shows the absolute value of the daily active power requested by the load at bus #18. Figure 10.3b shows the active power of the PV unit at bus #30. Figure 10.3c shows the electricity price [19].

10.5.1 Case 1: Single-Objective Optimization for the Minimization of Microgrid Costs

The day was divided into $N_C = 5$ time intervals (T1 = [0.00–8.00], T2 = [8.00–12.00], T3 = [12.00–14.00], T4 = [14.00–18.00] and T5 = [18.00–24.00]). Each interval was characterized by a specified energy request from aggregators. At each time interval, the upper and lower limits of the aggregators active powers during the whole day are also specified (Table 10.1). The specified values reported in the table derive from the forecasted requirements of the vehicles and are supposed to be treated and provided by the aggregator.

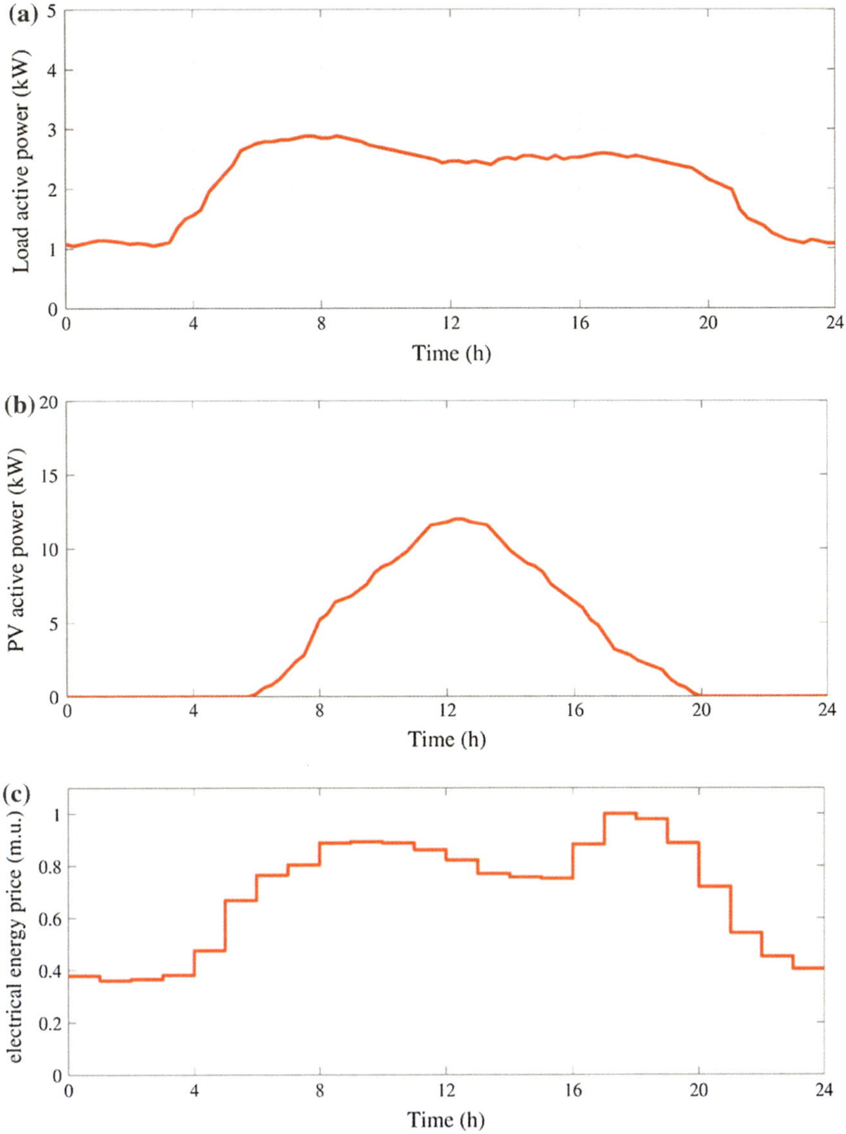

Fig. 10.3 **a** Absolute value of the power requested by the load at busbars #18. **b** PV unit power production (busbar #30). **c** Electricity price

As shown in Table 10.1, the PEV aggregator connected to bus #17 refers to a fleet of vehicles parked in a home parking area, while the PEV aggregator in bus #25 refers to a fleet of vehicles parked in an office parking area; PEVs for both of the two aggregators are allowed to perform the V2G service. In this application, for the sake of simplicity, ΔT_t is assumed to be 60 min. Consequently, since the energy

Table 10.1 Aggregators' energy requests and active power limits

Time period	Aggregator at bus #17		Aggregator at bus #25	
	Energy requested (kW h)	Upper and lower limits of the active power (kW)	Energy requested (kW h)	Upper and lower limits of the active power (kW)
T1 = [0.00–8.00]	144	+25; −25	40	+10; −10
T2 = [8.00–12.00]	32	+10; −10	100	+30; −30
T3 = [12.00–14.00]	20	+15; −15	40	+25; −25
T4 = [14.00–18.00]	40	+20; −20	32	+15; −15
T5 = [18.00–24.00]	120	+30; −30	30	+10; −10

price is usually an hourly price, EC_t is assumed to be constant at each time-slot, ΔT_t. To perform the peak-shaving service, the power that the μG is allowed to request to the upstream grid cannot exceed the maximum value of 50 kW.

Figures 10.4a–c report the profile of the active power, energy and reactive power of the aggregator at busbar #17; in Fig. 10.4a, the upper and lower limits of the aggregator active power during the whole day are also reported. Figure 10.4d reports the active power at the interconnection bus (i.e., #1). In Fig. 10.4d, the active power profile is reported with reference to the case in which the peak-shaving service was not performed.

Figures 10.4a, b show that the control actions allow for the satisfaction of the requirements of the aggregator in terms of power limits and energy requests. It also appears that when the power is positive in Fig. 10.4a, that means that the aggregator supplies power to the grid, the total energy stored in the PEVs' battery, reported in Fig. 10.4b, decreases. On the other hand, when the power is negative, the aggregator is absorbing power from the grid to charge the batteries on board of the vehicles and the energy stored increases. Figure 10.4a shows that the aggregator supplies power to the grid when the price is higher, if it is allowed by the energy requirements of the aggregators. This happens before 8 a.m. and after 5 p.m. In the other periods the aggregator is required to charge in order to satisfy its energy needs. In particular, by analyzing its behavior at each of the specified intervals T1 = [0.00–8.00], T2 = [8.00–12.00], T3 = [12.00–14.00], T4 = [14.00–18.00] and T5 = [18.00–24.00], it appears that the power requested by the aggregator for charging is always larger in the periods of lower energy prices coherently with the rationale of cost minimization.

With reference to the reactive power provided by the aggregators, in Fig. 10.4c the high contribution in terms of the reactive power of the aggregator is clearly highlighted. Always the aggregator supplies capacitive-reactive power close to the maximum value imposed by the interfacing converter size (40 kVA) and the absolute value of the aggregator's active power never reaches the maximum value. By the simulations, it emerged that also other DERs have similar behavior. For the sake of brevity, their reactive power profiles are not reported here.

The analysis of Fig. 10.4d reveals the smoothing effect of the peak-shaving service on the active power profile at the interconnection bus. The peak-shaving is

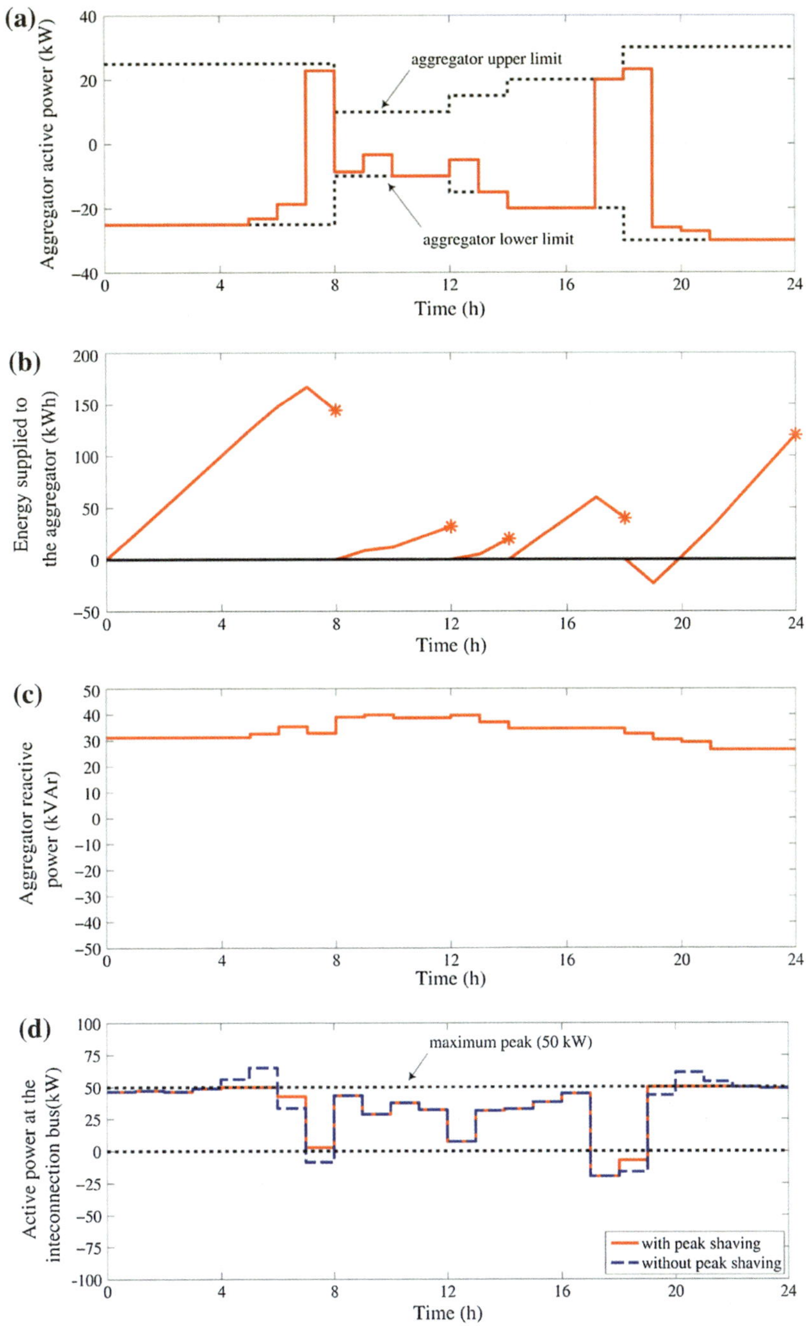

Fig. 10.4 **a** Active powers of the aggregator at busbar #17, **b** energy supplied to the aggregator at busbar #17, **c** reactive power of aggregator at bus #17, **d** profile of the active power at the interconnection bus

obtained thanks to the ability of the control actions to regulate the active power of the aggregators (e.g., during the interval [4.00–6.00 a.m.]). By comparing the profiles of the active power with and without peak shaving, it can be observed that, when the peak shaving is applied, the μG imports a higher value of energy in the interval [6.00–7.00 a.m.] whereas in the interval [7.00–8.00 a.m.] the imported energy is close to zero; without peak shaving, in the last interval, the μG is able to export energy to the upstream grid. During the interval [5.00–7.00 p.m.] the μG exports energy in both cases (with and without peak shaving) but when the peak shaving is applied, the energy exported to the upstream grid is lower. As a consequence, when the peak shaving service is performed, the objective function (i.e., the total cost suffered by the μG for energy) is higher than the value it assumes in the case without peak shaving. In particular, the increment of the objective function value resulted to be about 3 %. To justify this, it has to be considered that, although the control actions are aimed at the minimization of the total cost sustained by the μG for energy consumption, in order to satisfy the peak shaving constraint, the μG is forced to import power even in higher energy price periods.

10.5.2 Case 2: Multi-objective Optimization

In this application, a time interval of 60 s was simulated. The proposed MO optimization model was applied at each four second time step. During the 60 s time interval, the CCS sends the dispatch command every 4 s (i.e., 15 times), as illustrated in Fig. 10.5a. As a further example of the input data, Fig. 10.5b reports the load active power at busbar #18 and Fig. 10.5c reports the power production of the PV unit located at busbar #30.

A constant value was assumed for the voltage at busbar #1 (1.05 p.u.). The specified value of the voltage in the objective function (10.24) is 1.0 p.u.

Both the *weighted sum approaches* and the *objective sum method* (Appendix 1) were applied. With reference to the weighted sum approaches, the values of the weights were chosen according to different methods: the rank centroid and rank order. As discussed in the Appendix, Sect. 1.1, both of these two methods need to assume a specific rank order which, in this case, is:

f_1 = *squared voltage deviations,*
f_2 = *power losses,*
f_3 = *security margin.*

With this order, by applying relationships (10.29) and (10.30), the values of the weights in the *rank order centroid* (ROC) method were found to be $w_1 = 0.611$, $w_2 = 0.278$ and $w_3 = 0.111$, while in the *rank sum* (RS) method, they were $w_1 = 0.500$, $w_2 = 0.333$ and $w_3 = 0.167$.

With reference to the normalization of the objective functions, relationship (10.28) is applied. In this case, the maximum admissible value that each function can assume has to be evaluated. In the considered case, to calculate the maximum

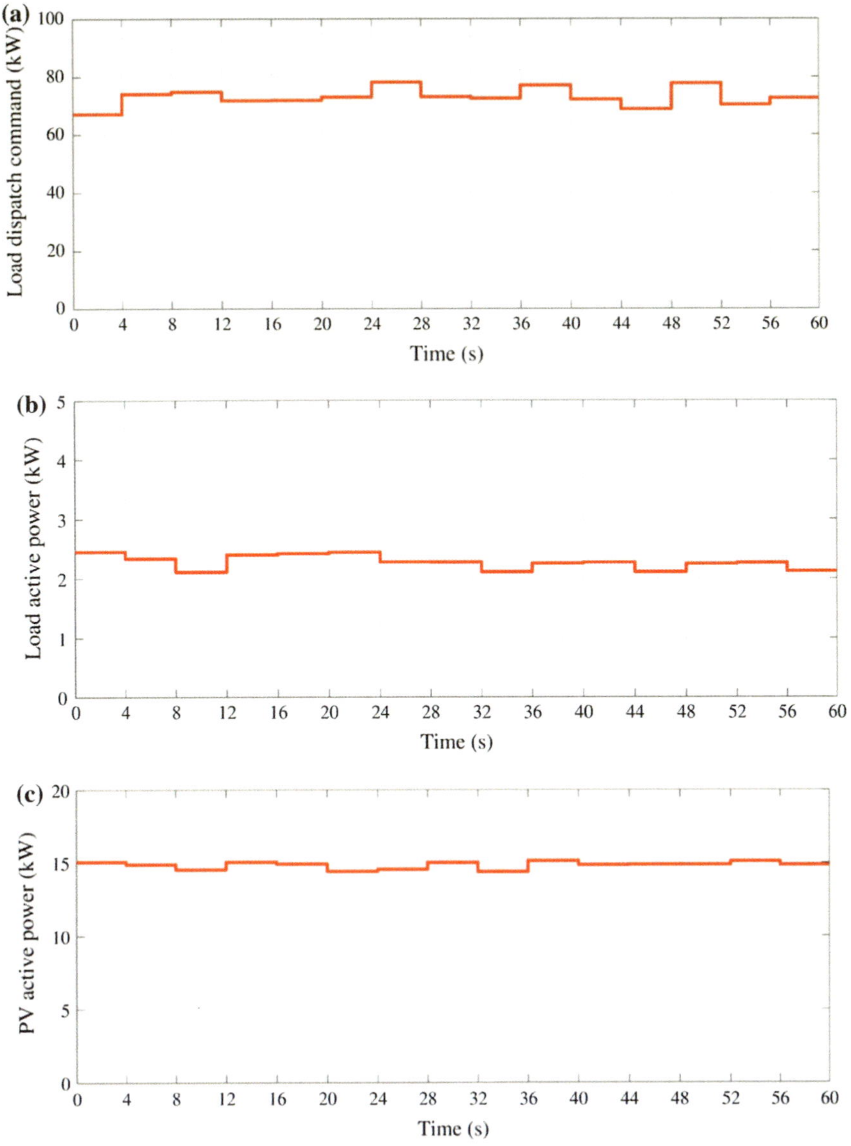

Fig. 10.5 Input data with reference to a 60 s time interval: **a** Load dispatch command sent by the grid operator to the CCS. **b** Load power at busbar #18. **c** PV unit power production (busbar #30)

value of the squared voltage deviation ($f_1^{\max}$) and that of the power losses ($f_2^{\max}$), two off-line power flows were performed in which the nominal values for the loads and a zero value for the reactive power of the DG units and aggregators were assumed. Moreover, for the calculation of $f_1^{\max}$, the power flow was performed assuming the nominal value for the DG unit's active powers. For the calculation of $f_2^{\max}$, the power flow was performed without considering any DG unit. Regarding the security margin, its maximum value ($f_3^{\max}$) was assumed to have the value 1.

As a first example of the obtainable results, the outputs of the proposed control strategy applying the *objective sum method* are shown in Figs. 10.6a, b. In Fig. 10.6a the active power signals sent from the CCS to the aggregators are reported. Figure 10.6b reports the reactive power requested to the aggregator at busbar #25 and to the PV unit at busbar #30.

As a further example of the output results, Figs. 10.7a, b show the voltage profiles at the aggregators' busbars and the line currents flowing in the lines connected to the aggregator located at busbar #25 and the PV unit located at busbar #30.

To study the impact of each objective function on the control procedure, some results are reported in Fig. 10.8, where the values assumed by each objective function in the case when the MO optimization is effected is compared to those

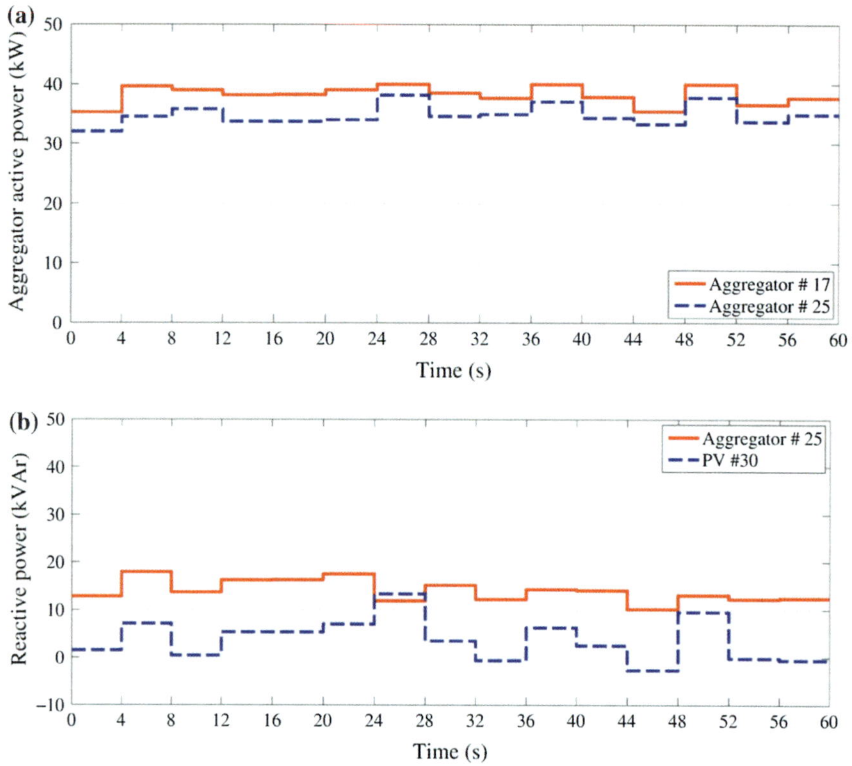

Fig. 10.6 Outputs of the control strategy applying the *objective sum method* method: **a** Active power of the aggregators. **b** Reactive power of the aggregator at node #25 and reactive power of the PV unit at node #30

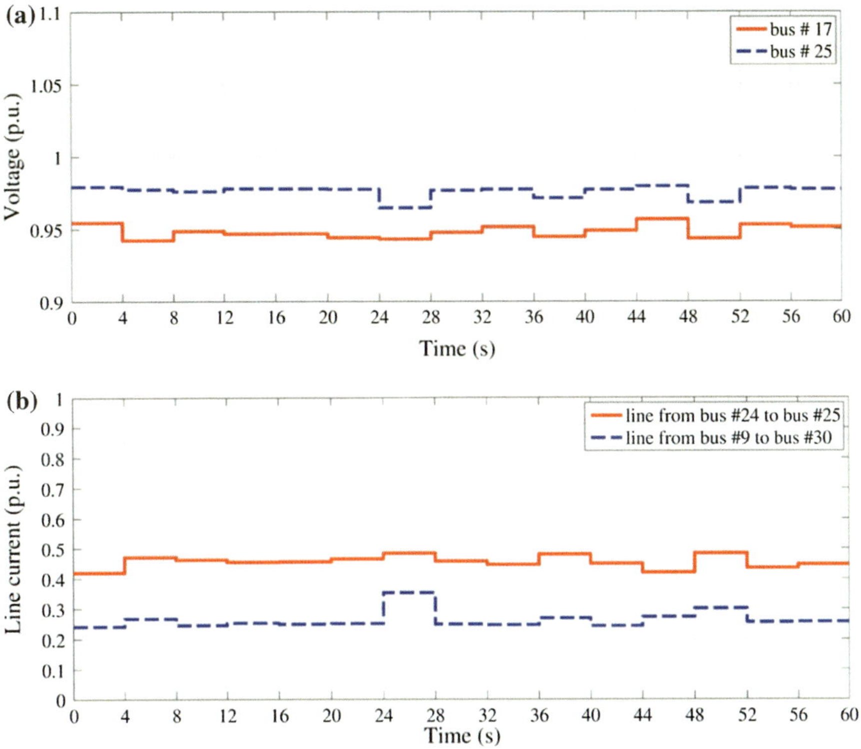

Fig. 10.7 Outputs of the control strategy applying the *objective sum method* method: **a** Voltage profile at busbars #17 and #25. **b** Current profiles at lines 24–25 and 9–30

obtained in the cases of SO optimizations (i.e., SO squared voltage deviation minimization, SO power losses minimization, and SO security margin maximization).

Figure 10.6a shows that the active power of the aggregator at bus #17 is always greater than that of the aggregator at bus #25. Figure 10.6b clearly reveals the significant contribution of the aggregator at bus #25 in terms of reactive power. It is interesting to note that the reactive power of the PV unit at bus #30 assumes, in some cases, negative values (i.e., during the intervals of 32–36 s and 44–48 s). This is probably due to the need, during those intervals, to decrease the voltages at bus #30.

Figure 10.7 confirms that the constraints on currents and voltages are fully satisfied. In particular, Fig. 10.7a shows that the voltage observed in the considered time interval is always greater than the lower limits imposed, even when the aggregator power requests are high. The voltage profile is also quite close to 1 p.u., as required by the first objective function. This is more marked in the busbar #25 due to the fact that the power supplied to the aggregator at bus #25 is lower than that supplied to the other aggregator and also because the PV and WT units located close to bus #25 supply a power greater than that supplied by the PV unit located

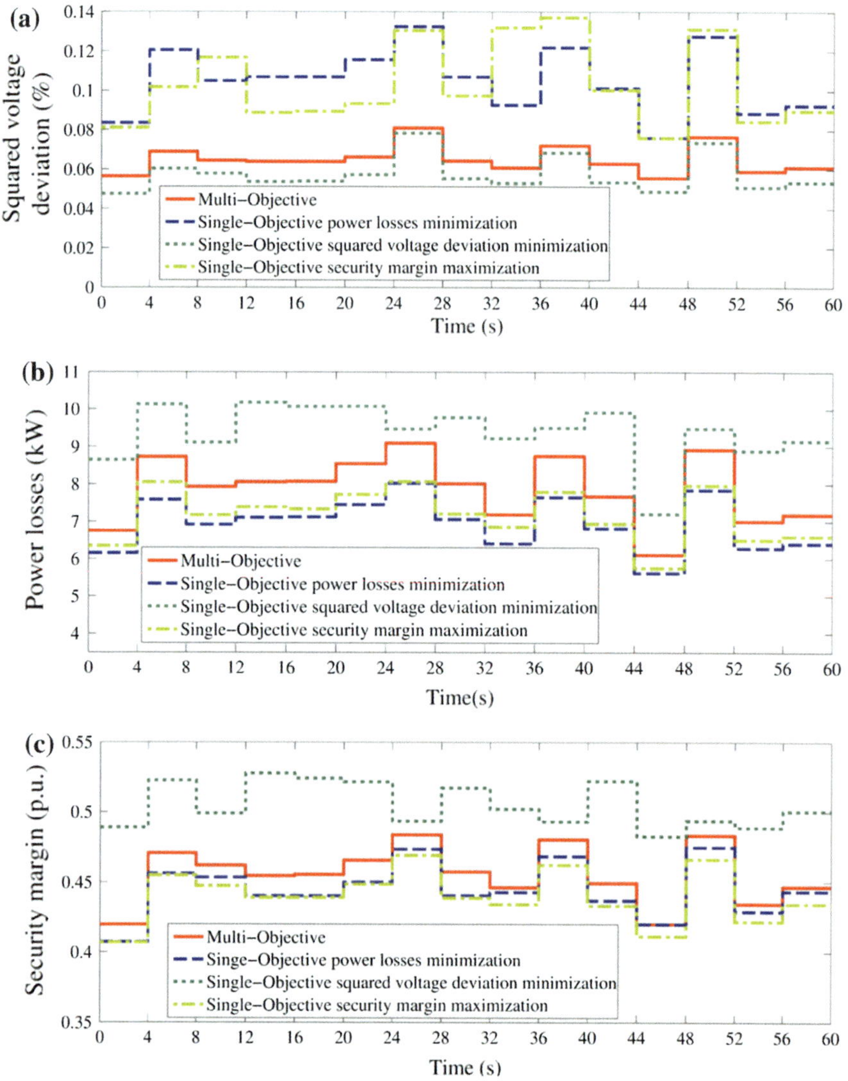

Fig. 10.8 Outputs of the control strategy applying different optimization methods: values assumed by **a** squared voltage deviation, **b** security margin, **c** power losses

close to bus #17. Figure 10.7b shows that the upper limits imposed on the values of the line currents are satisfied. In fact, in the figure, the ratio between the line current magnitude and the line ampacity is reported and these profiles are always lower than 1 p.u. This is obtained thanks to the contribution of the security margin objective function that is aimed at reducing the value of the current flowing into the line. The same result is obtained also by the minimization of the power losses objective function.

By observing Figs. 10.8a–c, it appears that the values of the objective functions in the MO approach are always included in the ranges bounded by the best and worst values of the equivalent functions in the SO approaches.

To study also the impact of the different solving algorithms on the control actions resulting from the MO approach, in Fig. 10.9, the values assumed by each

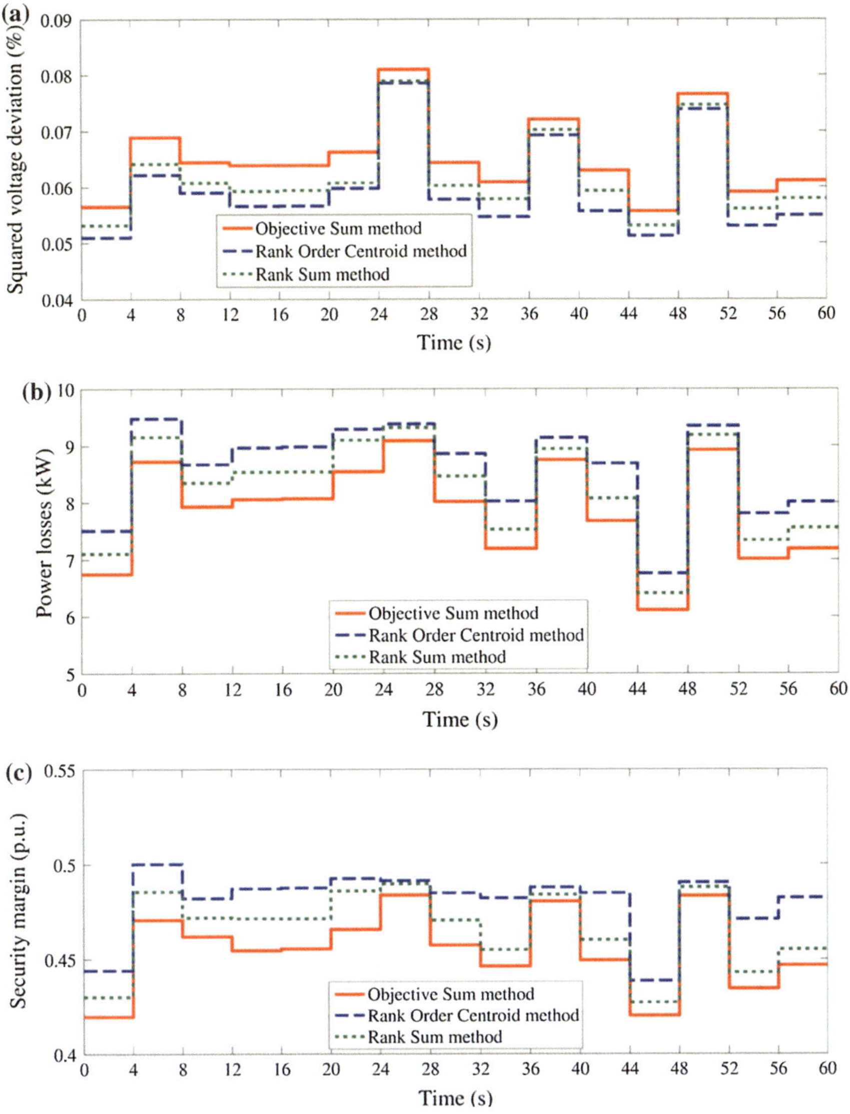

Fig. 10.9 **a** Squared voltage deviations for *objective sum*, *rank order centroid* and *rank sum* methods. **b** Power losses for *objective sum*, *rank order centroid* and *rank sum* methods. **c** Security margin for *objective sum*, *rank order centroid* and *rank sum* methods

objective function are reported by applying the *weighted sum* ROC and RS methods. In the same figure, the results obtained with the *objective sum method* approach are also reported, for comparison purposes. In more detail, Fig. 10.9a shows the values assumed by the squared voltage deviation, Fig. 10.9b reports the values assumed by the power losses, and the values obtained by the security margin are reported in Fig. 10.9c.

From the analysis of Fig. 10.9, it appears that the objective sum method yields the best result in terms of power losses and security margins, whereas the ROC method yields the best result in terms of the squared voltage deviation. The RS method, instead, yields results closer to the best solution, even if it doesn't provide the absolute optimum for any objective function. In fact, as an example, in terms of the squared voltage deviation, the RS method provides values included between the best results (obtained by using the ROC method) and worse results (obtained by using the OS method). The same considerations arise with reference to the other objective functions. It should be noted, however, that the previously mentioned results are strictly related to the rank order of preferences assumed for this application (i.e., squared voltage weight > power losses weight > security margin weight).

The mean values of each objective function over the 60 s time interval are reported in Table 10.2 whereas, with reference to each method, Table 10.3 reports the minimum and maximum deviations (with reference to the same time interval) of each objective function from their best values (in %). The results reported in the tables are coherent with the consideration that the ROC method emphasizes the first objective function, whereas the RS method provides flatter results due to the values assumed for the weights associated with each objective function in the two methods (w_1 = 0.611, w_2 = 0.278 and w_3 = 0.111 in the ROC method; w_1 = 0.500, w_2 = 0.333 and w_3 = 0.167 in the RS method) (Table 10.3).

Table 10.2 Mean values of the objective functions over the 60 s time interval

Method	Squared voltage deviation (%)	Power losses (kW)	Security margin (p.u.)
Objective Sum	0.0648	7.822	0.4548
Rank Order Centroid	0.0592	8.558	0.4805
Rank Sum	0.0615	8.197	0.4652

Table 10.3 Objective functions deviation ranges over the 60 s time interval

Method	Squared voltage deviation (%)		Power losses (%)		Security margin (%)	
	min	max	min	max	min	max
Objective Sum	2.99	13.00	0	0	0	0
Rank Order Centroid	0	0	3.28	13.15	1.48	8.41
Rank Sum	0.44	6.54	2.30	6.66	0.73	4.41

10.6 Conclusions

This chapter focused on the possible strategies available to optimally manage PEV fleets, considering the grid operator perspective. Two different approaches were considered: single-objective and multi-objective optimizations. Two examples of the possible strategies were presented to manage the PEV fleets together with other distributed resources included in a low voltage microgrid. The peculiarities of the two approaches used for the management of the PEV fleets were evidenced by implementing computer simulations. The examples reported in this chapter evidenced the importance that PEVs assume in the framework of modern power systems and smart grids. It has also been evidenced that an optimal management of PEV fleets, as well as their integration with other distributed resources, can provide both economical and technical benefits. It is worth to note that forecasting errors affecting the proposed control strategies would decrease the achievable economic benefits. When using procedures based on very short time forecasts as that reported in Sect. 10.4, less significant forecasting errors should be expected.

Appendix 1

A.1.1 Weighted Sum Approaches

The weighted sum approach consists of solving the following single-objective optimization problem [20]:

$$\min \sum_{i=1}^{m} w_i F_i(\mathbf{X}, \mathbf{C}) \tag{10.27}$$

subject to constraints (10.2) and (10.3). In (10.27), w_i are positive weights, whose values reflect the relative importance of the objectives. Typically, the weights satisfy the conditions $\sum_{i=1}^{m} w_i = 1$.

Before running the optimization algorithm, the DM indicates the preferences by setting the weights in (10.27). Regarding the relationship between the preferences and weights, even if the DM has a good knowledge of the objective functions and a precise idea of his own preferences, the solution obtained by minimizing (10.27) may not necessarily reflect the preferences [25]. The value of a weight must be relative to the other weights and relative to its corresponding objective function as outlined in [20]. Thus, the objective functions have to be transformed so that they all have similar magnitudes. As such, in this chapter, the following transformation in (10.27) is applied:

$$F_i = \frac{f_i}{f_i^{\max}}, \tag{10.28}$$

with $f_i^{\max}$ the maximum value of the ith objective function. In this transformation F_i should result in a non-dimensional objective function with an upper limit of one and a lower limit of zero. To have a fast computation of $f_i^{\max}$, for each objective function, the maximum value is approximated on the basis of an engineering intuition and computed in particularly critical grid operative conditions.

Several methods can be applied for the choice of the weights in (10.27). The type of method depends on the kind of information available from the DM. As a general rule, when the ratio scale properties of the DM evaluation of the objectives are available, the *ratio weights methods* should be used. When only the ordinal properties of the DM judgments are known, the *rank order methods* could be considered.

Ratio weights methods can be more effective than rank order weights methods but require extra time or effort to assess the ratio weights. Rank order methods are typically adopted when DMs are not able to provide more than ordinal information about objective function importance or when the decision process depends on the choice of a group, rather than one DM, whose members are able to agree on the ranking of the objectives, but not on the precise weights. Moreover, when choices require tradeoffs between objectives, the uncertainty regarding the preferences is even stronger [20, 25].

In this chapter, only the rank order methods were considered as criteria for choosing weights in the weighted sum approaches; in particular, the methods adopted were the *rank order centroid* (ROC) and *rank sum* (RS) methods [25].

In the ROC method a uniform distribution of the weights can be assumed on the simplex of the rank-order weights. Given that $w_{(1)} > w_{(2)} > \cdots > w_{(\mathrm{m})}$, if (i) is the rank position of $w_{(i)}$ and m is the number of objective functions, the expected values of the true weights are given by:

$$w_{(i)} = \frac{1}{m}\sum_{k=i}^{m}\frac{1}{k} \quad i = 1, \ldots, m. \tag{10.29}$$

In the RS method, the objective functions are ranked in terms of their relative importance. Each objective is then ranked in the proportion of its position in the rank order:

$$w_{(i)} = \frac{2(m+1-i)}{m(m+1)} \quad i = 1, \ldots, m. \tag{10.30}$$

where: (i) is the rank position of the objective function i, $0 < w_{(i)} < 1$ and $\sum_i w_{(i)} \leq 1$.

It should be noted that, both the two approaches analyzed in this chapter (ROC and RS methods) are sharable and provide quite accurate results. The choice between the methods primarily depends on the particular circumstances. In more detail, as evidenced in [25], the RS weights are flatter than the ROC weights, then, the choice between these two approaches could depend also on the steepness of the true weights from which the DM's decisions derive. The greater the value of the first few weights, the more attractive the ROC method.

It has to be observed that when dealing with a larger number of objectives, the differences among the methods are more pronounced.

A.1.2 Objective Sum Method

The *objective sum* (OS) method has no articulation of preferences. It can be stated that OS is a particular case of the weighted sum method with all weights being equal. In this way, it does not require any DM preference input. This is not necessarily a limit, because, in some cases, the DM has incomplete information and does not have a clear idea about their desires. At worst, the DM may have no preferences at all. The primary advantages of this method are its easy application and the reduced subjective requirements.

Appendix 2

The line parameters are reported in Table 10.4. The nominal values of active and reactive power requested by the loads are reported in Table 10.5.

Table 10.4 Line parameters

Line		R [Ω]	*X* [Ω]	Ampacity [A]
From bus	To bus			
2	3	0.0195	0.0070	249
3	4	0.2603	0.0137	60
4	5	0.3192	0.0168	60
5	6	0.0190	0.0010	60
3	7	0.2033	0.0107	60
7	8	0.1938	0.0102	60
3	9	0.0620	0.0164	210
9	10	0.0310	0.0082	210
10	11	0.0268	0.0071	210
11	12	0.0807	0.0073	120
12	13	0.1944	0.0713	91
12	14	0.0573	0.0051	120
14	15	0.0746	0.0067	120

(continued)

Table 10.4 (continued)

Line		R [Ω]	*X* [Ω]	Ampacity [A]
From bus	To bus			
15	16	0.1502	0.0135	120
15	17	0.0903	0.0081	120
11	18	0.0379	0.0255	150
18	19	0.0618	0.0416	150
19	20	0.1664	0.0611	91
20	21	0.0779	0.0041	65
19	22	0.0348	0.0235	150
22	23	0.0628	0.0424	150
23	24	0.0618	0.0416	150
24	25	0.0312	0.0028	120
24	26	0.0868	0.0078	120
23	27	0.1664	0.0611	91
19	28	0.0574	0.0201	150
9	30	0.1596	0.0084	60
10	29	0.0665	0.0035	60

Table 10.5 Load active and reactive powers

Bus	P_L (kW)	Q_L (kVAr)
4	3.00	1.45
5	3.00	1.45
6	6.00	2.91
7	3.00	1.45
8	3.00	1.45
13	3.00	1.45
14	3.00	1.45
16	2.00	0.97
18	3.00	1.45
21	2.50	1.21
26	3.00	1.45
27	14.00	6.78
28	3.00	1.45
30	6.00	2.91

References

1. Kempton W, Tomic J (2005) Vehicle-to-grid power fundamentals: calculating capacity and net revenue. J Power Sources 144(1):268–279
2. Brooks A, Lu E, Reicher D, Spirakis C, Weihl B (2010) Demand Dispatch. IEEE Power Energ Mag 8(3):20–29

3. Campbell JB, King TJ, Ozpineci B, Rizy DT, Tolbert LM, Xu Y, Yu X (2005) Ancillary services provided from DER, Report ORNL/TM-2005/263, OAK Ridge National Laboratory. http://web.ornl.gov/sci/decc/Reports.htm. Accessed 21 July 2014
4. Wu D, Aliprantis DC, Ying L (2012) Load scheduling and dispatch for aggregators of plug-in electric vehicles. IEEE Trans Smart Grid 3(1):368–376
5. Wu C, Mohsenian-Rad H, Huang J (2012) Vehicle-to aggregator interaction game. IEEE Trans Smart Grid 3(1):434–442
6. Peças Lopes JA, Soares FJ, Rocha Almeida PM (2009) Identifying management procedures to deal with connection of electric vehicles in the grid. In: Proceedings of the IEEE powertech conference, Bucharest, Romania
7. Kempton W, Tomic J, Letendre S, Brooks A, Lipman T (2001) Vehicle-to-grid power, battery, hybrid and fuel cell vehicles as resources for distributed electric power in California. Davis, CA: Institute of Transportation Studies Report. http://www.udel.edu/V2G. Accessed 21 July 2014
8. Tomic J, Kempton W (2007) Using fleets of electric-drive vehicles for grid support. J Power Sources 168:459–468
9. Kempton W, Letendre S (1997) Electric vehicles as a new power source for electric utilities. Trans Res D 2(3):157–175
10. Andreotti A, Carpinelli G, Mottola F, Proto D (2012) Review of single-objective optimization models for plug-in vehicles operation in smart grids part I: theoretical aspects. In: Proceedings of the IEEE power and energy society general meeting, San Diego, California, 22–26 July 2012
11. Clement K, Haesen E, Driesen J (2010) The impact of vehicle-to-grid on the distribution grid. Electric Power System Research 81:185–192
12. Carradore L, Turri R, Cipcigan LM, Papadopoulos P (2010) Electric vehicle as flexible energy storage systems in power distribution networks. In: Proceedings of the ecologic vehicles and renewable energies conference, Monte-Carlo, Monaco, 25–28 March 2010
13. Andreotti A, Carpinelli G, Mottola F, Proto D (2012) Review of single-objective optimization models for plug-in vehicles operation in smart grids part II: numerical applications to vehicles fleets. In: Proceedings of the IEEE power and energy society general meeting, San Diego, California, 22–26 July 2012
14. Méndez Quezada VH, Abbad JR, San Román TG (2006) Assessment of energy distribution losses for increasing penetration of distributed generation. IEEE Trans Power Syst 21 (2):533–540
15. Caramia P, Carpinelli G, Proto D, Varilone P (2013) Deterministic approaches for the steady-state analysis of distribution systems with wind farms. In: Pardalos P, Rebennack S, Pereira MVF, Iliadis NA, Pappu V (eds) Handbook of wind power systems. Springer, Berlin, p 211
16. Bracale A, Carpinelli G, Di Fazio A, Khormali S (2014) Advanced, cost-based indices for forecasting the generation of photovoltaic power. Int J Emerg Electr Power Syst 15(1):77–91
17. Bracale A, Caramia P, Carpinelli G, Di Fazio AR, Varilone P (2013) A bayesian-based approach for a short-term steady-state forecast of a smart grid. IEEE Trans Smart Grids 4:1760–1771
18. Bracale A, Caramia P, Carpinelli G, Di Fazio AR, Ferruzzi G (2013) A bayesian method for short-term probabilistic forecasting of photovoltaic generation in smart grid operation and control. Energies 6:733–747
19. Carpinelli G, Mottola F, Proto D, Bracale A (2012) Single-objective optimal scheduling of a low voltage microgrid: a minimum-cost strategy with peak shaving issues. In: Proceedings of the international conference on environment and electrical engineering, Venice, Italy, 18–25 May 2012
20. Marler RT, Arora JS (2004) Survey of multi-objective optimization methods for engineering. Struct Multidiscip Optimiz 26:369–395
21. Proos KA, Steven GP, Querin OM, Xie YM (2001) Multicriterion evolutionary structural optimization using the weighting and the global criterion methods. AIAA J 39(10):2006–2012

22. Carpinelli G, Caramia P, Mottola F, Proto D (2014) Exponential weighted method and a compromise programming method for multi-objective operation of plug-in vehicle aggregators in microgrids. Int J Electr Power Energy Syst 56:374–384
23. Carpinelli G, Mottola F, Proto D (2012) Multi-objective approach for plug-in vehicle optimal operation in a smart grid. In: Proceedings of the IEEE international symposium on power electronics, electrical drives, automation and motion, Sorrento, Italy, 20–22 June 2012
24. Carpinelli G, Andreotti A, Mottola F, Proto D (2011) Optimal control of plug-in vehicles fleets in a smart grid including distributed generation. In: Proceedings of the Cigrè symposium, The electric power system of the future, Bologna, Italy, 13–15 Sept 2011
25. Jia J, Fischer GW, Dyer JS (1998) Attribute weighting methods and decision quality in the presence of response error: a simulation study. J Behav Decis Mak 11(2):85–160

Chapter 11
Energy Management in Microgrids with Plug-in Electric Vehicles, Distributed Energy Resources and Smart Home Appliances

Okan Arslan and Oya Ekin Karaşan

Abstract Smart Grid is transforming the way energy is being generated and distributed today, leading to the development of environment-friendly, economic and efficient technologies such as plug-in electric vehicles (PEVs), distributed energy resources and smart appliances at homes. Among these technologies, PEVs pose both a risk by increasing the peak load as well as an opportunity for the existing energy management systems by discharging electricity with the help of Vehicle-to-grid (V2G) technology. These complications, together with the PEV battery degradation, compound the challenge in the management of existing energy systems. In this context, microgrids are proposed as an aggregation unit to smartly manage the energy exchange of these different state-of-the-art technologies. In this chapter, we consider a microgrid with a high level of PEV penetration into the transportation system, widespread utilization of smart appliances at homes, distributed energy generation and community-level electricity storage units. We propose a mixed integer linear programming energy management optimization model to schedule the charging and discharging times of PEVs, electricity storage units, and running times of smart appliances. Our findings show that simultaneous charging and discharging of PEV batteries and electricity storage units do not occur in model solutions due to system energy losses.

Keywords Microgrids · Energy management · Optimization · DER · Smart grid · Smart appliances

O. Arslan (✉) · O.E. Karaşan
Department of Industrial Engineering, Bilkent University, Ankara, Turkey
e-mail: okan.arslan@bilkent.edu.tr

O.E. Karaşan
e-mail: karasan@bilkent.edu.tr

S. Rajakaruna et al. (eds.), *Plug In Electric Vehicles in Smart Grids*,
Power Systems, DOI 10.1007/978-981-287-302-6_11

11.1 Introduction

The level of information exchange and ease of communication we are witnessing today is transforming the way we interact with the world around us. In the context of energy management, this translates to what has become the "Smart Grid". This new technology, referred to as "internet of energy" [1], facilitates development of environment-friendly, economic and efficient technologies such as PEVs (plug-in electric vehicles), DERs (distributed energy resources) and smart appliances at homes. Furthermore, classical problems are being reevaluated to cope with these developments. One such problem is the way we manage our daily energy requirements. For decades, electricity generation has always been centralized to reduce the costs and increase the efficiency. Smart Grid is transforming the way energy is being generated and distributed. Today, in addition to reducing costs and increasing efficiency, decentralizing the generation units and making use of renewable, more environmental friendly technologies have become new objectives.

In this context, microgrids are proposed as an aggregation unit to smartly manage the energy exchange of these different state-of-the-art technologies. Microgrid is an electric power system operating in either autonomous mode or connected to an electricity grid. It is formed by DERs, storage units and loads without a large scale infrastructure setup requirement. Due to its independence from the grid, it offers increased reliability. Furthermore, higher efficiencies and increased flexibility can be achieved by microgrids with respect to conventional electricity grids. Due to the smart grid technology, small scale energy resources can efficiently be integrated into the energy management systems. A microgrid generally operates in a confined geography and therefore transmission losses are reduced. Furthermore, it offers reduced costs for its participants.

In this chapter, we consider a microgrid with a high level of PEV penetration into the transportation system, widespread utilization of smart appliances at homes, distributed energy generation and community-level electricity storage units. We propose a mixed integer linear programming (MILP) energy management optimization model to schedule the charging and discharging times of PEVs, electricity storage units, and running times of smart appliances. Furthermore, we show that simultaneous charging and discharging of PEV batteries and electricity storage units do not occur in model solutions due to system energy losses.

11.1.1 Literature Review

The problem of energy management is addressed from several different aspects in the recent literature: Morais et al. [2] search for the least-cost schedule for a number of power resources for the forecast load by using MILP and take into account generators and storage units. Their study is one of the first examples of microgrid energy scheduling. The total cost of generating the energy is minimized and the forecast load is satisfied. Khodr et al. [3] use the same model with additional power loss constraints and implement it for an experimental setting with in-place power resources. Kriett and Salani [4] consider thermal energy and electricity, and model energy scheduling with the objective of minimizing operating costs. Naraharisetti et al. [5] improve the energy scheduling model by adding a constraint to maintain diversity between several resources so that no resource is idle. Moghaddam et al. [6] report a multi-objective model, including pollutant emission minimization and operating cost minimization. They use a variant of the particle swarm optimization (PSO) heuristic to reach a near-optimal solution. Basu et al. [7] author a comprehensive survey about the advantages and disadvantages of the microgrid. The aforementioned microgrid scheduling literature addresses the resource scheduling problem for a forecast load usually with the objective of cost minimization, but does not include PEV battery charging and discharging.

Xiong et al. [8] analyze a home microgrid, comprised of smart appliances capable of responding to hourly changing electricity prices. Pedrasa and Spooner [9] and Rastegar et al. [10] include a PEV as a load in a smart home microgrid. Elma and Selamogullari [11] give another example of the home microgrid scheduling model, but the load is time-inelastic and therefore not schedulable. The home microgrid scheduling literature approaches the energy scheduling problem from the demand side with the perspective of the end-user. It assumes that energy is provided from a single source, determines the on/off status of home appliances to minimize the individual customer cost and regards the PEV as an "appliance" in the house.

Aside from the microgrid energy scheduling literature, several different papers address the energy scheduling problem in PEV-penetrated networks: Fernandes et al. [12] consider massive deployment of EVs (electric vehicles) and investigate the impact of V2G capability on the power system operation in terms of cost, but they do not explicitly provide their models. Studies such as Arslan and Karasan [13] and Sioshansi and Denholm [14] analyze the value of PEVs as grid resources and model the charge scheduling of PEV batteries. Sioshansi and Denholm [14] is a unit commitment model of the Electricity Reliability Council of Texas (ERCOT) electric power system, formulated as an MILP. The objective of the model is to minimize the total system cost, which consists of conventional generator costs and PEV operation costs. The model approaches the problem from the energy generation perspective and neglects the pricing of individual PEV owners. Even though the sum of PEV operation costs is included in the objective function, this does not necessarily imply that the cost for each PEV owner is the least possible cost individually. Arslan and Karasan [13] analyzes the value of PEVs in a virtual power plant (VPP) formation. Several sensitivity analyses are carried out to see the impacts of PEV penetration in different settings. However, smart devices are not considered in [13]. Furthermore, energy trade between the VPP and the national grid is only one sided, that is, the VPP cannot sell its excess energy to the grid. A later study by Sioshansi [15] uses the same model of [14] and includes another aspect to make driving and charging decisions for the PEV owner. As a result, the paper examines the incentives for individual drivers with different electricity tariffs.

Su and Chow [16] propose a PSO algorithm for scheduling PEV charging at a municipal parking station. They model the probabilistic nature of the problem, whose objective function is to maximize the average state of charge for all vehicles in the next time period. Saber and Venayagamoorthy [17] schedule PEV batteries under uncertainty using PSO by taking plug-in and plug-out times with associated probability density functions. Their objective is to minimize the expected cost and emissions. Kristoffersen et al. [18] model EV charging in a market environment from the aggregator's point of view for two cases: when the aggregator is a price-taker and when it has market power. Sousa et al. [19] address the problem of energy scheduling from the microgrid perspective, also considering technical constraints such as bus voltage magnitude and angle limits. In [20, 21], the authors discuss the impacts of PEV specifications, road network features and driver tolerances on the route selection to minimize the costs. In this chapter, we assume that drivers prefer to drive using electricity whenever possible. If not, gasoline is used as the source of energy for transportation.

11.1.2 Problem Definition

We consider a group of house owners coming together to form a microgrid to benefit from economies of scale (Fig. 11.1). Rather than providing the energy solely from the grid, each entity in the new formation is generating a certain level of energy with DERs such as small capacitated photovoltaics and/or wind turbines. Houses have

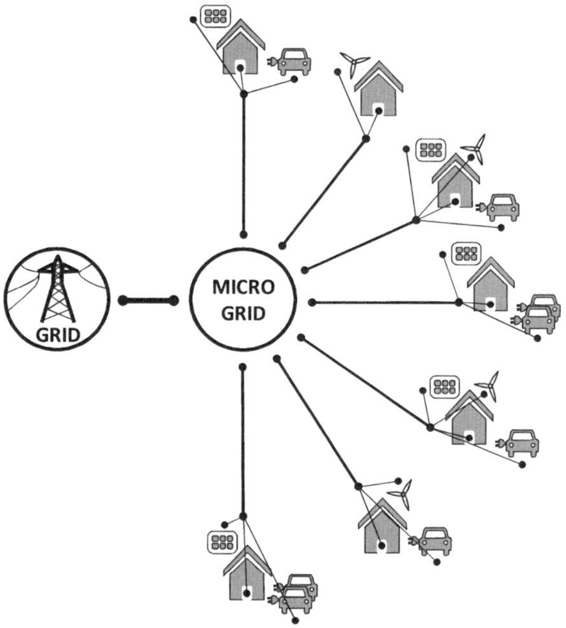

Fig. 11.1 The microgrid energy management model

time-inelastic (TIE) load demands, e.g. refrigeration or television, which need to be satisfied as soon as demanded. They also have time-elastic (TE) loads and these loads need to be satisfied within a given time frame. TE loads are non-preemptive; once started, the "smart" device must run non-stop for at least a certain amount of time. In addition to TIE and TE loads, the houses also have a number of PEVs with a certain driving profile in a given day. A PEV drives on charge depleting (CD) mode until a certain minimum state-of-charge (SoC) is reached, and switches to charge sustaining (CS) mode preserving the remaining charge in its battery and driving on gasoline. In order to drive on CD mode, PEVs needs to be charged when they are connected to the microgrid. The difference between TE loads and PEV charging loads is that the latter can be preemptively charged and does not require continuous energy flow. The PEVs can also supply the energy in their batteries to the microgrid via V2G technology. When charging or discharging PEV batteries, the battery degrades with respect to the level of discharge. Modeling battery degradation is another important aspect of the energy management models. In this respect, PEVs pose both a risk and an opportunity for the existing energy management systems. The microgrid as the energy management unit of the houses is responsible for satisfying the loads that are defined above. The energy generated by the DERs is at the discretion of the microgrid which "smartly" manages the total energy available within the microgrid. If a house is underutilizing DER capacities, the excess generated energy can be used to satisfy another house's demand. The microgrid also has a number of electricity storage units to balance the network loads and to postpone the usage of generated energy for a short period of time. There is also a trading mechanism between the national grid and the microgrid according to a pricing schema. If the energy generated is in excess of the required load and the storage limits, then the energy can be sold to the national grid. In this context, this work deals with the microgrid energy management model. In Sect. 1.2, we present a mixed integer linear programming model to handle the problem and discuss some insights related to the model. In Sect. 1.3, we provide a realistic case study and discuss the value of the microgrid for its participants. Sensitivity analyses of the results to gasoline prices, electricity prices and driving patterns are carried out in Sect. 1.4.

11.2 Methodology

In this section, we introduce the methodology for energy management of microgrids. First, we define the sets, parameters and variables to be used in the model.

11.2.1 Parameters and Variables

The following sets, parameters and variables are used in the microgrid energy management model. Note that our system of interest is the microgrid. Therefore, when defining the variables, we used a "+" superscript to indicate that the microgrid is *receiving* the energy, and a "–" superscript to indicate that the microgrid is *providing* the energy.

11.2.1.1 Sets

H	Set of microgrid participant homes
A_h	Set of appliance running tasks (non-preemptive) in home $h \in H$
D_h	Set of DERs in home $h \in H$
P_h	Set of PEVs in home $h \in H$
S	Set of electrical storage units
T	Set of time periods

11.2.1.2 Appliance-Related Parameters

α_{ah}^{app}	Period at which appliance running task $a \in A_h$ at home h can be started
β_{ah}^{app}	Period at which appliance running task $a \in A_h$ at home h must be finished
τ_{ah}^{app}	Non-preemptive running period of appliance task $a \in A_h$ at home h
ψ_{ah}^{app}	Energy requirement of appliance running task $a \in A_h$ at home h in one time period (kWh)

11.2.1.3 PEV-Related Parameters

φ_{ph}^{cs}	Gasoline consumption of PEV $p \in P_h$ at home h in CS mode (gallons /mile)
ε_{ph}^{cd}	Electricity consumption of PEV $p \in P_h$ at home h in CD mode (kWh/mile)
$\eta_{ph}^{pev^+}$	Discharging efficiency of PEV $p \in P_h$ at home h
$\eta_{ph}^{pev^-}$	Charging efficiency of PEV $p \in P_h$ at home h
$\rho_{ph}^{pev^+}$	Total transferable energy from PEV $p \in P_h$ at home h in one period
$\rho_{ph}^{pev^-}$	Total transferable energy to PEV $p \in P_h$ at home h in one period
I_{ph}^{pev}	Initial state of charge for PEV $p \in P_h$ at home h
o_{pht}^{pev}	1 if PEV $p \in P_h$ at home h is connected to the microgrid for charging/discharging during period $t \in T$, 0 otherwise
$\bar{K}_{ph}^{pev}$	Maximum capacity of PEV $p \in P_h$ at home h
$\underline{K}_{ph}^{pev}$	Minimum capacity of PEV $p \in P_h$ at home h

d_{pht}^{pev} Total travel distance during period t by PEV $p \in P_h$ at home h (miles)
c_{ph}^{deg} Battery degradation cost parameter for PEV $p \in P_h$ at home h (¢)

11.2.1.4 DER and Grid-Related Parameters

η_{dh}^{der} Generation efficiency of DER $d \in D_h$ at home h
c_{dh}^{der} Cost of energy generation for DER $d \in D_h$ at home h
K_{dht}^{der} Generation limit of DER $d \in D_h$ at home h during period $t \in T$ (kWh)
$c_t^{grid^+}$ Electricity price of buying from the grid during period $t \in T$ (¢)
$c_t^{grid^-}$ Electricity price of selling to the grid during period $t \in T$ (¢)

11.2.1.5 Storage Unit-Related Parameters

$\eta_s^{stor^+}$ Discharging efficiency of storage unit $s \in S$
$\eta_s^{stor^-}$ Charging efficiency of storage unit $s \in S$
c_s^{stor} Maintenance cost of storage unit $s \in S$ per period per kWh usage (¢)
I_s^{stor} Initial state of charge for storage unit $s \in S$ (kWh)
$\rho_s^{stor^+}$ Total transferable energy from the storage unit $s \in S$ in one period (kWh)
$\rho_s^{stor^-}$ Total transferable energy to the storage unit $s \in S$ in one period (kWh)
K_s^{stor} Capacity of storage unit $s \in S$ (kWh)
δ_{st}^{stor} Battery depth of discharge of storage unit $s \in S$ at the end of period $t \in T$

11.2.1.6 Other Parameters

l_{ht}^{tie} Time-inelastic load of home $h \in H$ during period $t \in T$ (kWh)
M Energy trade limit between the grid and the microgrid in one time period (kWh)
c_t^{gas} Price of gasoline during period $t \in T$ (¢)

11.2.1.7 Variables

e_{dht}^{der} Energy transfer from DER $d \in D_h$ at home h during period $t \in T$ (kWh)
$e_t^{grid^+}$ Energy import from the grid during period $t \in T$ (kWh)

$e_t^{grid^-}$	Energy export to the grid during period $t \in T$ (kWh)
w_t^{grid}	1 if electricity is purchased from the grid during period $t \in T$, 0 otherwise
$e_{pht}^{pev^+}$	Energy transfer from the PEV $p \in P_h$ at home h during period $t \in T$ (kWh)
$e_{pht}^{pev^-}$	Energy transfer to the PEV $p \in P_h$ at home h during period $t \in T$ (kWh)
e_{pht}^{pev}	State of energy of the PEV $p \in P_h$ at home h at the end of period $t \in T \cup \{0\}$ (kWh)
r_{pht}^{pev}	Required energy to run PEV $p \in P_h$ at home h in period $t \in T$ (kWh)
δ_{pht}^{pev}	Battery depth of discharge of PEV $p \in P_h$ at home h at the end of period $t \in T$
b_{pht}^{pev}	1 if PEV $p \in P_h$ at home h is charged during period $t \in T$, 0 otherwise
d_{pht}^{CD}	CD mode travel distance during period $t \in T$ by PEV $p \in P_h$ at home h (miles)
d_{pht}^{CS}	CS mode travel distance during period $t \in T$ by PEV $p \in P_h$ at home h (miles)
$e_{st}^{stor^+}$	Energy transfer from the storage unit s during period $t \in T$ (kWh)
$e_{st}^{stor^-}$	Energy transfer to the storage unit s during period $t \in T$ (kWh)
e_{st}^{stor}	State of energy of the storage unit s at the end of period $t \in T \cup \{0\}$ (kWh)
δ_{st}^{stor}	Battery depth of discharge of storage unit $s \in S$ at the end of period $t \in T$
y_{st}^{stor}	1 if storage unit s is charged during period $t \in T$, 0 otherwise
s_{aht}^{app}	1 if appliance task $a \in A_h$ at home h is started at the beginning of period $t \in T$, 0 otherwise
x_{aht}^{app}	1 if appliance task $a \in A_h$ at home h is running during period $t \in T$, 0 otherwise

11.2.2 Objective Function

The objective of the model is the cost minimization as depicted in Eq. 11.1. The cost is incurred due to energy generation, energy trade with the grid (buying and selling), maintenance of equipment, battery degradations and satisfying transportation requirements.

$$\text{minimize} \sum_{t \in T} \left[\begin{array}{r} c_t^{grid^+} \times e_t^{grid^+} - c_t^{grid^-} \times e_t^{grid^-} + \sum_{h \in H} \sum_{d \in D_h} (c_{dh}^{der} \times e_{dht}^{der}) + c_s^{stor} \times \sum_{s \in S} (e_{st}^{stor^+} + e_{st}^{stor^-}) \\ + \sum_{s \in S} \left[f(\delta_{st}^{stor}) - f(\delta_{s\,t-1}^{stor}) \right]^+ + \sum_{h \in H} \left(\sum_{p \in P_h} c_t^{gas} \times \varphi_{ph}^{cs} \times d_{pht}^{CS} + c_{ph}^{deg} \times \left[f(\delta_{pht}^{pev}) - f(\delta_{ph\,t-1}^{pev}) \right]^+ \right) \end{array} \right] \tag{11.1}$$

The first two terms in the objective function correspond to the price of buying energy from and selling energy to the main grid. If microgrid can satisfy the demand load by its own resources, then the additional energy generated by the DERs is sold back to grid which is an income. On the other hand, if the resources are not enough for the load at any given period, then the microgrid provides energy from the grid according to a pricing schema set by the grid. The third term in the objective function is for the energy generation cost by the DERs. Even if the resources of some DERs (such as sun or wind) do not incur any cost for the owners, there is a fixed operation and maintenance cost per each kWh of energy generation. Similarly, the fourth term is for the cost incurred due to storage unit maintenance per each kWh of usage. The fifth term corresponds to battery degradation of the storage units. In this term, operator $[\cdot]^+$ equals only a non-negative value. If the term in the parenthesis is less than zero, then the operator returns a zero value. Note that if the battery is charged in period *t*, then the term in the parenthesis might be a negative value. Thus, we consider the battery cost component when this cost value is non-negative by the $[\cdot]^+$ operator. The battery degradation cost accounts for the battery degradation of the storage units at each charging cycle similar to PEV batteries [13 and 14]. Peterson et al. [22] identifies that in real world applications, the life cycle of a battery is a linear function of the depth of discharge (DoD). Therefore, the degradation cost component in the objective, i.e. function *f*, is a linear function of DoD difference between periods [13]. The last term is related to the cost incurred by the PEVs. It has the gasoline cost component which is for traveling the distance in CS mode, and the battery degradation cost component similar to storage units degradation.

11.2.3 Energy Balance Constraint

The energy balance between the generation units and the electricity demanding units must be satisfied at every time period. The left hand side of Eq. 11.2 is the summation of the energy that is received by the microgrid. The electricity can be received from the grid, from the storage units, from the PEV batteries by discharging and from the DERs. The right hand side is the summation of the energy that is leaving the microgrid system: the electricity exported to the grid, to the storage units, to the PEV batteries by charging, to the smart appliances at homes and for satisfying the TIE loads.

$$\begin{aligned} & e_t^{grid^+} + \sum_{s \in S} e_{st}^{stor^+} + \sum_{h \in H} \left(\sum_{p \in P_h} \eta_{ph}^{pev^+} \times e_{pht}^{pev^+} + \sum_{d \in D_h} \eta_{dh}^{der} \times e_{dht}^{der} \right) \\ & = e_t^{grid^-} + \sum_{s \in S} e_{st}^{stor^-} + \sum_{h \in H} \left(\sum_{p \in P_h} \eta_{ph}^{pev^-} \times e_{pht}^{pev^-} + \sum_{a \in A_h} \psi_{ah}^{app} \times x_{aht}^{app} + l_{ht}^{tie} \right) \quad t \in T. \end{aligned} \tag{11.2}$$

11.2.4 PEV Modeling

PEV modeling is an important aspect of our formulation and in this section, we model the constraints related to the PEVs. Equation 11.3 sets the upper and lower bounds for the SoC of PEV at every time period. Observe that SoC of a PEV battery is bounded above by the capacity and bounded below by the technical requirements. Typically, a PEV battery is not discharged below 20–30 %.

$$\underline{K}_{ph}^{phev} \leq e_{pht}^{phev} \leq \bar{K}_{ph}^{phev} \quad h \in H, p \in P_h, t \in T \tag{11.3}$$

In Eq. 11.4, we require the energy balance of a PEV battery to hold between periods. The energy at the end of a given period t is equal to the summation of energy at the end of the previous period $t-1$ and the energy difference in period t. The energy difference might be due to charging or discharging (by the virtue of the V2G technology) when the PEV is connected to the grid, or it might be due to consumption in CD mode transportation. When energy is transferred to PEV, it can only receive a percentage of the transferred energy due to system losses. Similarly, when the energy is transferred from PEV to the microgrid, only a percentage of the energy can be provided to the microgrid. Thus, we make sure that more energy is discharged from the battery to ensure that the required level of energy is provided to the microgrid.

$$e_{pht}^{pev} = e_{ph\,t-1}^{pev} + \eta_{ph}^{pev^-} \times e_{pht}^{pev^-} - \frac{1}{\eta_{ph}^{pev^+}} e_{pht}^{pev^+} - r_{pht}^{pev} \quad h \in H, p \in P_h, t \in T \tag{11.4}$$

The initial SoC of PEV batteries are set by Eq. 11.5 in which e_{ph0}^{pev} corresponds to the SoC at the beginning of the planning horizon. We also require by Eq. 11.6 that the SoC at the end of the planning horizon is greater than or equal to the same SoC as the beginning state. This makes sure that the PEV is ready for the following day travels.

$$e_{ph0}^{pev} = I_{ph}^{pev} \quad h \in H, p \in P_h \tag{11.5}$$

$$e_{ph,|T|}^{pev} \geq I_{ph}^{pev} \quad h \in H, p \in P_h \tag{11.6}$$

Equations 11.7 and 11.8 make sure that energy cannot be charged to or discharged from PEV batteries if they are not connected to the microgrid. Observe that if the parameter o_{pht}^{phev} equals to zero, then charge and discharge levels are essentially set equal to zero. If this parameter equals to one, then there is a maximum limit on the charge and discharge levels. Therefore, the same sets of equations also make sure that there is an upper bound on the level of energy that can be charged or discharged in one period. Furthermore, by the use of binary variables, we ensure that the PEV batteries are not simultaneously charged and discharged.

$$e_{pht}^{pev^{+}} \leq \rho_{ph}^{pev^{+}} \times o_{pht}^{pev} \times b_{pht}^{pev} \quad h \in H, p \in P_h, t \in T \tag{11.7}$$

$$e_{pht}^{pev^{-}} \leq \rho_{ph}^{pev^{-}} \times o_{pht}^{pev} \times (1 - b_{pht}^{pev}) \quad h \in H, p \in P_h, t \in T \tag{11.8}$$

The distance to be travelled on CD mode is limited from above by the available energy in the PEV battery. The energy that can be used for transportation is the difference of the current SoC and the minimum SoC level. Dividing this value by the energy consumption of the PEV gives the mileage that can be traveled on CD mode which is ensured by Eq. 11.9.

$$d_{pht}^{CD} \leq \left(\frac{e_{pht-1}^{pev} - \underline{K}_{ph}^{pev}}{\varepsilon_{ph}^{cd}} \right) \quad h \in H, p \in P_h, t \in T \tag{11.9}$$

Note also that the CD mode trip distance is also bounded above by the total distance that will be traveled in a given time period. This constraint is realized by Eq. 11.10.

$$d_{pht}^{CD} \leq d_{pht}^{pev} \quad h \in H, p \in P_h, t \in T \tag{11.10}$$

If the total travel distance requirement of PEV cannot be traveled solely on CD mode, then the PEV travels the remaining distance on CS mode using gasoline as the source of energy for transportation which is ensured by Eq. 11.11.

$$d_{pht}^{CS} = d_{pht}^{pev} - d_{pht}^{CD} \quad h \in H, p \in P_h, t \in T \tag{11.11}$$

The energy requirement of PEV for transportation is set by Eq. 11.12. The total level of energy that PEV consumes in transportation is the mileage that is actually traveled in CD mode times the energy consumption of PEV per mile.

$$r_{pht}^{pev} = d_{pht}^{CD} \times \varepsilon_{ph}^{cd} \quad h \in H, p \in P_h, t \in T \tag{11.12}$$

DoD is used for obtaining the battery degradation cost in the objective function. Equation 11.13 sets the DoD for each period depending on the current SoC of PEV battery. Observe that it can assume values in [0, 1] range.

$$\delta_{pht}^{pev} = 1 - \frac{e_{pht}^{pev}}{\overline{K}_{ph}^{pev}} \quad h \in H, p \in P_h, t \in T \tag{11.13}$$

11.2.5 DERs and Grid Modeling

The energy supply is limited by an upper bound depending on the forecast sunshine, wind and/or generation capacity of the DER. The limit is enforced by Eq. 11.14.

$$e_{dht}^{der} \leq K_{dht}^{der} \quad h \in H, d \in D_h, t \in T \tag{11.14}$$

The energy trade between the microgrid and the grid is modeled by Eqs. 11.15 and 11.16. M is the limit of energy transfer. Practically, it can be set to a sufficiently large number. The binary variables in the equations ensure that the energy is not simultaneously transferred both ways in the same time period.

$$e_t^{grid^+} \leq M \times w_t^{grid} \quad t \in T \tag{11.15}$$

$$e_t^{grid^-} \leq M \times (1 - w_t^{grid}) \quad t \in T \tag{11.16}$$

11.2.6 Electricity Storage Units Modeling

Electricity storage units' modeling has a similar characteristic with the PEV constraints as presented in Sect. 11.2.4. Equation 11.17 models the capacity of the storage unit: the SoC of the storage unit can at most equal capacity.

$$e_{st}^{stor} \leq K_s^{stor} \quad s \in S, t \in T \tag{11.17}$$

Equations 11.18 and 11.19 ensure that charge and discharge is bounded at any given period. Furthermore, simultaneous charging and discharging is avoided by the use of binary variables.

$$e_{st}^{stor^+} \leq \rho_s^{stor^+} \times y_{st}^{stor} \quad s \in S, t \in T \tag{11.18}$$

$$e_{st}^{stor^-} \leq \rho_s^{stor^-} \times (1 - y_{st}^{stor}) \quad s \in S, t \in T \tag{11.19}$$

Similar to PEV batteries, the energy of the storage units between periods must be balanced. Equation 11.20 states that the energy level at the end of a given period t is the summation of the energy at the end of the previous period $t - 1$ and the energy change during period t. The energy change is due to charging or discharging. Furthermore, Eq. 11.20 takes into account the system losses. Only a percentage of the energy provided by the microgrid can actually be stored by the unit, and similarly, only a percentage of the energy discharged from the storage unit can actually be used by the microgrid.

$$e_{st}^{stor} = e_{st-1}^{stor} + \eta_s^{stor^-} \times e_{st}^{stor^-} - \frac{1}{\eta_s^{stor^+}} \times e_{st}^{stor^+} \quad s \in S, t \in T \tag{11.20}$$

Equation 11.21 sets the initial conditions for the storage units and Eq. 11.22 ensures that the SoC at the end of the planning horizon is at least as much as the starting SoC.

$$e_{s0}^{stor} = I_s^{stor} \quad s \in S \tag{11.21}$$

$$e_{s,|T|}^{stor} \geq I_s^{stor} \quad s \in S \tag{11.22}$$

Similar to Eqs. 11.13, and 11.23 sets the DoD for each period depending on the current SoC of storage units.

$$\delta_{st}^{stor} = 1 - \frac{e_{st}^{stor}}{K_s^{stor}} \quad s \in S, t \in T \tag{11.23}$$

11.2.7 Smart Appliances Modeling

The users of smart devices define a feasible time interval for running tasks. Smart devices need to run for a fixed number of periods non-stop. In this regard, Eq. 11.24 models the starting time of the smart devices. Beginning from time period α_{ah}^{app}, the appliance must be started before time period $\beta_{ah}^{app} - \tau_{ah}^{app}$ to be able to finish the task by time period β_{ah}^{app}. Note that $\beta_{ah}^{app} - \tau_{ah}^{app}$ might exceed the planning horizon, i.e. period $|T|$. Thus, we consider the minimum of these two terms when planning the starting time period of the appliance. Once the device is started, Eq. 11.25 ensures that the binary variables indicating that the device is running, x_{aht}^{app}, is set equal to 1 for the number of periods that the device needs to run. Equation ensures that the variable x_{aht}^{app} equals 1 if the appliance is started any time period in the previous τ_{ah}^{app} time periods. Note that the term $t - \tau_{ah} + 1$ might be less than 1. Thus, we only consider the planning horizon, starting from time period 1 until time period T.

$$\sum_{t=\alpha_{ah}}^{\min\{|\mathrm{T}|,\beta_{\mathrm{ah}}-\tau_{\mathrm{ah}}\}} s_{aht}^{app} = 1 \quad h \in H, a \in A_h \tag{11.24}$$

$$\sum_{i=\max\{1,\mathrm{t}-\tau_{\mathrm{ah}}+1\}}^{t} s_{ahi}^{app} \leq x_{aht}^{app} \quad h \in H, a \in A_h, t \in T \tag{11.25}$$

Lastly, Eqs. 11.26 and 11.27 are the domain restrictions of the model.

$$e_{dht}^{der}, e_t^{grid^+}, e_t^{grid^-}, e_{pht}^{pev^+}, e_{pht}^{pev^-}, e_{pht}^{pev}, r_{pht}^{pev}, \delta_{pht}^{pev}, e_{st}^{stor^+}, e_{st}^{stor^-}, e_{st}^{stor}, d_{pht}^{CD}, d_{pht}^{CS} \geq 0$$
$$t \in T, s \in S, h \in H, p \in P_h, d \in D_h, a \in A_h \qquad (11.26)$$

$$s_{aht}^{app}, x_{aht}^{app}, b_{pht}^{pev}, y_{st}^{stor}, w_t^{grid} \in \{0, 1\} \quad t \in T, s \in S, h \in H, p \in P_h, a \in A_h \quad (11.27)$$

11.2.8 Binary Variable Reduction

In the above energy management model, we have made use of binary variables in order to model that the results exclude simultaneous occurrence of energy transfer between microgrid and the other entities: PEVs, storage units and the grid. These variables that are traditionally included in the energy management models are actually not required due to system energy losses. To see, we consider two different settings. In the first one, simultaneous charging and discharging of a PEV v occurs in a given time period t. Let c be the energy provided by the microgrid to PEV v, d be the energy received by the microgrid from the PEV v and $e < 1$ be the efficiency of PEV v. In the second setting, the microgrid provides $(c - d)$ units of energy to a PEV v and does not receive any energy from it. Observe that in the second setting, PEV v can only charge $(c \times e)$ units of energy due to system losses, and need to send (d/e) units of energy to the microgrid to make sure that the microgrid receives d units of energy. The net energy stored in the PEV battery is then $(c \times e - d/e)$ units of energy, which is strictly less than $(c - d)$. In both of the scenarios, the net energy difference of the microgrid is the same, but PEV v stores more energy in the second setting. Since the objective function of the above energy management model is cost minimization, the first setting is always a suboptimal solution and the model never simultaneously charges and discharges a PEV in a given time period t.

The same result was also obtained in a different study [13] for a different setting. The same logic also proves that simultaneous charging and discharging of storage units, and simultaneous buying and selling of energy from the national grid is never profitable. Excluding these binary variables from the energy management model ensures speed and efficiency in reaching the optimal solutions.

11.3 Case Study

In this section, we present a case study to observe the model behavior in different settings. The data related to the case study is obtained from official sources as well as recent literature. We consider a group of 100 houses forming a microgrid in California. In addition to TIE loads, the houses have 0, 1 or 2 PEVs, 0, 1 or 2 DERs and/or 2, 3 or 4 smart devices. The total number of PEVs, DERs and smart appliances in the microgrid are 90, 100 and 300, respectively. The microgrid also has one storage unit. The energy management model above is used for scheduling. 365 days are simulated to observe the impacts in a one-year time period. In the following, we present the data and the experimental design details for the sensitivity analyses.

11.3.1 PEV Data

Four types of PEVs are considered in this study with all-electricity ranges of 7, 20, 40 and 60 miles. The features for each type of PEV are presented in Table 11.1. The capacities of the vehicles, electricity and gasoline consumption data are obtained from [23]. Transferrable energy per period is 7 kWh [24]. The minimum capacity is 30 % of the capacities. The initial SoC is assumed to be the minimum capacity. Cost of gasoline in a day changes between $4.00 and $4.10.

The driving data is obtained from the US Department of Transportation's National Household Travel Survey [25]. The number of PEVs travelling in any given period is presented in Fig. 11.2. The peak at 8 a.m. is due to morning commute and a similar peak occurs at close of business hours around 5 p.m.

Travel distances of PEVs are depicted in Fig. 11.3. The average number of travels per vehicle is 2.1 per day in the case study.

11.3.2 DER and Grid Data

We consider three types of DERs that can be used in smart homes: photovoltaics (PVs), wind turbines and natural gas engine with combined heat and power (CHP). The related data is presented in Table 11.2 [26].

Observe that the electricity generation is dependent on two natural factors: sun and wind. Figure 11.4 plots the availability of the sun and the wind as a percentage of its maximum capacity throughout the day [27].

Similarly, these two factors also follow different trends during different seasons of the year [28]. Since we are simulating a year in the case study, we also take into account the seasonal changes of the sun and the wind levels. Figure 11.5 plots the change of wind and sun in a year.

We assume that the grid electricity prices in a given day change in the range of 4.80¢–4.85¢ [29], and that the microgrid sells the excess energy to the grid for half the price that grid sells the electricity for.

Table 11.1 Technical data of the PEVs

PEV type	All electricity range (miles)	Electricity consumption (kWh/mile)	Gasoline consumption (gallons/mile)	Minimum capacity (kWh)	Maximum capacity (kWh)	Charging efficiency (%)	Discharging efficiency (%)
PEV-7	7	0.179	0.0194	0.6	3	90	80
PEV-20	20	0.183	0.02	1.64	8.2	90	80
PEV-40	40	0.188	0.0204	3.36	16.8	90	80
PEV-60	60	0.197	0.0209	5.28	26.4	90	80

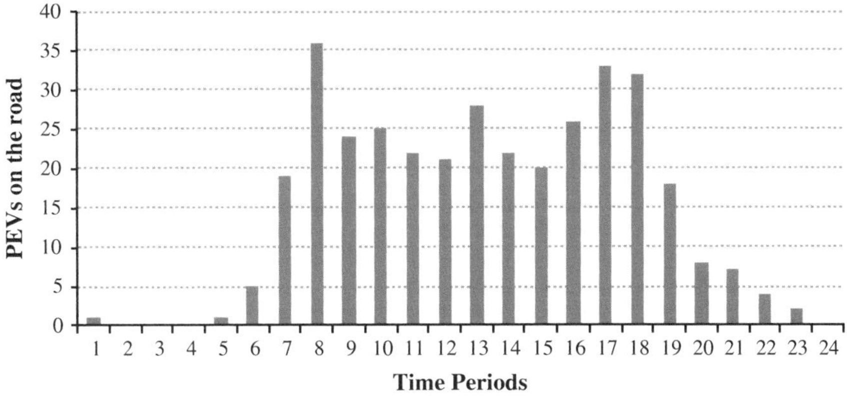

Fig. 11.2 Number of PEVs on the road according to the driving patterns

Fig. 11.3 Number of PEV trips for different daily travel distances

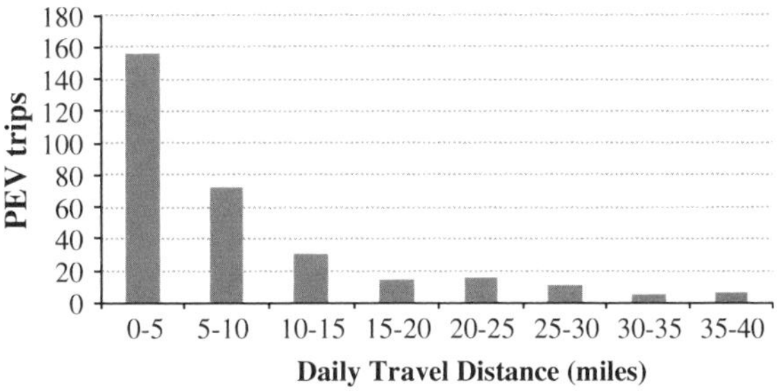

Table 11.2 Technical data of the DERs

DER type	Range of capacities (kWh)	Efficiency (%)	Operation and maintenance cost (¢/kWh)	Number in the case study
Photovoltaics	[1, 5]	N/A	0.2	30
Wind turbines	[1, 5]	N/A	1.0	30
Natural gas w/CHP	[1, 5]	85	2.7	30

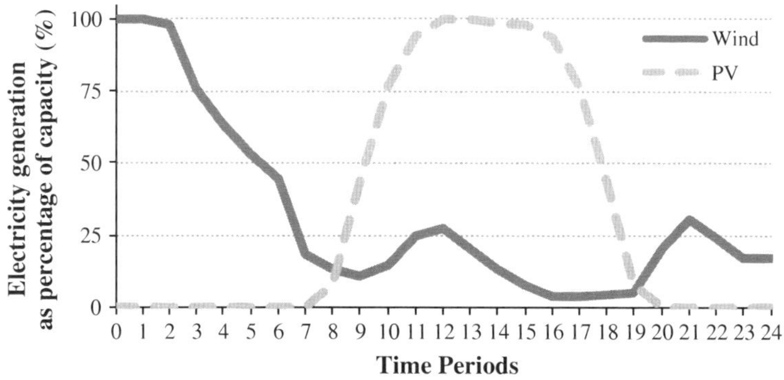

Fig. 11.4 Energy generation of PVs and wind turbines as percentage of total capacity

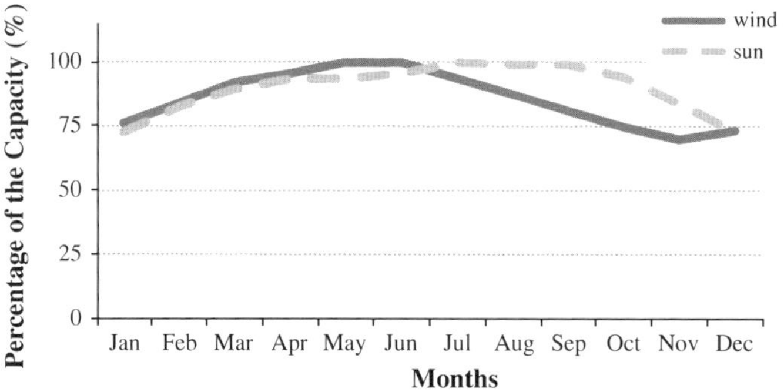

Fig. 11.5 Capacity of PVs and wind turbines as percentage of maximum, yearly change

Table 11.3 Technical data of the smart appliances

Appliance type	Energy requirement per period (kWh)	Running hours	Number in the case study
Dishwasher	2.8	2	75
Washer and dryer	2.5	3	75
Water heater	5	4	75
Air conditioner	0.75	5	75

11.3.3 Load and Smart Device Data

The smart appliances in the case study are presented in Table 11.3 [30], and the maximum and minimum limits on the TIE loads are depicted in Fig. 11.6 [31].

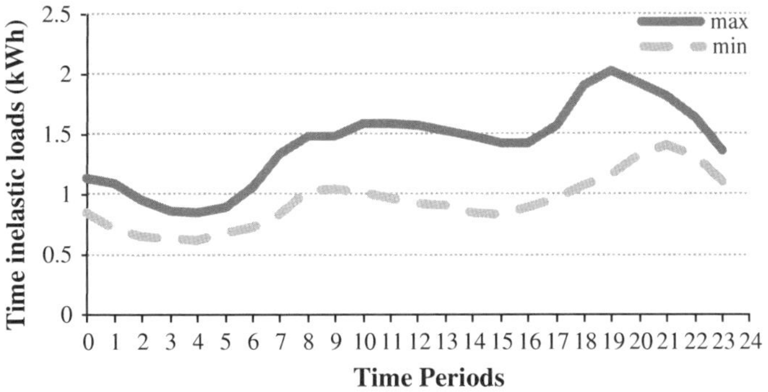

Fig. 11.6 Time-inelastic load range per house for a given day

11.3.4 Storage Unit Data

In the case study, we consider a single storage unit with a capacity of 85 kWh and a charging efficiency of 93 % [32]. The cost of battery maintenance per kWh of use is 1¢ [26]. The transfer rate is 19.2 kWh when charging and 14.2 kWh when discharging [33, 34].

11.3.5 Emission Data

In the case study, two types of greenhouse gas (GHG) emissions are considered: NO_x and CO_2. The CHP, grid and gasoline usage are the main sources of GHG emission. The emission for the DERs [26] and PEVs on CS mode [35] is shown in Table 11.4.

11.3.6 Experimental Design

We constructed 365 different datasets with the above baseline data. A total of 100 driving profiles are randomly distributed among PEVs in the scenarios. Every day,

Table 11.4 Emission data of DERs and PEVs (in CS mode)

GHG emission source	NO_x emission	CO_2 emission
Natural Gas w/CHP (lb/kWh)	0.0059	0.97
National Grid (lb/kWh)	0.005	1.2
PEV7/PEV20 (on CS mode) (g/mile)	0.693	368.4
PEV40/PEV60 (on CS mode) (g/mile)	0.95	513.5

Table 11.5 Scenario settings

Scenario	PEVs	DERs	DER capacities	Gasoline pricing	Electricity pricing	Driving patterns	Smart appliances
1	+	+	Baseline	Baseline	Baseline	Baseline	Baseline
2	−	+	Baseline	Baseline	Baseline	Baseline	Baseline
3	+	−	Baseline	Baseline	Baseline	Baseline	Baseline
4	−	−	Baseline	Baseline	Baseline	Baseline	Baseline
5	+	+	×1.5	Baseline	Baseline	Baseline	Baseline
6	+	+	×2.0	Baseline	Baseline	Baseline	Baseline
7	+	+	Baseline	$3.50–$3.60	Baseline	Baseline	Baseline
8	+	+	Baseline	$3.00–$3.10	Baseline	Baseline	Baseline
9	+	+	Baseline	$2.50–$2.60	Baseline	Baseline	Baseline
10	+	+	Baseline	$2.00–$2.10	Baseline	Baseline	Baseline
11	+	+	Baseline	Baseline	9.60¢–9.70¢	Baseline	Baseline
12	+	+	Baseline	Baseline	19.20¢–19.40¢	Baseline	Baseline
13	+	+	Baseline	Baseline	Baseline	×2.0	Baseline
14	+	+	Baseline	Baseline	Baseline	×4.0	Baseline
15	+	+	Baseline	Baseline	Baseline	Baseline	0
16	+	+	Baseline	Baseline	Baseline	Baseline	×2.0

randomly selected smart appliances are required to run. We also randomly assign the TIE loads to the microgrid participant houses. In order to test the impacts of Smart Grid, PEVs, DERs, gasoline and electricity pricing, we built 16 different scenarios as presented in Table 11.5. "Baseline" in the table refers to the data settings presented in Sect. 1.3 above. When the scenario excludes PEVs (i.e. Scenarios 2 and 4), an equivalent performance vehicle is assumed to be in the scenario to travel the required driving patterns.

11.4 Results and Discussion

In this section, we present the results of 16 scenarios, each of which are run for 365 days. For solving the models, IBM ILOG CPLEX Optimization Studio 12.5 was used on a computer with Intel®Core™2 Duo CPU at 2.00 GHz and 2.00 GB RAM. The average solution time for each model is 2.25 s. Daily average energy distribution among generation (grid and DERs) and consumption (PEVs, smart appliances, TIE loads) units are listed in Table 11.6. The corresponding cost and emission values are shown in Table 11.7. In the following part, we shall refer to these tables when discussing the results and providing insights.

Table 11.6 Daily average energy distribution of the microgrid in kWh

Scenario	Grid	DER	PEV	Appliances	TIE loads
1	2926.66	3427.88	395.1	2955.9	2747.7
2	2584.59	3412.44	0.0	2955.9	2747.7
3	6059.13	0.00	395.1	2955.9	2747.7
4	5703.55	0.00	0.0	2955.9	2747.7
5	1360.80	5141.38	395.1	2955.9	2747.7
6	202.84	6388.72	395.1	2955.9	2747.7
7	2926.66	3427.88	395.1	2955.9	2747.7
8	2926.66	3427.88	395.1	2955.9	2747.7
9	2919.24	3427.88	386.8	2955.9	2747.7
10	2584.78	3412.63	0.4	2955.9	2747.7
11	2926.66	3427.88	395.1	2955.9	2747.7
12	2926.66	3427.88	395.1	2955.9	2747.7
13	3198.12	3427.88	696.7	2955.9	2747.7
14	3546.85	3427.88	1084.2	2955.9	2747.7
15	327.13	3020.60	395.1	0.0	2747.7
16	5882.56	3427.88	395.1	5911.8	2747.7

Table 11.7 Daily average cost and emission impacts of the microgrid

Scenario	Cost (¢)	NO_x emission (kg)	CO_2 emission (kg)
1	30325.48	15.82	3107.68
2	36450.71	15.03	2927.99
3	38370.51	13.75	3304.50
4	44472.55	12.97	3124.25
5	26303.93	16.86	3009.28
6	22989.24	17.57	2927.80
7	30146.85	15.82	3107.68
8	29968.21	15.82	3107.68
9	29789.45	15.81	3104.04
10	27963.66	15.04	2932.15
11	44440.09	15.82	3107.68
12	72669.29	15.82	3107.68
13	42651.78	16.46	3265.40
14	70645.14	17.30	3484.59
15	16691.44	8.84	1513.53
16	44581.47	22.53	4716.61

11.4.1 Value of PEVs and DERs

The base scenario (i.e. Scenario 1) has an average electricity generation of 6,354.6 kWh and consumption of 6,098.7 kWh. The difference between generation and consumption is due to system losses which accounts to 4 % of the generation. When the PEV is excluded from the base scenario, less energy is purchased from the grid. In particular, the amount of reduction is as much as the energy that is used by the PEVs. On the other hand, the cost increases around 20 % when PEVs are excluded. The reason for this effect is that rather than utilizing electricity as the source of energy for transportation, gasoline is used. Thus, the cost increases in Scenario 2. However, counter-intuitively, the emissions are decreasing when more gasoline is used. Observe that the energy that is used by the PEVs to drive on CD mode is solely generated by the grid. The average emissions of the grid (as presented in Sect. 1.3) are higher than that of gasoline. Therefore, we do not observe a reduction in emissions when more electricity is used in transportation. One critical insight is that in order to observe the emission reduction when PEVs penetrate the transportation network, the source of electricity generation for the PEVs will be crucial.

The benefits of DERs can be observed by analyzing the results for Scenarios 1, 3, 4, 5, and 6. Observe that when the DERs are not considered in the microgrid in Scenario 3, the level of electricity to be obtained from the grid is the maximum among the scenarios considered in this study. The cost also increases drastically when compared to the base scenario. When both DERs and PEVs are excluded, i.e. Scenario 4, the cost increase is more than 45 %. When the DER capacities are increased by 1.5 and 2 times in Scenarios 5 and 6, respectively, the cost benefits are between 13 and 25 %. The CO_2 emission also reduces by increasing DER capacities. However NO_x is increasing. This increase is due to the extra NO_x generation by the natural gas with CHP as presented in Table 11.4. Considering a PEV with an average electricity usage of 0.352 kWh per mile, the natural gas with CHP generates 0.96 g of NO_x per mile, the highest amount of emission in our experiments. Therefore, NO_x is increasing in Scenarios 5 and 6. However this increase rests on the assumption that there is no limit on the level of NO_x emission. If the governments put certain limits on the emission levels, then the cost benefits of the microgrid might decrease.

In the base scenario, different PEV types perform differently in terms of electricity utilization in transportation (Fig. 11.7). CD mode usage can be considered as

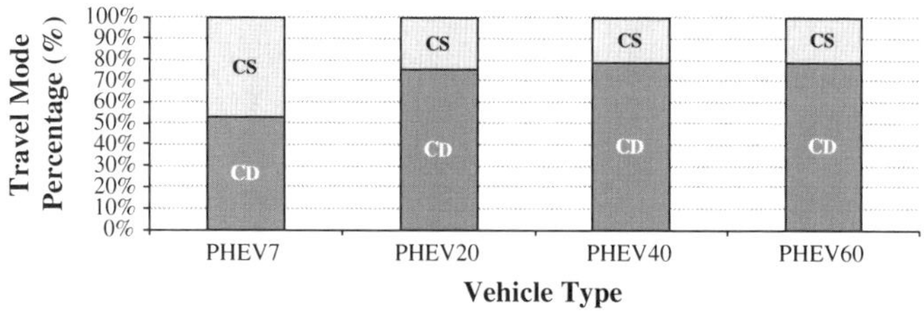

Fig. 11.7 Travel mode percentage for different vehicle types

the level of advantage that a PEV owner benefits from driving a hybrid vehicle. Thus, the benefits are more for higher all-electricity range vehicles. However, note that the CD more drive percentages almost do not change for PHEV20, PHEV40 and PHEV60. This indicates that after a basic all-electric range is attained (20 miles in our scenarios), the benefits are almost similar for all longer all-electric range vehicles.

11.4.2 Gasoline Pricing Sensitivity

Scenarios 7–10 are dedicated to analyzing the results' sensitivity for gasoline prices. Observe that decreasing the costs does not affect the PEV electricity requirement to a great extent in Scenarios 7, 8, and 9 (Table 11.6). However, in Scenario 10, the energy requirement for PEVs is practically zero. This indicates that there is a critical gasoline pricing between \$2.0 and \$2.5 beyond which drivers prefer gasoline over electricity drive. This is another critical insight of this study.

11.4.3 Electricity Pricing Sensitivity

In scenarios 11 and 12, we consider electricity pricing by the national grid twice and four times more than the baseline pricing, respectively. Even though costs are increased in both scenarios, the electricity purchase from the grid has not changed. The reason is that DERs are utilized at full capacity in even the baseline scenario so that increasing the prices does not significantly affect the amount of electricity purchase. The microgrid still needs to satisfy the loads. Therefore the microgrid purchases the energy that it requires in excess of the generation capacity of the DERs from the grid regardless of the price.

11.4.4 Driving Patterns

Increasing the driving mileage by two and four times in Scenarios 13 and 14, we observe a gradual increase in the cost and the electricity purchase from the grid. Note that the impact of traveling longer distances is significant and might result in doubling the total costs, however impact on emissions is not that significant.

11.4.5 Smart Appliances

In the last two scenarios, we consider excluding the smart devices from the scenarios and doubling the number of devices. Both of the scenarios give the same effect: the cost is halved or doubled as expected. The reason is the change in the amount of electricity purchase from the grid.

11.5 Conclusions

In this study, we consider a microgrid that manages the DERs, PEVs, and smart devices with the objective of cost minimization. The opportunities of microgrids over classical national grid are investigated. Smart management of loads is a way to tackle the excess energy requirement of PEVs as well as peak load increase. High level of PEV penetration into the transportation system, widespread utilization of smart appliances at homes, distributed energy generation and community-level electricity storage units all complicate the energy management problem, however if smartly managed, these complications can be regarded as strengths and opportunities for the next generation energy management units: the microgrids.

In this scope, we propose a mixed integer linear programming energy management optimization model to schedule the charging and discharging times of PEVs, electricity storage units, and running times of smart appliances. Our findings show that simultaneous charging and discharging of PEV batteries and electricity storage units do not occur in model solutions due to system energy losses.

Critical insights are also presented in this study. First of all, in order to observe the emission reduction when PEVs penetrate the transportation network, the source of electricity generation for the PEVs is crucial. If charged from renewable, more benefits can be attained. However charging from the national grid reduces the benefits of PEVs and might even increase the emission level. Another important result is that after a basic all-electric range, cost and emission benefits are almost similar for all longer all-electric range vehicles. This range is 20 miles in this study. Lastly, there is a critical gasoline pricing between \$2.0 and \$2.5 beyond which drivers prefer gasoline over electricity drive.

References

1. Asmus P (2010) Microgrids, virtual power plants and our distributed energy future. Energ J 23 (10):72–82
2. Morais H, Kàdàr P, Faria P, Vale ZA, Khodr H (2010) Optimal scheduling of a renewable micro-grid in an isolated load area using mixed-integer linear programming. Renew Energ 35 (1):151–156

3. Khodr H, Halabi NE, García-Gracia M (2010) Intelligent renewable microgrid scheduling controlled by a virtual power producer: a laboratory experience. Renew Energ 48:269–275
4. Kriett PO, Salani M (2012) Optimal control of a residential microgrid. Energy 42(1):321–330
5. Naraharisetti PK, Karimi I, Anand A, Lee DY (2011) A linear diversity constraint application to scheduling in microgrids. Energy 36(7):4235–4243
6. Moghaddam AA, Seifi A, Niknam T, Pahlavani MRA (2011) Multi-objective operation management of a renewable MG (micro-grid) with back-up micro-turbine/fuel cell/battery hybrid power source. Energy 36(11):6490–6507
7. Basu AK, Chowdhury S, Chowdhury S, Paul S (2011) Microgrids: energy management by strategic deployment of DERs—a comprehensive survey. Renew Sustain Energ Rev 15 (9):4348–4356
8. Xiong G, Chen C, Kishore S, Yener A (2011) Smart (in-home) power scheduling for demand response on the smart grid. In: Innovative smart grid technologies (ISGT), IEEE PES, pp 1–7
9. Pedrasa MAA, Spooner TD, MacGill IF (2011) A novel energy service model and optimal scheduling algorithm for residential distributed energy resources. Electr Power Syst Res 81 (12):2155–2163
10. Rastegar M, Fotuhi-Firuzabad M, Aminifar F (2012) Load commitment in a smart home. Appl Energ 96:45–54
11. Elma O, Selamogullari US (2012) A comparative sizing analysis of a renewable energy supplied stand-alone house considering both demand side and source side dynamics. Appl Energ 96:400–408
12. Fernandes C, Frías P, Latorre JM (2012) Impact of vehicle-to-grid on power system operation costs: the Spanish case study. Appl Energ 96:194–202
13. Arslan O, Karasan OE (2013) Cost and emission impacts of virtual power plant formation in plug-in hybrid electric vehicle penetrated networks. Energy 60:116–124
14. Sioshansi R, Denholm P (2010) The value of plug-in hybrid electric vehicles as grid resources. Energ J 31(3):1–24
15. Sioshansi R (2012) Modeling the impacts of electricity tariffs on plug-in hybrid electric vehicle charging, costs, and emissions. Oper Res 43(4):1199–1204
16. Su W, Chow MY (2012) Computational intelligence-based energy management for a large-scale PHEV/PEV enabled municipal parking deck. Appl Energ 96:171–182
17. Saber A, Venayagamoorthy G (2012) Resource scheduling under uncertainty in a smart grid with renewables and plug-in vehicles. IEEE Syst J 6(1):103–109
18. Kristoffersen TK, Capion K, Meibom P (2011) Optimal charging of electric drive vehicles in a market environment. Appl Energ 88(5):1940–1948
19. Sousa T, Morais H, Soares J, Vale Z (2012) Day-ahead resource scheduling in smart grids considering vehicle-to-grid and network constraints. Appl Energ 96:183–193
20. Arslan O, Yildiz B, Karasan OE (2014) Impacts of battery characteristics, driver preferences and road network features on travel costs of a plug-in hybrid electric vehicle (PHEV) for long-distance trips. In: Energy policy, http://dx.doi.org/10.1016/j.enpol.2014.08.015
21. Arslan O, Yildiz B, Karasan OE (2014) Minimum cost path problem for plug-in hybrid electric vehicles. In: Technical report, Bilkent University, Department of Industrial Engineering
22. Peterson SB, Apt J, Whitacre J (2010) Lithium-ion battery cell degradation resulting from realistic vehicle and vehicle-to-grid utilization. J Power Sources 195:2385–2392
23. Shiau CSN, Samaras C, Hauffe R, Michalek JJ (2009) Impact of battery weight and charging patterns on the economic and environmental benefits of plug-in hybrid vehicles. Energ Policy 37:2653–2663
24. Motors T (2014) Supercharger. http://www.teslamotors.com/supercharger. Accessed 10 Jun 2014
25. U.S. Department of Transportation (2011) 2009 National household travel survey version 2.1
26. U.S. Department of Energy, Energy Efficiency and Renewable Energy (2013) Federal energy management program. http://www1.eere.energy.gov/femp/technologies/derchp_derbasics.html. Accessed 01 Aug 2013

27. California ISO (2014) Daily renewable watch, hourly breakdown of renewable resources. http://content.caiso.com/green/renewrpt/20140315_DailyRenewablesWatch.txt. Accessed 10 May 2014
28. National Oceanic and Atmospheric Administration (2014) Comparative climatic data. http://ols.nndc.noaa.gov/plolstore/plsql/olstore.prodspecific?prodnum=C00095-PUB-A0001. Accessed 30 May 2014
29. Pacific Gas and Electric (PG&E) (2013) Hourly electric commodity prices. http://www.pge.com/nots/rates/tariffs/pxdy0212.html. Accessed 04 May 2013
30. U.S. Department of Energy (2014) Estimating appliance and home electronic energy use. http://www.energy.gov/energysaver/articles/estimating-appliance-and-home-electronic-energy-use Accessed 24 May 2014
31. Pasific Gas and Electric (PG&E) (2002) Residential load profiles. http://www.pge.com/nots/rates/2002_static.shtml. Accessed 01 Jun 2014
32. Motors T (2014) Model S specs. http://www.teslamotors.com/models/design. Accessed 16 May 2014
33. Motors T (2014) Features and specs. http://www.teslamotors.com/models/features#/battery. Accessed 02 Jun 2014
34. Motors T (2014) How long does charging take? http://www.teslamotors.com/goelectric#charging. Accessed 02 Jun 2014
35. U.S. Environmental Protection Agency (EPA) (2012) Average annual emissions and fuel consumption for gasoline-fueled passenger cars and light trucks report